Stress Field of the Earth's Crust

Stress Field of the Earth's Crust

Arno Zang
German Research Centre for Geosciences (GFZ)
Potsdam, Germany

and

Ove Stephansson
GFZ and KTH Stockholm, Sweden

With Foreword by Bezalel Haimson

Springer

Priv.-Doz. Dr. Arno Zang
GFZ German Research Centre
for Geosciences
Telegrafenberg
14473 Potsdam
Germany
zang@gfz-potsdam.de

Prof. Ove Stephansson
GFZ German Research Centre
for Geosciences
Telegrafenberg
14473 Potsdam
Germany
ove@gfz-potsdam.de

Additional material to this book can be downloaded from http://extras.springer.com

ISBN 978-94-007-9515-0 ISBN 978-1-4020-8444-7 (eBook)
DOI 10.1007/978-1-4020-8444-7
Springer Dordrecht Heidelberg London New York

Cover illustrations: i) Cube of rock mass in the Earth's crust with open cracks at top, progressively closing cracks in the middle and residual pores at bottom. Principal stresses are indicated by arrows ($S_V > S_H > S_h$) forcing a favourable oriented shear crack to slide. ii) Section of the World Stress Map (Heidbach et al. 2007) shows orientations of the maximum horizontal principal stress (S_H). For detailed description see Chapter 11.

Cover design: deblik

Printed on acid-free paper

Springer is part of Springer Science+Business Media (www.springer.com)

For my wife Corinna and our children Felix, Charlotte and Marlene.

I am grateful to my academic teachers Hans Berckhemer and Gerhard Müller (1940–2002) both at Goethe-University Frankfurt/Main

Arno Zang

I like to thank my wife Almut for her support and our children Samuel and Naemi for bringing in a good balance between daily life and work.

Ove Stephansson

Foreword

Stress Field of the Earth's Crust is based on lecture notes prepared for a course offered to graduate students in the Earth sciences and engineering at University of Potsdam. In my opinion, it will undoubtedly also become a standard reference book on the desk of most scientists working with rocks, such as geophysicists, structural geologists, rock mechanics experts, as well as geotechnical and petroleum engineers. That is because this book is concerned with what is probably the most peculiar characteristic of rock – its initial stress condition. Rock is always under a natural state of stress, primarily a result of the gravitational and tectonic forces to which it is subjected. Crustal stresses can vary regionally and locally and can reach in places considerable magnitudes, leading to natural or man-made mechanical failure. Preexisting stress distinguishes rock from most other materials and is at the core of the discipline of "Rock Mechanics", which has been developed over the last century.

Knowledge of rock stress is fundamental to understanding faulting mechanisms and earthquake triggering, to designing stable underground caverns and productive oil fields, and to improving mining methods and geothermal energy extraction, among others. Several books have been written on the subject, but none has attempted to be as all-encompassing as the one by Zang and Stephansson. The present book does not limit itself to just providing a detailed description of the known methods of measuring stress in-situ, and of the different stress fields around the globe. It first presents a detailed and thorough description of the concept of stress, the sources of in-situ stress, and rock failure criteria. These first three chapters are the very foundation of rock mechanics, and could be used as a text for an introductory course in this field. The last three chapters go beyond stress measurements, to describe stress profiles through the Earth's crust, regional stress fields, the World Stress Map, and three recent international field projects in which scientists from many countries collaborated, and in which one of the first priorities was the determination of the state of in-situ stress. These are the KTB, SAFOD, and Olkiluoto projects. The book provides excellent summaries of these major projects, the results of which are otherwise scattered over many publications. These case histories are an invaluable resource to researchers, teachers, and students in the Earth sciences.

The chapters dealing directly with stress measurements are encyclopedic. Each method listed is presented in some detail, accompanied by an exhaustive list of references, so that the reader could dig deep into any of the techniques at the level he/she chooses. The variety of stress measuring methods in practice today is divided logically into borehole- and core-based. Naturally, the more commonly accepted methods, hydrofracturing, overcoring, and borehole breakouts, are given added emphasis, but all methods are treated as equals even if their use at this time is rather rare.

PD Dr. Zang and Prof. Stephansson, who are among the top echelon of in-situ stress researchers and consultants, have produced a much needed book on the state of stress in the Earth's crust, one that complements previous texts, which were considerably more restricted in scope. The book thoroughly and convincingly integrates in-situ stress, its sources, measurement, and applications, into the fields of geophysics, geology, geomechanics, and geoengineering. It is therefore that I enthusiastically recommend *Stress Field of the Earth's Crust*.

University of Wisconsin Bezalel Haimson
Madison, USA
June 2009

Preface

Rock stress is a key parameter in solid Earth sciences and technology. Long-term geological processes like plate tectonics are driven by mechanisms that generate different types of stresses in the Earth's crust. These stresses are acting as we extract raw materials from the crust and deposit human altered materials into the crust in boreholes, mines and underground constructions. To better use and save our resources there is an obvious need for a greater understanding of mechanical stresses in the Earth's crust. This book is directed toward graduate students, teachers and practitioners in geology, geophysics and civil, mining, petroleum and rock engineering. The book aims to fill the gap in the existing literature between principles in rock mechanics (Jaeger, Cook & Zimmerman 2007), rock stress measurements (Amadei & Stephansson 1997) and stress regimes in the lithosphere (Engelder 1993).

Mechanical stress and rock stress are fictitious terms as stress can never be directly measured. Stresses in rock originate from gravity and tectonic forces and can only be inferred by disturbing the rock by drilling a borehole, making a slot and coring the rock. The drill core can be brought to the laboratory and stresses determined by different physical methods. The complex nature of rocks prevents us from exactly determinating the magnitudes and orientation of the components of the stress tensor and often we have to accept large variability and uncertainties. Stress in rock is usually described in the context of continuum mechanics. To introduce the *Stress Field of the Earth's Crust* to students in geosciences, one has to adopt methods from a number of otherwise self-supporting disciplines like the theory of elasticity, continuum mechanics, fracture mechanics, structural geology, geophysics, geodesy, experimental physics, rock and petroleum engineering.

This book starts by introducing the physical *Concept of Stress* from continuum mechanics, and continues to describe *Rock Failure* from classical to strength of material and modern fracture mechanics approaches. The chapter on *Rock Stress Terminology*, presented from a material sciences and rock engineering viewpoint, is followed by simple physical models describing the variation of stress magnitudes with depth in the Earth's crust. Then a chapter on the *Physics of Stress Measurements* is presented, where techniques from experimental physics are applied to determine residual stresses using material sciences standard. Methods for determining crustal stresses are separated into *Borehole Techniques* (e.g., overcoring, hydraulic fracturing, borehole breakouts) and *Core-Based Methods* (e.g., anelastic strain recovery, Kaiser effect). The focus of the chapter *Local Stress Data* is to demonstrate facets of integrated stress measuring strategies as applied to scientific deep wells (*KTB,* Germany), to nuclear waste repository (*Olkiluoto,* Finland), and to monitor a seismically active fault zone (*SAFOD,* USA). The chapter *Generic Stress Data* reports on general trends in stress-magnitude profiles and stress-orientation maps, where stress-state scaling relations are used to find the best estimated stress model. In the last chapter the European and *World Stress Map* is interpreted in terms of *Plate Tectonics* by a thermally self-balancing planet Earth.

We wrote this as a classical *black and white textbook* (apart from 8 color figures) taking into account our experience and expertise in the topic. The reader will benefit from the presentation of the material as a textbook combined with *DVD movies.* Where a movie symbol is found in the margin of the body text of the book, the reader is able to click on the corresponding movie-file of the DVD. The movie material explains complex *scientific relationships,* demonstrates sophisticated experimental apparatus or *field testing equipment* for stress determination, and allows filing *interviews* with experts in rock stress and its measurements. After listening and watching the digital information, the reader returns to the textbook letters. In the case of getting lost while reading or watching, the reader can go directly to the *Note-Box* at the end of each section. From there, one can work backwards to obtain the full knowledge.

The content presented in this book is based on the many years of research and practical work of the two authors. In writing the book, we have made ample references to key publications in related fields and have tried to bring the reader up-to-date about theory, experiments, field tests and stress data compilations and analysis. In doing this, we may have omitted some references and hope the reader will forgive us. The material of this book was first compiled as lecture notes in 1999 when the first author started to teach a course entitled *Stress Field of the Earth's Crust* at the University of Potsdam for students in geosciences.

Potsdam
May 2009

Arno Zang and Ove Stephansson

Acknowledgements

The writing of this book could not have been possible without using the excellent research environment of the Helmholtz Centre Potsdam, GFZ German Research Centre for Geosciences. We thank the Chairman of the Executive Board, Prof. Dr. Dr. h.c. *Reinhard Hüttl* and the Director of Department 3 Geodynamics and Geomaterials, Prof. Dr. *Onno Oncken,* for agreeing to publish this book and using GFZ resources. We are thankful to the Head of the Sect. 3.2 Geomechanics and Rheology, Prof. Dr. *Georg Dresen* for supporting this work. We profited from the stimulating discussions and the helpful and open atmosphere among the colleagues of Sect. 3.2. In addition, the writing of the book strongly benefited from the comments and questions asked by Geosciences students at the University of Potsdam while the first author was lecturing in the decade 1999 to 2009.

The quality of this manuscript was substantially improved by a review process taking into account comments and suggestions of 16 colleagues and one graduate student. We gratefully acknowledge the highly recognized experts in their field who commented on earlier versions of single chapters of the manuscript, namely Prof. *Frank Roth,* GFZ (Chap. 2); Prof. *Teng-fong Wong,* State University of New York at Stony Brook, USA (Chap. 3); Prof. *John A. Hudson,* Imperial College, London, UK (Chap. 4); Prof. *Robert W. Zimmerman,* Imperial College, UK and Dr. *Erik Rybacki,* GFZ (Chap. 5); Prof. *Hans-Rudolf Wenk,* University of California at Berkeley, USA and Dr. *Christian Scheffzük,* GFZ (Chap. 6); Prof. *Fritz Rummel,* Bochum University and MeSy GmbH, Prof. *François Cornet*, Institute Physique du Globe, Paris, France and PD Dr. *Sebastian Hainzl*, GFZ (Chap. 7); Prof. *Rune Holt* and Dr. *Alexandre Lavrov,* SINTEF Petroleum Research, Trondheim, Norway (Chap. 8); Prof. *Jörg Erzinger;* GFZ and Dr. *Martin Brudy,* Baker Hughes, Houston, Texas, USA (Chap. 9), Dr. *Oliver Heidbach,* GFZ and World Stress Map project (Chaps. 10 and 11). Among these experts we asked one eminent authority to comment on the revised book script as a whole. We are grateful to Prof. *Bezalel C. Haimson,* University of Wisconsin-Madison (USA) for his thorough reading and valuable comments, and for providing the foreword of this book. Last, we thank *Katrin Plenkers,* GFZ and University of Potsdam for her useful comments from a student's point of view.

We very much appreciate the expertise and patience of *Andreas Hendrich* who created and modified most of the viewgraphs. Some illustrations including the draft of the book cover was made available by *Manuela Dziggel. Stephanie Teteris* helped in finalizing the viewgraphs of Chap. 5. We gratefully acknowledge the excellent work of these three draftsmen located at GFZ.

We would like to thank Dr. *Diedrich Fritzsche* (Alfred-Wegener Institut für Meeres- und Polarforschung Potsdam) for his help in obtaining the copyright permission of the portrait of *Alfred Wegener* in Fig. 1.7. Dr. Fritzsche helped us in locating *Käthe Schönharting* (second daughter of Alfred Wegener with his wife *Else Wegener;* born *Köppen*) and Käthe Schönharting's son Dr. *Günther Schönharting* (grandson of Alfred Wegener), who own the portrait of Alfred Wegener sketched by the painter Achton Friis (1871–1939). Friis accompanied Alfred Wegener during his first Greenland (Denmark) expedition in 1907. Günther Schönharting provided us the address of the Arctic Institute at Copenhagen, Denmark, from where we received a high-resolution scan of the original drawing and the permission to reproduce the portrait.

Ulrich Harms and *Thomas Wöhrl,* International Deep Drilling Project ICDP, located at GFZ provided two photographs from the German Continental Deep Drilling Program, one cross-section in Fig. 9.1b and the aerial photograph of the drill site in Fig. 9.2.

We apologize for the quality of some of the movies. All 17 movies are self-made documentaries, and therefore the viewer cannot expect Hollywood quality pictures. Raw movie material was processed with the software package MAGIX Video deluxe 2008 Premium by *Wolfgang Kunert* (wksyspro.de). The DVD movie material was examined in terms of content by the head of GFZ Sect. 3.2 Prof Dr. Georg Dresen and in terms of copyright queries by GFZ public relation manager *Franz Ossing.* The help of Mr. Otto Grabe in video related queries and using GFZ equipment is grateful acknowledged.

Movie M1 contains the introduction of the first author to Stresses in a Body (Book Chap. 1). This movie was recorded at www.wksyspro.de. M2 contains the introduction of the second author to rock stress as a key parameter to Solid Earth Sciences (Chap. 2). This movie was recorded at GFZ. Movie M3.1 shows a laboratory uniaxial compression test and the result of a laboratory triaxial compression test. The licence to use 27 seconds of ZDF Enterprises GmbH material is grateful acknowledged. M3.2 shows laboratory equipment used to perform triaxial testing. Both Chap. 3 movies were recorded at GFZ Sect. 3.2 facilities using an MTS (Mechanical Test Systems GmbH) 815 loading frame. Special thanks go to Prof. Dr. *Georg Dresen,* our colleagues Dipl.-Ing. *Andreas Reinicke,* Dr. *Sergei Stanchits* and our former colleague Dr. *Christian Wagner.* Movie M4 captures a discussion of both authors about the Stress Terminology shown in diagram 4.1 (Chap. 4). In movie M5.1 an introduction to the high-pressure-high-temperature Paterson Laboratory at GFZ Sect. 3.2 facility is given. In movie M5.2 our colleague Dr. *Erik Rybacki* is interviewed about the stress depth variation in terms of extrapolating laboratory creep data to nature. Both Chap. 5 movies were recorded at GFZ Sect. 3.2 facilities. Special thanks go to Prof. Dr. *Georg Dresen* and our colleagues Dipl.-Ing. *Michael*

Naumann and Dr. *Erik Rybacki.* The two movies M6.1 Overcoring Borre Probe and M6.2 Overcoring Data Analysis were recorded at Vattenfall Power Consultant AB, Lulea, Sweden (Chap. 6). We are grateful to MSc Civil Eng. *Ulf Lindfors,* Dr. *Daniel Ask* and Mrs. *Sofi Berg.* The movie M7.0 Interview with Prof. *Bezalel C. Haimson* (Chap. 7) was recorded at GFZ Potsdam during the World Stress Map Conference in October 2008. Movie M7.1 is split into Conventional Hydraulic Fracturing and the latest development of the Quadruple Packer system at Vattenfall Power Consultant AB, Lulea, Sweden. Movies M7.2 Sleeve Fracturing and M7.3 Hydraulic Test on Pre-existing Fractures (HTPF) all are recorded at Vattenfall Power Consultant AB, Lulea, Sweden. We are grateful to Dr. *Daniel Ask* for the final interview in movie M7.4 related to Chap. 7 topics. Movie M8 Anelastic Strain Recovery was recorded at DCM GmbH Walluf, Germany (Chap. 8). We are grateful to geologist Dr. *Klaus E. Wolter.* The KTB drill core 202A1c used for demonstration purpose in this movie was kindly provided by *Ulrich Harms* and *Thomas Wöhrl* both located at ICDP, GFZ. Movie M9.1 Interview with Prof. *Mark D. Zoback* (Chap. 9) was recorded at GFZ in January 2008. We apologize for digital zooming and low resolution pictures, but the student will find all important information in the spoken words. Finally, Movie M9.2 Interview with Dr. *Jonny Sjöberg* (Chap. 9) was recorded at Vattenfall Power Consultant AB, Lulea, Sweden.

The *World Stress Map* (WSM) release 2008 data set was kindly provided by Dr. *Oliver Heidbach.* He also contributed to Fig. 11.2 by sending WSM data released 2005 and smoothed stress orientations from Heidbach et al. (2007) publication via electronic mail. We gratefully acknowledge the permission of the *Heidelberg Academy of Sciences and Humanities* to store and distribute the World Stress Map data set on the accompanying DVD. We also thank the WSM Editors for their agreement to do so, namely *O. Heidbach, M. Tingay, J. Reinecker, A. Barth, D. Kurfeß, B. Müller, K. Fuchs and F. Wenzel.* The initial help of Prof. emeritus Karl Fuchs in getting this permission is appreciated.

We thank *Martin H. Trauth* who encouraged the first author to contact Springer about the book manuscript for the very first time at European Geosciences Union 2006 meeting on a flight from Berlin to Vienna. Discussing the book project with Martin at the University of Potsdam was very helpful in design, advertising and short course related queries.

Our final thank goes to the publisher Springer SBM NL. We acknowledge the excellent support by *Petra D. van Steenbergen* (Senior Publishing Editor) and *Hermine Vloemans* (Publishing Assistant Geosciences) while working together on the book manuscript.

Permissions to Reproduce Figures

We would like to thank the following publishers, organizations and individuals for permission to reproduce previously published material:

American Geophysical Union (Figs. 1.3b, 3.7a, 3.8b, 4.3a, 4.6, 5.8a, 5.12a, 7.10, 7.13, 9.1b, 9.6c, 9.7b, 9.13, 9.15–9.18, 10.3, 10.9, 11.2, 11.3, 11.5, 11.7a, b, 11.9a, 11.10–11.12); American Institute of Physics (Fig. 3.9); American Society of Mechanical Engineers (Fig. 8.2b); Arctic Institute, Copenhagen (Fig. 1.7); ASM International® (Fig. 8.10); Prof. Dr. H. Berckhemer (Fig. 1.3a); Blackwell Publishing Ltd. (Figs. 4.9d, 6.7); Cambridge University Press (Figs. 5.8b, 5.9, Tables 3.3, 5.2); CSIRO Publishing and Australian Society of Exploration Geophysicists (Fig. 4.3b); Elsevier (Figs. 3.3, 3.5a, b, 3.6b, 4.5, 4.9c, 4.11, 5.4, 5.7, 5.10, 7.11b, d, f, 8.14, 8.16, 9.7a, 10.7, 11.4); E. Schweizerbart'sche Verlagsbuchhandlung (Fig. 7.6a); Geological Society of America (Fig. 4.4c, Table 4.2); GFZ German Research Centre for Geosciences, Sect. 3.2 (Table 3.3); GFZ German Research Centre for Geosciences, ICDP (Fig. 9.2); Dr. O. Heidbach, GFZ German Research Centre for Geosciences (Cover, DVD World Stress Map database and software CASMI for the World Stress Map database); Japan Agency for Marine-Earth Science and Technology (Fig. 1.8); John Wiley & Sons Ltd (Fig. 11.9b); Dr. B. Müller (Figs. 10.2, 10.4); Nature Publishing Group (Figs. 9.1c, 11.9c); Niedersächsisches Landesamt für Bodenforschung, Hannover (Fig. 9.1a); CRC Press Balkema (Taylor and Francis Group) (Figs. 3.2a, b, 8.13); Posiva Oy, Finland (Figs. 9.8–9.12); Princeton University Press (Fig. 8.4); Prof. Dr.-Ing. A. Peiter (Figs. 6.4, 6.5, 6.9, 6.10); Rudarsko-Geolosko-Naftni Zbornik, Kroatia (Fig. 7.11a, c, e); Prof. Dr. F. Rummel (Fig. 9.4); Society of Petroleum Engineers (Fig. 8.17); Springer (Figs. 3.6a, 5.12c, 6.6, 6.8, 7.14, 8.6–8.8, 8.11, 9.5c, 9.7a, c, 10.la, b, 10.10, Tables 3.la, b, 3.2, 4.2, 5.1); SINTEF, Norway (Fig. 7.7); Swedish Nuclear Fuel and Waste Management Company (Fig. 10.11); Taylor & Francis Group (Figs. 8.15, 11.1); The American Association for the Advancement of Science (Fig. 9.14); Wissenschaftliche Verlagsgesellschaft mbH, Stuttgart (Fig. 11.8).

Contents

DVD-ROM included inside back cover

Chapter 1
Introduction

This book is about the stress field of the Earth's crust and consists of three major parts. Part I is dedicated to the definition and terminology of rock stress (Chaps. 2–4) resulting in simple Earth stress models (Chap. 5). Part II is an overview of the various stress-measuring methods from a physical point of view (Chap. 6) and a rock mechanics point of view (Chaps. 7 and 8). Rock mechanics and rock engineering techniques for determining stress are divided into borehole (Chap. 7) and core-based methods (Chap. 8). In Part III, stress data are interpreted at the local scale (Chap. 9) and at the global scale in terms of plate tectonics (Chap. 11). A generic stress approach relates local to global data through stress-scaling relations and the best estimated rock stress model (Chap. 10).

In the theory of elasticity, the term *mechanical stress* is introduced as the average internal force acting across a cross-sectional area. This descriptive definition of stress is addressed in Sect. 1.1 and introduces stress as a fictitious quantity. In rock mechanics, stress is known as an abstract concept (Chap. 2) originating from the study of strength and failure of solids (Chap. 3). Both approaches (abstract and descriptive) result in two important consequences of rock stress. Firstly, since both force and area are described by vectors, then the stress turns out to be a second-rank tensor quantity (Chap. 2). Secondly, since the area over which averaging is performed is not specified, stress can be regarded at different scales (Chap. 10). In this book, we introduce *rock stress* as force per area resulting in the unit of a pressure ($Nm^{-2}=Pa$).

The way the state of stress varies through space is called the *stress field* which is defined as the distribution of internal forces in a body that balance a given set of external forces. The stress field is visualized by the magnitudes and orientations of stress components (Sect. 2.4). At any point of a deformable body, three stress components (principal stresses (S_1, S_2, S_3)) and three associated stress directions (so-called principal stress axes ($\bar{n}_1, \bar{n}_2, \bar{n}_3$)) can be found, which describe the state of stress at this point completely and unequivocally with respect to a global coordinate system (x, y, z) (Fig. 1.1). Compared to a vector defined by two quantities (magnitude and orientation such as wind velocity) and a scalar quantity defined by a single value (e.g., temperature at a point), the stress field is a more complex quantity requiring six quantities to be determined, see Sect. 2.2.

A. Zang, O. Stephansson, *Stress Field of the Earth's Crust*,
DOI 10.1007/978-1-4020-8444-7_1, © Springer Science+Business Media B.V. 2010

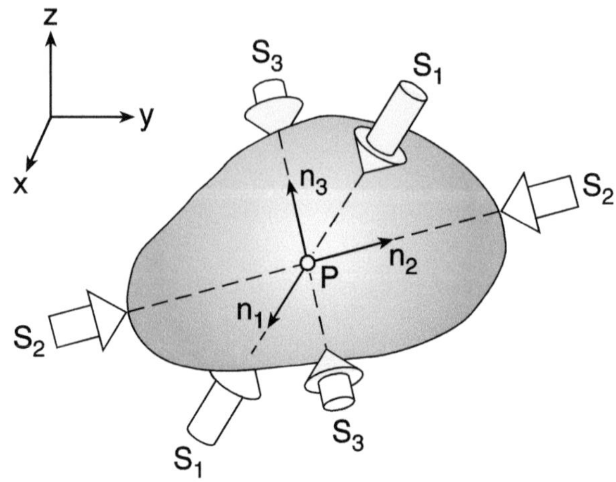

Fig. 1.1 State of stress at an arbitrary point *P* within a deformable body. Three principal stress components and their directions are needed to unequivocally define mechanical stress at a point

In general, stresses within the Earth's crust are typically defined in terms of principal stresses. Intuitively, we understand that one of the principal stresses results from the Earth's gravity field (Sect. 4.1) and therefore must be oriented vertically. This principal component of the crustal stress is called *vertical stress* S_V. The two other principal components required to unequivocally define the crustal stress field are the *minimum horizontal stress* S_h and the *maximum horizontal stress* S_H. Any state of stress in which all principal stresses are equal (Fig. 1.1, $S_1 = S_2 = S_3$) is referred to in mechanics as a *hydrostatic stress* state (Chap. 2). In the rock mechanics context, however, to avoid confusion with stresses caused by pore pressure (Sect. 5.4), the stress state governed by the overlying rock density and equal in all directions ($S_V = S_H = S_h$) is referred to as *lithostatic stress* (Chap. 5).

The assumption of a lithostatic stress state as we penetrate deeper into the Earth's crust is known as *Heim's rule*. The Swiss geologist Albert Heim (1849–1937) postulated that the rock under conditions of constant, persistent loads at depth will compensate for internal differential stresses (differences in principal stresses) by creep processes, given sufficient (geological) time. In Fig. 1.2, the rule of Heim (1878) is demonstrated by a shift in principal stress relations from a scenario where

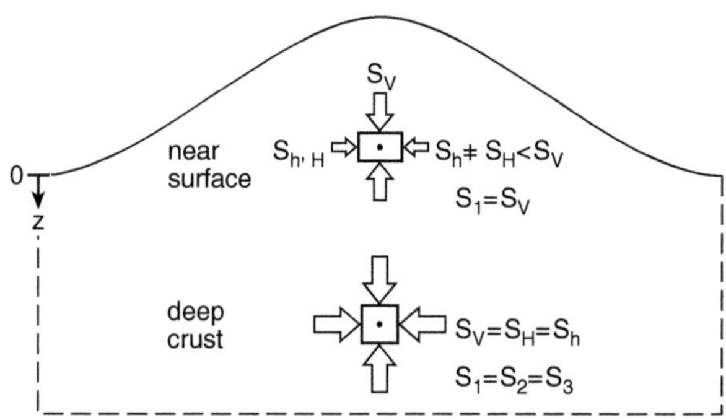

Fig. 1.2 Heim's idea of stress at depth in the deeper Earth's crust (1878)

the vertical stress is maximum and the two other horizontal principal stresses are smaller in magnitude ($S_V > S_H \neq S_h$) near the surface of the Earth to a lithostatic state of stress where all principal stresses are equal in magnitude ($S_V = S_H = S_h$) at greater depth in the Earth's crust.

The last term used in the book title to be defined is the Earth's crust. The *Earth's crust* is defined by the rock mass between the Earth's surface and a demarcation called the *Mohorovičić* discontinuity (*Moho*) named after the Croatian seismologist Andrija Mohorovičić, who discovered it in 1909. The Moho separates the Earth's crust from the Earth's mantle. The term crust refers to that region of the Earth above the Moho (Fig. 1.3a, $z < z_M$), which represents about 0.4% of the Earth's mass. Geophysically, the Moho is identified by the rise of the compressional wave velocity from approximately 7 km/s in the lower Earth's crust to approximately 8 km/s in the

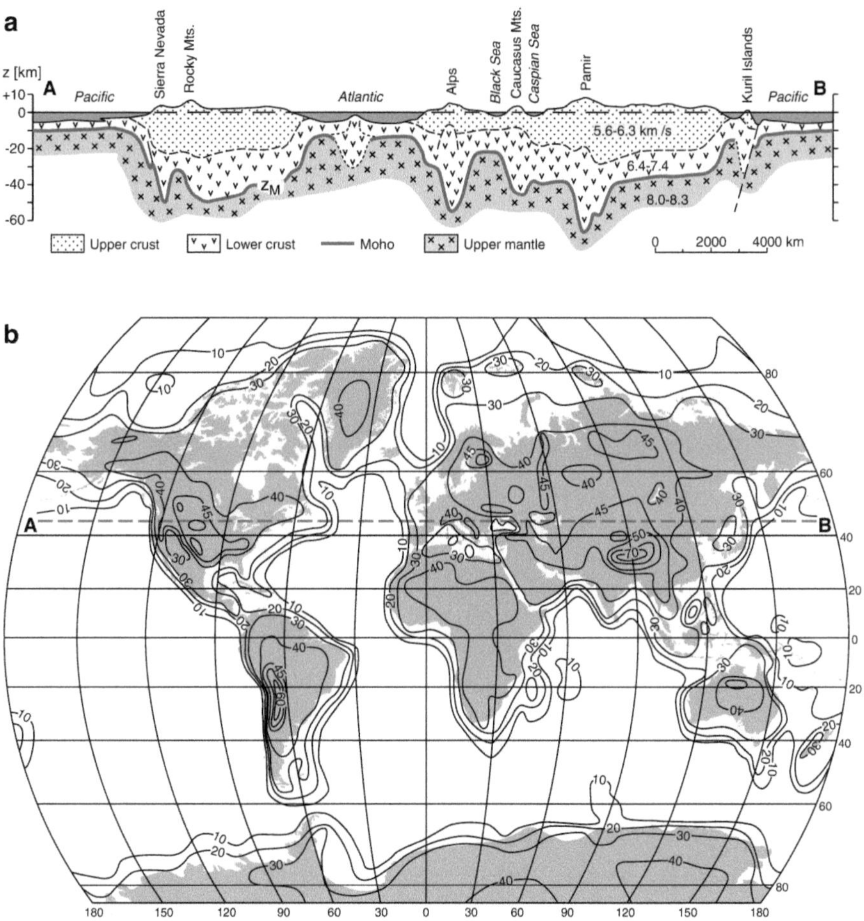

Fig. 1.3 **a** Depth section of the Earth's crust at latitude 45° North as derived from seismic data; after Berckhemer (1968). **b** Contour plot of crustal thickness based on seismic refraction data published 1948–1995; after Mooney et al. (1998)

upper Earth's mantle (Fig. 1.3a, depth z_M). The thickness of the continental crust ranges from about 25 to 70 km beneath mountain roots (Fig. 1.3b) corresponding to about 1% of the Earth's radius. The upper crust consists of quartz-rich crystalline rocks, the lower crust of intermediate to basic plutonic rocks. Physicochemically, the Moho represents a composition boundary within a tectonic plate where the underlying ultramafic, peridotitic upper mantle is stronger than rocks of the lower crust. The oceanic crust is about 5 to 15 km thick (Fig. 1.3b) corresponding to about 1‰ of the Earth's radius and is mainly composed of basaltic rocks. In this book, we limit ourselves to stress-determination techniques applied to the Earth's crust (Chaps. 7 and 8) and stress data obtained from the Earth's crust (Chaps. 9 and 10), but have to take into consideration plate tectonic aspects when discussing the physical origin of the crustal stress field (Chap. 11).

Why is the stress field of the Earth's crust of interest? (See also Sect. 1.2). The knowledge of the present-day stress field of the Earth's crust is of local interest, when (1) civil engineers are planning and constructing underground excavations and carrying out safety analyses of tunnels, (2) mining engineers are planning and exploiting underground mines, and (3) petroleum engineers are planning and extracting oil and gas from a petroleum field (reservoir). It is of global interest when (4) tectonic deformations near faults and related earthquake hazard needs to be quantified, (5) mantle flow models and (6) plate movements in a global reference frame need to be calibrated. Among the more notable reviews of stress measurements in the Earth's crust are those of Hast (1969), Herget (1974), Haimson (1975), McGarr & Gay (1978), Rummel (1979), Zoback & Zoback (1980), Rummel (1986), Gough & Gough (1987), Hickman (1991), Zoback (1992). In monographs many of these data sets have been given, e.g., by Engelder (1993), Amadei & Stephansson (1997), Rummel (2005), and Zoback (2007).

Information stored in current stress data bases, e.g., the World Stress Map (Fuchs and Müller (2001), Zoback & Zoback (2002a, b), Sperner et al. (2003), Wenzel et al. (2004), Reinecker et al. (2006), Heidbach et al. (2007) show that none of these simple hypotheses about the stress field of the Earth's crust are valid in general. Instead, near the Earth's surface one or both of the horizontal principal stress components are larger in magnitude compared to the vertical stress, S_V (Sect. 10.2). It has been argued that stress measurement results recorded close to the Earth's surface (less than 500 m) are not necessarily representative of the stresses prevailing in the deeper Earth's crust (Chap. 10), and that stress data recorded in the deeper crust (less than 10 km) are not necessarily representative of the stresses prevailing at depth in a tectonic plate (less than 100 km), see Chap. 11.

In zero-order approximation, stress magnitudes in the Earth's crust show a linear increase with depth (Sect. 10.1). One method to express stress as a function of depth is to plot measured principal stress components (e.g., $S_i(z)$ with $i=1,2,3$, Haimson 1978). Another widespread method is to use the difference in principal stress components (e.g., S_1–S_3). McGarr (1980) concludes that on average the maximum shear stresses, (S_1–S_3) have a gradient of 3.8 MPa/km for "soft" rocks and 6.6 MPa/km for "hard" rocks in the upper 5 km of the crust, with no suggestion of diminishing gradient at deeper levels. Thus, a linear extrapolation to crustal depth of 15 km would result in a value of maximum shear stress of about 100 MPa in hard crystalline rock (6.6 MPa/km × 15 km = 99 MPa).

Recent regional and global stress maps based on results from borehole methods (e.g., breakouts, hydraulic fracturing, drilling induced fractures) and core-related measurements (e.g., overcoring), as well as field indicators (e.g., earthquake focal mechanisms, geologic faults) show a consistent orientation of crustal stresses with respect to the movement of tectonic plates (Chaps. 10 and 11). The intraplate maximum horizontal stress directions are aligned parallel to ridge-push torque and absolute velocity azimuths for stable North America, Western Europe and South America (Richardson 1992). In plate spreading regions tensile stresses are expected (Fig. 1.4a), whereas compressive stresses should dominate in convergent regions of plates (Fig. 1.4b). Even if we are not aware of any location where tensile stresses have been measured, including near-plate spreading such as in Iceland (Haimson & Voight 1977, Haimson & Rummel 1982), the schematic sketch in Fig. 1.4a holds for large-scale intraplate stress patterns. Explanation for this is the separation of rock mass under tension into blocks of rock under compression. Due to the self-gravitation, large blocks of rocks near the spreading area generate and often contain frozen-in compressive stresses. At a local scale, stress orientations near mountain ranges, postglacial uplift regions and salt domes may be perturbed significantly (Sect. 4.4, near-field and far-field stresses).

The schematic plate margins shown in Fig. 1.4 are realized in nature at the Mid-Atlantic Ridge of Iceland (Fig. 1.5a) and at the Andes subduction zone of South America (Fig. 1.5b). A detailed present-day, kinematic model of plate-boundary deformation in southwest Iceland is discussed by Arnadottir et al. (2006). Plate velocity vectors are obtained from global positioning systems, GPS, recorded on the surface (Kreemer et al. 2003, Feissel-Vernier et al. 2007). Often, the orientation of the maximum horizontal stress coincides with the velocity vectors of the plate-motion model used. Direction and magnitudes of stresses, however, can be perturbed by irregularities like weak plate boundaries and fault zones. In the southern part of South America, the orientation of maximum horizontal stress, S_H west of and close to the plate boundary align parallel to the plate margin, while orientations of

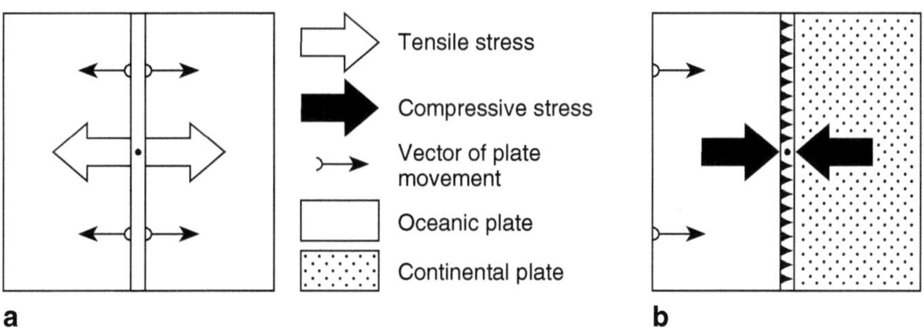

Fig. 1.4 Schematic sketch of stress field and plate movement **a** in a divergent plate tectonic setting, and **b** in a convergent plate tectonic setting

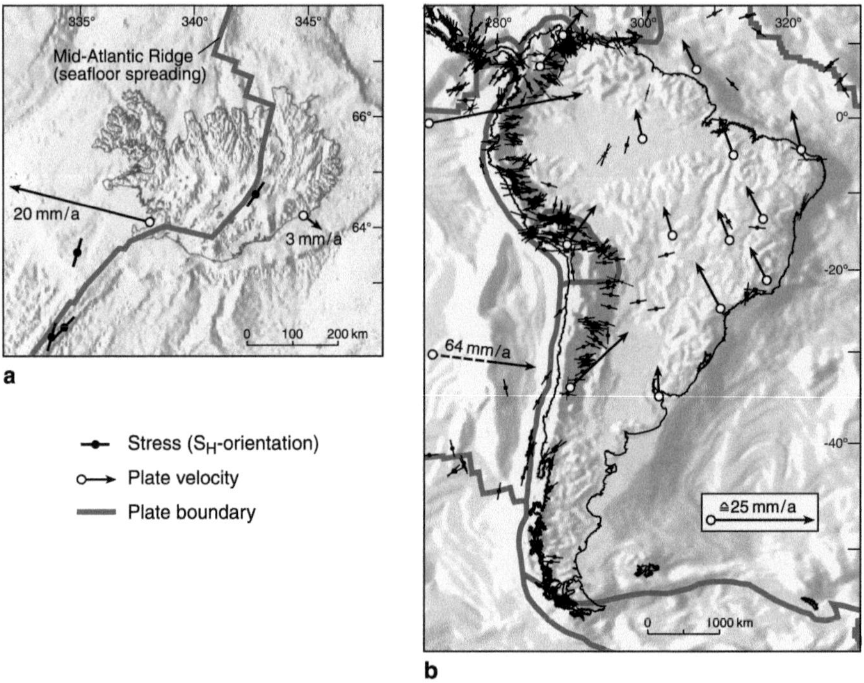

Fig. 1.5 Measured plate movement vectors and stress orientations for **a** Iceland, and **b** South America. Stress data are taken from the World Stress Map. Plate velocity vectors in **a** are inserted from La Femina et al. (2005). Plate velocity vectors in **b** are inserted 2007 from UNAVCO at Boulder, Colorado

S_H east of the plate boundary in the Andes plateau (Chile) align perpendicular to the plate margin (Fig. 1.5b). Maps of plate velocity vectors (Fig. 1.5b, open circles with arrows) can be overprinted by local geologic structures and may not reflect the averaged, overall plate velocity field trajectories at the lower bound of the tectonic plate (Zoback 1992). A generalized stress map together with absolute velocity trajectories can be found in Amadei and Stephansson (1997).

In contrast to the Earth's crust, *lithospheric plates* are about 100 km thick and comprise the crust and uppermost mantle. If the Earth is viewed in purely mechanical terms on a geologic time-scale (Cathles 1975, Brown & Musset 1981), the mechanically strong lithosphere (*rigidity* 5×10^{24} Nm) floats on the mechanically weak asthenosphere (*kinematic viscosity* $4\times10^{19}\,m^2\,s^{-1}$). In the concept of *plate tectonics*, the lithosphere is divided into a number of rigid plates (like ice floes), which are moving over the asthenosphere (like water). Generation of new plate material occurs by seafloor spreading along active midocean ridges (Iceland Fig. 1.5a), so-called divergent plate boundaries (Fig. 1.4a). Seafloor spreading must be balanced by consumption of plates in subduction zones (Andes, Fig. 1.5b), so-called convergent plate boundaries (Fig. 1.4b). Downgoing plates penetrate the lower mantle (viscosity $10^{21}\,m^2\,s^{-1}$) to a depth of about 650 km as manifested by the occurrence of deep earthquakes along the subducted plate. Lithospheric plates are capable of transmitting stresses over large horizontal distances (Fowler 1990). The pattern of intraplate stresses in the crust can be used to assess forces acting on plate boundaries. Stress orientation can be correlated with absolute plate motions (Minster & Jordan 1978, DeMets et al. 1990) to obtain correct directions of individual stress components. Construction of a global reference frame for plate tectonics is based on true polar wander taking into account a hotspot track system and a paleomagnetic absolute reference frame (Torsvik et al. 2008). The maximum speed of true polar wander caused by mantle convection is about 1 degree per 1 million years (Steinberger & Torsvik 2008).

In Chap. 11, the modern concept of plate tectonics based on Alfred Wegener's idea of continental drift is presented. Here, we evaluate the stress field of the Earth's crust as a present-day snap shot of a geodynamical planet Earth by matching observations from intraplate stresses given by the World Stress Map (Sect. 11.2) with calculated stresses induced by mantle flow, crustal heterogeneity and topography (Sect. 11.3).

1.1 Stresses in a Body

M1

To introduce *mechanical stress* as a fictitious term, consider the sketch in Fig. 1.6. Imagine an elastic body in equilibrium with a *fictitious slicing plane* through the body. The plane is entirely internal to the body (Fig. 1.6a, plane *A*). In Fig. 1.6b the same body is shown with a *realistic slicing plane* separating the body into parts *1* and *2*. To compensate for the new equilibrium, the realistic slicing plane splits into two free surfaces by finite body displacements (Fig. 1.6b, planes A^+ *and* A^-). In order to prevent re-equilibrium deformations of the realistic slicing plane, opposing forces must be applied at the fictitious slicing plane. Hence, it follows that before

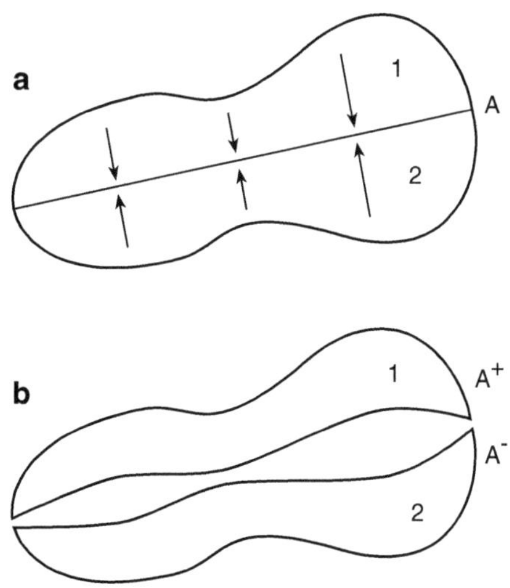

Fig. 1.6 Elastic body in equilibrium **a** with fictitious slicing plane *A*, and **b** with realistic slicing plane forming two stress-free surfaces A^+ and A^- by re-equilibrium deformation of the two newly created body parts. Mechanical stress is the result of internal body forces (*arrows*) per slicing area (*straight line*, *A*)

cutting the body open along the fictitious plane, internal forces must have prevailed. As we see in Chap. 2, these forces obey *Newton's law of action and reaction*. They counterbalance each other along the fictitious plane (Fig. 1.6a, arrows), since both resulting forces and resulting moments are equal to zero. Intuitively, a body in the state of equilibrium seems to have no internal forces acting inside it. In reality, however, internal stresses are always there. This fact is of great practical importance in industry, when stress-free materials or materials with controlled frozen-in stresses are manufactured (Sect. 4.3, residual stresses).

1.2 Importance of Rock Stress

Knowledge of the state of stress in the Earth's crust is very important when dealing with rocks in civil, mining, and energy (e.g., petroleum, geothermal) engineering, as well as in geology and geophysics. In civil and mining engineering, rock stresses and stress changes have to be known as precisely as possible when underground openings like tunnels, mines, shafts or caverns are designed and excavated into the pre-stressed rock mass. Stress concentrations at openings may be high enough to reach the limit of rock strength resulting in rock failure, bursting or collapse (worst-case scenario). In general, stress generated stability problems increase with depth in the Earth's crust. A positive aspect of high lithostatic stress (e.g. in nuclear waste deposites), however, is that the rock mass under consideration is tighter, implying less permeability and shorter pathways of contaminant transport.

One primary goal in underground design is to find the easiest and safest way through the rock mass by excavating through low stressed, high strength rock in order to minimize stress concentration problems and related failure. In general, large underground caverns are oriented with their long axes parallel to the maximum horizontal in-situ stress S_H (Mimaki & Matsuo 1986) avoiding the re-opening of fractures crossing the tunnel and generating the optimum stress distribution around the walls of the cavern.

The acceptance of rock mechanics methods and techniques in underground mining is due to the use of rock mass classification systems (Barton et al. 1974, Bieniawski 1989), reliable methods for rock stress measurements (Amadei & Stephansson 1997), and rapid advancement in numerical modelling (Jing 2003). The state-of-the-art in rock engineering for underground mining is presented by Brady and Brown (2004). Virgin stresses in the Earth's crust generate the loading system for underground excavations. Stress redistribution due to mining causes deformation and sometimes generates failure of the rock mass. Mining methods applied, the depth and geometry of the mine, as well as the rate and the volume of rock extracted (apart from geology and tectonic setting), determine the redistribution of stresses, the occurrence of rock bursts and mine seismicity (McGarr & Wiebols 1977). An *a priori* knowledge of pre-existing zones of weakness in rock masses (e.g., excavated damage zone, Sect. 4.4) and the virgin stress state can assist mining engineers in identifying areas where mining-induced failure and seismic events are likely to occur (Wong 1993). Some mining methods, like block caving, make use of the virgin rock stress for fragmentation of the ore prior to transportation.

Rock stress measurements are often performed to provide input to numerical modelling of mine design and give information about deformability, strength and rock support of mine openings. A common practice is to locate measuring spots remote from the mining stopes. In this way, the measured stress field is unaffected by the mining activity. In the final stress compilation, however, stress data from stopes together with stress measurements in undisturbed areas are used as input for the best stress estimate by numerical modelling for the design and stability predictions of mining (Sect. 10.4).

Knowledge of virgin stresses is also important for the storage of nuclear waste in rock (Sect. 9.2). In-situ stress should be measured at the site, before, during and after the construction of a repository (Kim 1992). One of the best documented case studies on rock-stress measurements was conducted at the Underground Research Laboratory (URL) for radioactive waste in the Lac du Bonnet granite batholith in the province of Manitoba on the western edge of the Canadian Shield. This research facility was used by Atomic Energy of Canada Limited to investigate the disposal of spent nuclear fuel in plutonic rock. The field experiments conducted at URL answered many questions with regard to stress in hard rock (Martin & Simmons 1993).

Rock-stress measurements in deep water-filled boreholes drilled from the surface and measurements from tunnels in the Äspö Hard Rock Laboratory in Sweden are important components in the Swedish and Finnish programs for deep geological final disposal of spent nuclear fuel. New measurement techniques of overcoring in water-filled boreholes and analysis of stresses of anisotropic rocks have been developed (Sjöberg et al. 2003, Hakala & Sjöberg 2006). Gunzburger et al. (2006) conducted deep borehole stress measurements with hydraulic methods in the Jurrassic shales of the Paris Basin for the French underground nuclear waste research laboratory at Bure.

Knowledge of the virgin stress field is also important for the fracturing of formations of oil and gas fields (Sect. 7.2) to stimulate production (Teufel 1986). Proper reservoir management requires that stresses and stress changes are known before and during reservoir depletion. In the North Sea Ekofisk field, Teufel et al. (1991) reported a minimum of 20 MPa reduction in reservoir pore pressure as a result of a follow-up study of 20 years of petroleum production (1 MPa pore pressure reduction per production year). As the pore pressure in the chalk formation decreases, the overburden load has to be transferred into the weak chalk matrix material, which in turn may cause reservoir compaction and seafloor subsidence. Such secondary effects sometimes require oil and gas platforms to be jacked up in order to prevent breaking by sea waves.

In order to select the direction of a wellbore and reduce borehole deviations and possible breakouts (Sect. 7.3) or collapse, a clear understanding of in-situ stresses is necessary. Borehole stability is governed by stress concentrations along borehole walls. Overbreaks due to mobilization of the rock strength may create sever borehole stability problems (Maury 1987, 1991).

To understand the stress field of the Earth's crust, the origin of lithospheric stresses has to be understood. Stresses are caused by the relative motion between mantle flow and plate motion. In this way, the crustal stress field and its variation is intimately related to the driving mechanism of plate tectonics (Chap. 11). Plate tectonics and mantle convection are one system, because oceanic plates are the cold upper thermal boundary layer of convection (Tackley 2000). *Alfred Wegener* (1880–1930) launched his hypothesis of drifting continents in 1915. He died in 1930 during an expedition in Greenland when he tried to confirm his theory. A portrait of Alfred Wegener sketched in 1907 by the painter Achton Friis (1871–1939) during his first Greenland expedition lead by Mylius Erichsen (Danmark Expedition 1906–1908) is shown in Fig. 1.7. After half a century of quiescence, the first sketch of relative motion of tectonic plates and complementary flow in the Earth's mantle was published by Isacks et al. (1968). After accepting Wegener's new paradigm, the plate tectonic picture was further refined by Forsyth and Uyeda (1975) by classifying the forces involved. Finally, Zoback et al. (1989) polished our present-day image of sources of tectonic stress in the Earth's crust and mantle. According to this present-day *plate tectonic stress model*, one has to understand the mechanisms of (1) plate genesis ("birth of plate") by *sea-floor spreading* involving forces like *ridge push* and *transform fault* friction, (2) plate movement indicated by geodetic velocity vectors involving forces like basal drag or drive at the base of the lithosphere ("life time of plate"), and (3) plate collision ("burial of plate") involving collision forces, slab pull

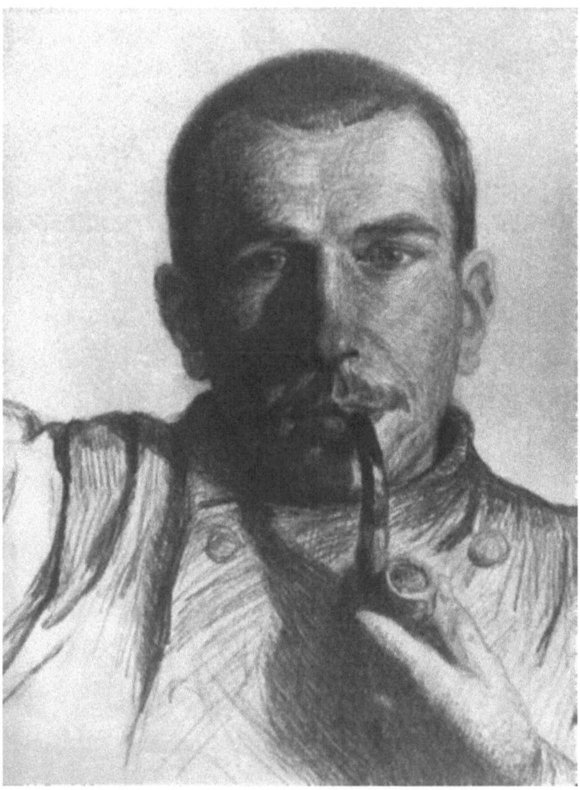

Fig. 1.7 Pencil drawing portrait of Alfred Wegener sketched 1907 by the painter Achton Friis (1871–1939) who accompanied the inventor of the continental drift during his first Greenland expedition lead by Mylius Erichsen. © Arktic Institute, Copenhagen, Denmark

and mantle resistance forces. Analysis of 15,969 stress data entries into the World Stress Map has revealed major stress patterns and stress regimes in the Earth's crust (Reinecker et al. 2006), which can be interpreted in terms of first-order stress patterns (plate stress), second-order stress patterns (isostacy of continental roots) and third-order stress patterns (faults) as indicated by Heidbach et al. (2007).

Compared to the thickness of the Earth's crust, however, the human efforts in drilling the Earth's crust are very limited (Chap. 9). Within the International Continental Deep Drilling Program (ICDP) and the Ocean Drilling Program (OPD), only the outermost part of the Earth's crust is penetrated. The depth record in ultra-deep continental drilling is held by the Russian well bore at the Kola peninsula with a final depth of 12.3 km, whereas one of the deepest open holes (meaning partially cased but not refilled) is located at Windischeschenbach, Germany (Sect. 9.1). At the KTB drilling site with a final depth of 9.1 km, stress measurements were conducted using hydraulic methods until the bottom of the hole, which is one of the deepest hydraulic measurements ever made (Chap. 7). The deepest hole in the oceanic crust drilled from the scientific research ship JOIDES (Joint Oceanographic Institutions for Deep Earth Sampling) Resolution (hole 504B), is located in the Central Pacific close to the Costa Rica shoreline and reached a final depth of 2.1 km. In 2005, the riser drilling ship "*Chikyu*", meaning "Earth", was built in Japan (Fig. 1.8). In 2006,

Fig. 1.8 Chikyu, the Japanese deep-sea drilling ship: length 210 m, heights 130 m, gross tonnage 57,087 t, living space for 150 staff and scientists, as well as helicopter deck. Hopefully this vessel will refine our understanding of the state of stress in the oceanic crust

57,087 t of steel and a 515 million US-Dollar expenditure started off-shore drilling of the oceanic crust in a maximum water depth of 2.5 km with an ultimate length of drill string of 10 km. The unique *riser drilling technique* on board with a double pipe system allows drilled mud to circulate and stabilize the borehole while drilling. One aim of Chikyu is to collect in-situ core material of the Earth's mantle for the very first time (apart from mantle xenoliths). A so-called *Mohole* has to be drilled, a hole which penetrates the oceanic crust into the upper mantle below the Moho.

1.3 History of Interest in Rock Stress

Rock mechanics engineers, structural geologists and geophysicists have to be well acquainted with the basics of rock stress measurements (Chaps. 6–8) and rock stress data (Chaps. 9–11). The need to understand the state of stress in the Earth's crust has been recognized for a long time and techniques to determine stresses have been proposed since the early 1930s. The growing interest in the stress field of the Earth's crust and stress-measurement methods is reflected in the number of meetings dealing with the subject. In Table 1.1, five decades of rock-stress meetings are summarized indicating the year, title, location and type of meeting, as well as the editors of the proceedings.

Table 1.1 History of rock stress meetings

Year	Title and location	Type	Editor and proceedings
1963	State of Stress in the Earth's Crust, Santa Monica, CA	C	WR Judd, Elsevier Publishing Company, New York, 1964
1966	In-situ Stress, Residual Stress, Lisbon, Portugal	CS	1st Congress of the International Society of Rock Mechanics (ISRM), 1966
1969	Determination of Stresses in Rock Masses, Lisbon, Portugal	S	Laboratorio Nacional de Engenharia Civil (LNEC), Lisbon, 1971
1976	Investigation of Stress in Rock, Sydney, Australia	S	The Institution of Engineers, Australia, 1976
1981	Hydraulic Fracturing Stress Measurements, Monterey, CA	WS	MD Zoback & B Haimson, US National Commission on Rock Mechanics, National Academic Press, 1983
1986	Rock Stress and Rock Stress Measurements, Stockholm, Sweden	σ_1	O Stephansson, Centek Publishers, Lulea, Sweden, 1986
1988	Hydraulic Fracturing Stress Measurements, Minneapolis, Minnesota	WS	B Haimson, Pergamon Press, International Journal of Rock Mechanics and Mining Sciences & Geomechanic Abstracts 26(6), 1989
1990	Stresses in Underground Structures, Ottawa, Canada	C	Canada Center for Mineral and Energy Technology (CANMET), Ottawa, Canada, 1990

Table 1.1 (continued)

Year	Title and location	Type	Editor and proceedings
1991	Stresses in the Earth's Crust, Aachen, Germany	WS	O Stephansson, Proceedings of 7th ISRM Congress, Balkema, Rotterdam, 1993
1995	Rock Stress Measurements at Great Depth, Tokyo, Japan	WS	Matsuki et al., Proceedings of 8th ISRM Congress, Balkema, Rotterdam, 1997
1997	Rock Stress, Kumamoto, Japan	σ_2	K Sugawara & Y Obara, Proceedings of the International Symposium on Rock Stress, AA Balkema, Rotterdam, 1997
1998	Earth Stress and Industry, Heidelberg, Germany	WSM	MD Zoback & K Heffers, Proceedings of the 1st EuroConference (1st World Stress Map Conference)
1999	Deformation and Stress in the Earth's Crust, Äspö, Sweden	WSM	T Batchelor & J Kakkuri, Proceedings 2nd EuroConference (2nd World Stress Map Conference)
2003	Rock Stress, Kumamoto, Japan	σ_3	K Sugawara, Y Obara & A Sato, Proceedings of the International Symposium on Rock Stress, Swets & Zeitlinger, Lisse, The Netherlands, 2003
2006	In-situ Rock Stress, Trondheim, Norway	S	M Lu, CC Li, H Kjorholt & H Dahle, Proceedings of the International Symposium on In-situ Rock Stress, Taylor & Francis Group plc, London, UK
2008	Frontiers of Stress Research, GFZ German Research Centre for Geosciences, Potsdam, Germany	WSM	O Heidbach, M Tingay, F Wenzel, Proceedings 3rd World Stress Map Conference, Tectonophysics 2009

C conference, *CS* conference session, σ_i sigma-conference, *S* symposium, *WS* workshop, *WSM* world stress map conference

Part I
Definition and Terminology

Chapter 2
Stress Definition

"Stress tensor" is a tautology since the term tensor (latin tensio) means stress. The pleonasm, however, became naturalized in books dealing with mechanical stress (Hahn 1985, p. 20, fn. 1). Referring to other second-rank tensors in physics like the inertial tensor, this tautology disappears

This chapter presents the fundamental concept of stress as it is defined from a mathematical, physical and continuum mechanics point of view. The stress tensor defining the state of stress at a point is introduced using the continuum concept of a stress vector (traction) defining the state of stress on a plane (Sect. 2.1). Principal stresses and their orientations are deduced from solving the eigenvalue problem (Sect. 2.2). The Mohr circle of stress is a way of visualizing normal and shear stress components for traction vectors associated with all possible planes through one point (Sect. 2.3). Since elastic stress is a fictitious term, the display of stress involves some mathematical gimmicks (Sect. 2.4).

2.1 Stress Tensor

In this section, mechanical stress is quantified mathematically as a second-order tensor and physically by its tensor invariants. In analogy to continuum mechanics (Fung 1965; Timoshenko and Goodier 1970; Hahn 1985), consider a deform-
12 able body subjected to some arbitrary sets of loads in equilibrium (Fig. 2.1). At any given point $P(\bar{x}) = P(x_1, x_2, x_3)$ within this body, we imagine a plane A slicing through the body at an angle with respect to the Cartesian coordinate system with unit vectors $(\bar{e}_1, \bar{e}_2, \bar{e}_3)$. The fictitious slicing plane (Sect. 1.1) divides the body into volumes V_1 and V_2, and has a normal $\bar{n} = (n_1, n_2, n_3)$ which points towards V_1. The action that V_1 exerts on V_2 is denoted by a resultant force $\bar{F} = (F_1, F_2, F_3)$. The *traction vector* $\bar{\sigma}$ is defined as the ratio of the resultant force $\bar{F}$ to the surface area A (Fig. 2.1). In order to define the traction that acts over a specific point $P(\bar{x})$ in the body, the area A is now allowed to contract to a point ($dA \to 0$), so that the magnitude A goes to zero.

A. Zang, O. Stephansson, *Stress Field of the Earth's Crust*,
DOI 10.1007/978-1-4020-8444-7_2, © Springer Science+Business Media B.V. 2010

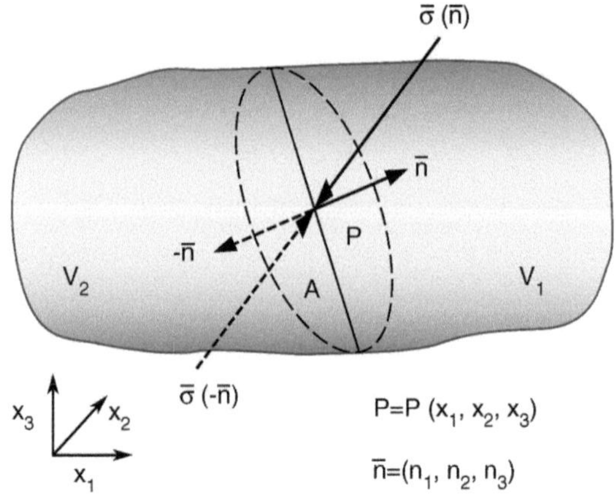

Fig. 2.1 Traction vector $\bar{\sigma}$ acting on a hypothetical (fictitious) slicing plane A with surface normal $\bar{n}$ within a deformable body

$$\bar{\sigma}(P(\bar{x}), \bar{n}) = \lim_{\Delta A \to 0} \frac{d\overline{F}}{dA} \tag{2.1}$$

In general, the traction vector $\bar{\sigma}$ can vary from point to point, and is therefore a function of the location of the point $P(\bar{x})$. However, at any given point, the traction will also, in general, be different on different planes that pass through the point. Therefore, $\bar{\sigma}$ will also be a function of $\bar{n}$, the outward unit normal vector of the slicing plane. In summary, $\bar{\sigma}$ is a function of two vectors, the position vector $\bar{x}$ and the normal vector of the slicing plane $\bar{n}$. In 1823, the French mathematician *Augustin Baron Cauchy* (1789–1857) introduced the concept of stress by eliminating the difficulty that $\bar{\sigma}$ is a function of two vectors, $\bar{\sigma}(\bar{x}, \bar{n})$ at the price that stress became a second-order tensor (Jaeger et al. 2007).

We have three remarks about Eq. (2.1). First, Eq. (2.1) is an empirical formula, i.e. is confirmed by experimental findings. Second, there are obvious practical limitations in reducing the size of a small area to zero, but it is important that, formally, *the stress* is defined in this way as a *point property*. Third, the magnitude of the total traction vector is

$$|\bar{\sigma}(P(\bar{x}), \bar{n})| = \frac{dF}{dA}. \tag{2.2}$$

To uniquely identify stress as a second-order tensor, Cauchy verified two laws. *Cauchy's first law* is visualized in Fig. 2.1 and reads

$$\bar{\sigma}(-\bar{n}) = -\bar{\sigma}(\bar{n}). \tag{2.3}$$

Equation (2.3) is a version of Newton's third law "*actio* = *reactio*" we know from Sect. 1.1. If material to the right of the slicing plane (Fig. 2.1, volume V_1) exerts a traction $\bar{\sigma}$ on the material to the left (Fig. 2.1, volume V_2), then the material to the left will exert a traction $-\bar{\sigma}$ on the material to the right. The Cartesian component of the traction vector in any given direction is considered to be positive if the inner product (dot product) is negative. This is called the *rock mechanics sign convention* (compression positive), and is inconsistent with most areas of mechanics where tension positive convention is used.

Cauchy's second law states that all possible traction vectors at a point (infinite number) corresponding to all possible slicing planes passing through that point (infinite number), can be found from the knowledge of traction vector on three mutually orthogonal planes in 3D. To derive this relationship for the traction on an arbitrary plane, Cauchy introduced an infinitesimal tetrahedron (Fig. 2.2). In the *Cauchy tetrahedron*, the arbitrary slicing plane dA is chosen as a small inclined triangle close to the point $P(\bar{x}) = P(0,0,0)$ at which the state of stress needs to be known. Cauchy's second law can be derived from balancing forces at the tetrahedron (Fig. 2.2). Three faces of the tetrahedron have outward unit normal vectors that coincide with the negative Cartesian coordinate directions $(-\bar{e}_1 = (-1,0,0)$, $-\bar{e}_2 = (0,-1,0)$, $-\bar{e}_3 = (0,0,-1))$. The inclined face of the tetrahedron has an outward unit normal vector of

$$\bar{n} = (n_1, n_2, n_3) = \cos(\bar{n}, \bar{e}_i). \tag{2.4}$$

The components of the vector $\bar{n}$ are given by the direction cosines that the outward unit normal vector of the fourth face makes with the three Cartesian coordinate axes. As the length of any unit vector is unity, $n_1^2 + n_2^2 + n_3^2 = 1$ applies. The area of the face with unit vector $\bar{n}$ is taken to be dA. The areas of the three other faces with outward unit normal vectors $\bar{n} = -\bar{e}_i$ equal

$$dA_i = n_i dA. \tag{2.5}$$

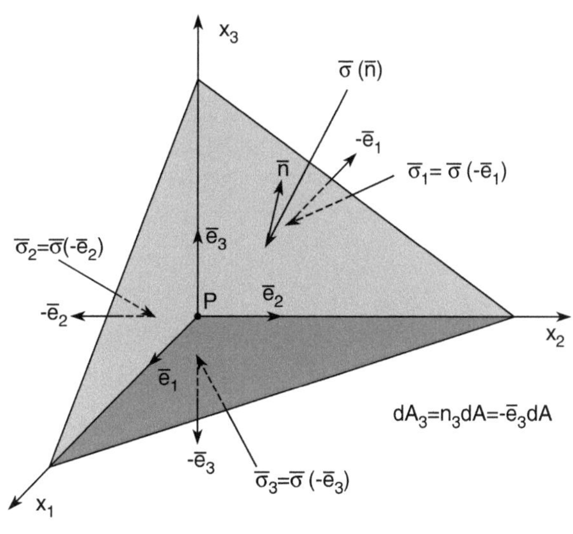

Fig. 2.2 Cauchy tetrahedron with traction vector balance on three Cartesian planes and the inclined slicing plane close to the origin of the infinitesimal tetrahedron

The traction vectors on these faces are denoted by (Fig. 2.2)

$$\bar{\sigma}_i = \bar{\sigma}(-\bar{e}_i) \tag{2.6}$$

and so the total force acting on these face are

$$dA_i\bar{\sigma}_i = n_i dA\bar{\sigma}(-\bar{e}_i). \tag{2.7}$$

Balancing forces on the inclined face of the tetrahedron leads to

$$\bar{\sigma}(\bar{n})dA + n_1 dA\bar{\sigma}(-\bar{e}_1) + n_2 dA\bar{\sigma}(-\bar{e}_2) + n_3 dA\bar{\sigma}(-\bar{e}_3) = 0. \tag{2.8}$$

Cancelling out the common area dA, and utilizing Cauchy's first law, for example $\bar{\sigma}(-\bar{e}_1) = -\bar{\sigma}(\bar{e}_1)$, leads to *Cauchy's second law*:

$$\begin{aligned} \bar{\sigma}(\bar{n}) &= n_1\bar{\sigma}(\bar{e}_1) + n_2\bar{\sigma}(\bar{e}_2) + n_3\bar{\sigma}(\bar{e}_3) \\ \sigma_i(n_j) &= \sigma_{ij}^T n_j \end{aligned} . \tag{2.9}$$

It is written both in vector notation ((2.9), upper half) and index notation ((2.9), lower half). The components of the three traction vectors that act on planes whose outward unit normals are in the three coordinate directions are denoted by

$$\begin{aligned} \bar{\sigma}(\bar{e}_1) &= (\sigma_{11}, \sigma_{12}, \sigma_{13})^T \\ \bar{\sigma}(\bar{e}_2) &= (\sigma_{21}, \sigma_{22}, \sigma_{23})^T, \\ \bar{\sigma}(\bar{e}_3) &= (\sigma_{31}, \sigma_{32}, \sigma_{33})^T \end{aligned} \tag{2.10}$$

where T stands for the transpose of a row vector since traction vector components appear as columns in Eq. (2.9). The components of each traction vector are denoted by two indices. The first refers to the direction of the outward unit normal vector $\bar{e}_i$ and the second refers to the component of the traction vector $\bar{\sigma}$. Substituting (2.10) into (2.9) leads to

$$\begin{aligned} \bar{\sigma}_1(\bar{n}) &= \sigma_{11}n_1 + \sigma_{21}n_2 + \sigma_{31}n_3 \\ \bar{\sigma}_2(\bar{n}) &= \sigma_{12}n_1 + \sigma_{22}n_2 + \sigma_{32}n_3. \\ \bar{\sigma}_3(\bar{n}) &= \sigma_{13}n_1 + \sigma_{23}n_2 + \sigma_{33}n_3 \end{aligned} \tag{2.11}$$

If one uses the standard matrix algebraic convention that the first subscript of the matrix components denotes the row and the second subscript denotes the column, it follows that

$$\begin{pmatrix} \sigma_1(\bar{n}) \\ \sigma_2(\bar{n}) \\ \sigma_3(\bar{n}) \end{pmatrix} = \begin{pmatrix} \sigma_{11} & \sigma_{21} & \sigma_{31} \\ \sigma_{12} & \sigma_{22} & \sigma_{32} \\ \sigma_{13} & \sigma_{23} & \sigma_{33} \end{pmatrix} \begin{pmatrix} n_1 \\ n_2 \\ n_3 \end{pmatrix}. \tag{2.12}$$

The matrix appearing in (2.12) is the transpose of the stress matrix. The *stress tensor* mathematically given by the stress matrix unequivocally defines the state of stress at an arbitrary point within a deformable body.

$$\sigma_{ij} = \begin{pmatrix} \sigma_{11} & \sigma_{12} & \sigma_{13} \\ \sigma_{21} & \sigma_{22} & \sigma_{23} \\ \sigma_{31} & \sigma_{32} & \sigma_{33} \end{pmatrix} \tag{2.13}$$

In 3D the stress tensor has nine components. The rows of the stress tensor are the traction vectors along the coordinate axes. Equation (2.12) is usually written with the transpose matrix defined in Eq. (2.13), because the stress matrix is always symmetric

$$\begin{aligned} \bar{\sigma} &= \sigma_{ij}\bar{n} = \sigma_{ij}^{T}\bar{n} \\ \sigma_i &= \sigma_{ij}n_j \equiv \sigma_{ji}n_j \end{aligned}. \tag{2.14}$$

Again, the equation is written in both vector notation ((2.14), upper half) and index notation ((2.14), lower half). The symmetry of the stress tensor can be proven by applying the mechanical law of conservation of angular momentum. The symmetry of the stress tensor reduces the components from nine to six in Eq. (2.13). Based on this property of the stress tensor, the first subscript i in (2.13), can be specified as normal to the actual slicing surface, while the second subscript j in (2.13), can be identified with the direction of the force. Depending upon the orientation of the slicing surface being normal and the force we can distinguish *normal stress* ($i=j$, force perpendicular slicing plane) with components (σ_{11}, σ_{22}, σ_{33}) pointing towards Cartesian axes and *shear stress* ($i \neq j$, force parallel slicing plane) with components ($\sigma_{12}=\sigma_{21}$, $\sigma_{13}=\sigma_{31}$, $\sigma_{23}=\sigma_{32}$) effective within Cartesian planes.

The physical significance of the stress tensor is illustrated by a 2D square element of an elastic body in Fig. 2.3. The *rock mechanics sign convention* is illustrated in Fig. 2.3a. The traction vector that acts on the face whose outward unit normal vector is in the x_1 direction, has components (σ_{11}, σ_{12}). As the traction components are considered positive if they are oriented in the directions opposite to the outward unit normal vector (Fig. 2.3a), we see that the traction σ_{11} is a positive number if it is compressive (Fig. 2.3a). In Fig. 2.3b, the *engineering mechanics sign convention* is illustrated with tensile normal stresses treated as positive. The direction of positive shear stresses is as shown in both sign conventions. Since compressive stresses are much more common for rocks in the Earth's crust, the rock mechanics sign convention (compression positive) is more appropriate in order to avoid frequent occurrence of negative signs in calculations involving stresses.

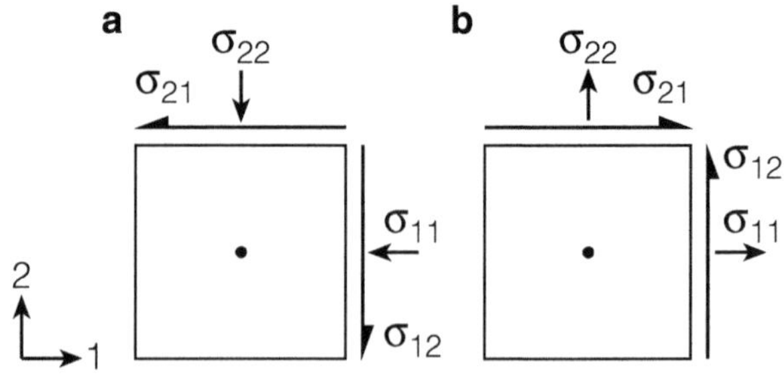

Fig. 2.3 Visualization of stress components in 2D. Sign convention in **a** rock mechanics or geosciences (compression positive), and **b** engineering mechanics or material sciences (tension positive)

We have three final remarks about the stress tensor. Firstly, the stress tensor can be calculated from the inner product (dot product) of the traction vectors with the unit vectors of the Cartesian reference frame (see (2.13) and (2.14), $\sigma_{ij} = \bar{e}_j \cdot \bar{\sigma}_i$). Secondly, the stress tensor can be written as a matrix, but a tensor has specific physical properties that are more important compared to that of a regular matrix. These properties relate to the manner in which the components of a tensor transform when the coordinate system is changed (Sect. 2.2). Thirdly, the fact that the state of stress at a point in 3D is completely specified by six independent components is important for the stress measuring techniques discussed in Part II of the book (Chaps. 6–8).

The unit of stress is pascal $Pa = Nm^{-2}$. The stress magnitude 1 Pa is produced by the force 1 N which acts normal or parallel to a square metre large surface. As 1 N is a small force and 1 m^2 a large surface, 1 Pa is a very small stress (= force/area). The "old" unit atmospheric pressure (1 bar) corresponds to 100 kPa. Crustal stresses in the Earth are usually measured in mega pascal whereby 1 MPa ("new unit") equals 10 bars ("old unit"). The stress magnitude 1 MPa is equal to the pressure p at a depth z of about 100 m in water, or about 37 m in rock using the relationship $p = \rho g z$, where ρ is the density of material (1000 kg m^{-3} for water, 2700 kg m^{-3} for rock) and g is the acceleration due to gravity, 9.81 m s^{-2}. Note that stress is not the same as pressure. Pressure is reserved for a specific stress state in which there are no shear components and all normal components are equal (e.g., in a fluid).

Note-Box A specific stress component acting on a specific slicing plane inside a deformable body can be described by a stress vector (traction). Three traction vectors are needed to unequivocally define the state of stress at a point inside the body resulting in nine components of the physical quantity stress tensor. Mathematically, the stress tensor can be written as stress matrix representing all stress components acting on three orthogonal slicing planes through a single, arbitrarily chosen body point. Due to the symmetry, only six stress components remain independent in the stress tensor (three normal and three shear stresses). The stress unit is force per area N m^{-2} = Pa (pascal), whereby 1 MPa equals 10 bar.

2.2 Principal Stresses

The principal stresses and principal directions can be found by asking whether or not there are planes on which the traction vector is purely normal, with no shear component. On such planes, the traction vector is aligned parallel to the outward unit vector, and can therefore be expressed using Cauchy's second law (see Eq. (2.14)) as

$$\begin{aligned} \bar{\sigma} &= \sigma \bar{n} \\ \sigma_i &= \sigma_{ij} n_j = \sigma n_j \end{aligned}, \tag{2.15}$$

where σ is a yet unknown scalar quantity. With help of *Kronecker's delta* ($\delta_{ij}=1$ for $i=j$ and $\delta_{ij}=0$ for $i\neq j$) it follows that

$$(\sigma_{ij} - \sigma \delta_{ij}) n_j = 0. \tag{2.16}$$

This is the fundamental equation for determining *eigenvalues* (principal stress magnitudes) and *eigenvectors* (principal stress directions) of the stress matrix. Splitting it into components results in the following set of equations

$$\begin{aligned} (\sigma_{11} - \sigma) n_1 + \sigma_{12} n_2 + \sigma_{13} n_3 &= 0 \\ \sigma_{21} n_1 + (\sigma_{22} - \sigma) n_2 + \sigma_{23} n_3 &= 0. \\ \sigma_{31} n_1 + \sigma_{32} n_2 + (\sigma_{33} - \sigma) n_3 &= 0 \end{aligned} \tag{2.17}$$

Admissible solutions of this linear, homogeneous set of equations can be found only if the determinant of the matrix coefficients equals zero, i.e.

$$\begin{vmatrix} \sigma_{11} - \sigma & \sigma_{12} & \sigma_{13} \\ \sigma_{21} & \sigma_{22} - \sigma & \sigma_{23} \\ \sigma_{31} & \sigma_{32} & \sigma_{33} - \sigma \end{vmatrix} \stackrel{!}{=} 0. \tag{2.18}$$

When the determinant is expanded out, it takes the form of a cubic equation in σ

$$\sigma^3 - I_1 \sigma^2 + I_2 \sigma - I_3 = 0, \tag{2.19}$$

where (I_1, I_2, I_3) are called *tensor invariants*. Values of stress invariants are independent of the coordinate system used. The physical content of a stress tensor is reflected exclusively in the stress invariants. For example, pressure in all directions, as is the case in the hydrostatic state of stress, results from I_1. The three invariants of the stress tensor are given by

$$I_1 = \sigma_{ii} = \sigma_{11} + \sigma_{22} + \sigma_{33} = trace(\sigma_{ij})$$

$$I_2 = \begin{vmatrix} \sigma_{11} & \sigma_{12} \\ \sigma_{21} & \sigma_{22} \end{vmatrix} + \begin{vmatrix} \sigma_{11} & \sigma_{13} \\ \sigma_{31} & \sigma_{33} \end{vmatrix} + \begin{vmatrix} \sigma_{22} & \sigma_{23} \\ \sigma_{32} & \sigma_{33} \end{vmatrix} = \frac{1}{2}(\sigma_{ii}\sigma_{jj} - \sigma_{ij}\sigma_{ij}) \quad (2.20)$$

$$I_3 = \begin{vmatrix} \sigma_{11} & \sigma_{12} & \sigma_{13} \\ \sigma_{21} & \sigma_{22} & \sigma_{23} \\ \sigma_{31} & \sigma_{32} & \sigma_{33} \end{vmatrix} = \frac{1}{6}(\sigma_{ii}\sigma_{jj}\sigma_{kk} + 2\sigma_{ij}\sigma_{ik}\sigma_{ki} - 3\sigma_{ij}\sigma_{ij}\sigma_{kk})$$

I_1 is called the *trace* of the stress tensor and $I_{2,3}$ can be computed from the coefficient determinants. The three solutions of the characteristic Eq. (2.19) are called *principal normal stresses* (σ_1, σ_2, σ_3). The double subscripts of the normal stress components can be reduced to single suffixes, since the shear components per definition become zero. With the help of the fundamental theorem of algebra, Eq. (2.19) can be written as

$$(\sigma - \sigma_1)(\sigma - \sigma_2)(\sigma - \sigma_3) = 0. \quad (2.21)$$

Then the tensor invariants follow from principal normal stresses

$$\begin{aligned} I_1 &= \sigma_1 + \sigma_2 + \sigma_3 \\ I_2 &= \sigma_1\sigma_2 + \sigma_2\sigma_3 + \sigma_3\sigma_1. \\ I_3 &= \sigma_1\sigma_2\sigma_3 \end{aligned} \quad (2.22)$$

After the transformation, the stress matrix has the following (diagonal) form:

$$\sigma_{ij} = \begin{pmatrix} \sigma_1 & 0 & 0 \\ 0 & \sigma_2 & 0 \\ 0 & 0 & \sigma_3 \end{pmatrix}, \quad (2.23)$$

whereby the principal axes are chosen in a way that magnitudes $\sigma_1 \geq \sigma_2 \geq \sigma_3$ apply for the principal stress. Using the principal stress magnitudes, the direction cosines of the principal axes can be deduced from Eq. (2.4).

The principal normal stress components can be visualized in 3D using a rotated cube (Fig. 2.4). Principal stress directions (x', y', z') are rotated with respect to the global (space-fixed) coordinate system (x, y, z). While six components of stress (taking into account the symmetry of stress tensor) are necessary to define the state of stress in an arbitrary oriented cube (Fig. 2.4a), only three components of stress (σ_1, σ_2, σ_3) are required in the rotated cube of principal stresses (Fig. 2.4b). Due to the fact that besides the magnitudes of the three principal stresses (Fig. 2.4b, eigenvalues σ_1, σ_2, σ_3) also the directions of the three principal stresses (Fig. 2.4b, eigenvectors

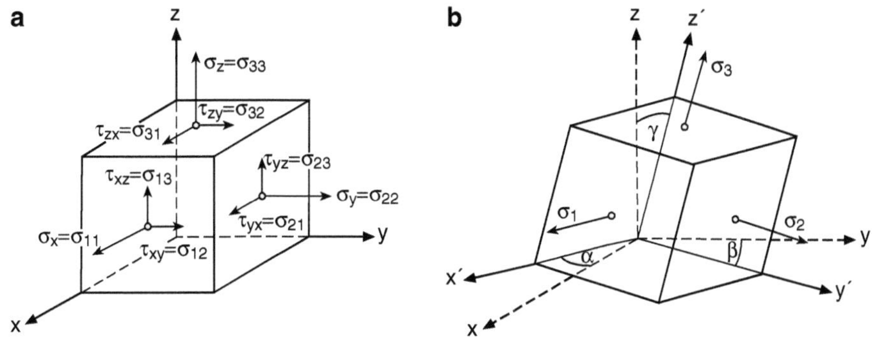

Fig. 2.4 Visualization of stress components in 3D on a cube **a** before and **b** after solving the eigenvalue problem of the stress matrix; engineering mechanics notation (tension positive convention)

α, β, γ) are necessary, six pieces of information are required for full determination of the state of stress at any point in any coordinate system. Principal normal stresses act perpendicular to the faces of the cube, the surfaces of which are free of shear stresses. An additional exercise for the reader is to redraw Fig. 2.4 so that it describes the situation when using rock mechanics (compression positive convention).

It is useful to have a way of presenting the stress tensor that clearly shows whether or not there are any shear stresses acting at the point in question. To do so, the stress tensor is decomposed into an isotropic (= hydrostatic) and a deviatoric part. The *isotropic part* of the stress tensor is defined as

$$\begin{aligned} \sigma_{ij}^{iso} &= \frac{1}{3} I_1 I = \sigma_{\mathrm{m}} I \\ \sigma_m &= \frac{1}{3}(\sigma_{11} + \sigma_{22} + \sigma_{33}) \end{aligned}, \tag{2.24}$$

where I is the *identity tensor* ($I\bar{a} = \bar{a}$) and σ_m is the *mean normal stress*. The *deviatoric stress* is obtained by subtracting the isotropic part of the stress tensor from the full stress tensor.

$$\sigma_{ij}^{dev} = \sigma_{\mathrm{ij}} - \sigma_{ij}^{iso} = \begin{pmatrix} \sigma_{11} - \sigma_m & \sigma_{12} & \sigma_{13} \\ \sigma_{21} & \sigma_{22} - \sigma_m & \sigma_{23} \\ \sigma_{31} & \sigma_{32} & \sigma_{33} - \sigma_m \end{pmatrix}. \tag{2.25}$$

The usefulness of this decomposition arises from the fact that, in the elastic range of deformation, the isotropic stress controls the volumetric change of a body, whereas the deviatoric stress controls the distortion. Even at very high levels of stress, no plastic flow is caused by a hydrostatic stress, because there are no shear stresses on any plane, since all planes are principal planes. The deviatoric stress, however, produces shear stress and can therefore lead to plastic flow if the elastic limit of material is exceeded. A deviatoric stress causes no dilatation because the sum of its components is always zero. Rock failure criteria (Chap. 3) are concerned primarily with distortion, in which case these criteria are most conveniently expressed in terms of the invariants of the stress deviation. J_2 is the *invariant of the stress deviator* that appears most often in rock failure criteria.

$$J_2 = \frac{1}{2}(\sigma_{11}^2 + \sigma_{22}^2 + \sigma_{33}^2) + \sigma_{12}^2 + \sigma_{23}^2 + \sigma_{13}^2$$
$$J_2 = \frac{1}{6}\left[(\sigma_1 - \sigma_2)^2 + (\sigma_2 - \sigma_3)^2 + (\sigma_3 - \sigma_1)^2\right]. \tag{2.26}$$
$$J_2 = 3\sigma_m^2 + I_2 = \frac{3}{2}\tau_{OCT}^2$$

Apart from principal normal stresses, it should be mentioned that also *principal shear stresses* exist which act in planes which are parallel to one principal axis and form an angle of 45° with the two other principal axes. To the magnitudes of principal shear stresses applies

$$\tau_1 + \tau_2 + \tau_3 = \frac{\sigma_2 - \sigma_3}{2} + \frac{\sigma_3 - \sigma_1}{2} + \frac{\sigma_1 - \sigma_2}{2} = 0. \tag{2.27}$$

With $\sigma_1 \geq \sigma_2 \geq \sigma_3$, where σ_1 is the maximum and σ_3 is the minimum (least) principal normal stress component, the *maximum shear stress* results in

$$\tau_{\max} = \frac{\sigma_1 - \sigma_3}{2} = |\tau_2|. \tag{2.28}$$

The maximum shear stress acts in a plane which cuts the angle between maximum and minimum principal normal stress into half. The planes in which the principal shear stresses act are not perpendicular to each other. They form a normal dodecahedron (Fig. 2.5). The planes of the dodecahedron are not free of normal stresses. Shear stress values are dictated by (2.27).

Exercise 2.1 Transformation of a (3×3) stress matrix. Consider the stress matrix

$$\sigma_{ij} = \begin{pmatrix} \left(\frac{13}{4}\right) & \left(\frac{\sqrt{3}}{4}\right) & 0 \\ \left(\frac{\sqrt{3}}{4}\right) & \left(\frac{15}{4}\right) & 0 \\ 0 & 0 & 3 \end{pmatrix}.$$

(a) Find the three principal stresses (σ_1, σ_2, σ_3) by solving the eigenvalue problem.
(b) Compute the stress invariants (I_1, I_2, I_3) from principal stresses.
(c) What are the direction cosines of the planes on which the principal stresses act?
(d) The matrix formed by the nine components of the three eigenvectors describes what kind of geometrical operation in space?

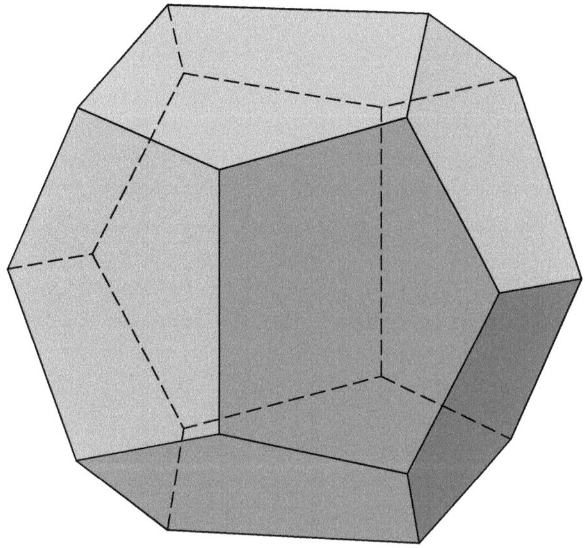

Fig. 2.5 Visualization of planes in which principal shear stresses are effective on a pentagon dodecahedron (3D arrangement of 12 surfaces with 5 corners)

We have two final remarks about principal stresses. Firstly, the fact that the stresses transform in an eigenvalue problem according to $\sigma'_{ij} = R\sigma_{ij}R^T$ when the coordinate system is rotated is the defining property that makes the stress a *second-order tensor*. The traction vector transforms according to $\bar{\sigma}' = R\bar{\sigma}$. The appearance of only one rotation matrix, R in this transformation law is the reason that vectors are referred to as *first-order tensors*. A *zero-order tensor* is a scalar quantity with magnitude only (e.g. temperature). Secondly, principal stresses have particular significance for rock engineering. The process of creating a new surface in a rock mass by excavation causes principal stresses to be locally oriented perpendicular and parallel to the free surface. The principal stress perpendicular to the free surface is zero. The other two principal stresses, i.e. the maximum and minimum value, occur in a direction parallel to the rock-free surface. Therefore, any excavation plane within the Earth's crust is a principal stress plane.

Note-Box Each stress matrix can be transformed from an arbitrary reference frame into the frame of principal axes. The state of stress in the new system is then defined by three principal stresses (stress magnitudes) and three principal axes (stress orientations). The physical meaning of the stress tensor is captured in tensor invariants, which are independent of the reference frame used. Principal normal stresses are visualized on a cube the faces of which are free of shear stresses. Principal shear stresses exist and act on a dodecahedron, which planes are not free of normal stresses. All unsupported rock excavation surfaces within the Earth's crust are principal normal stress planes.

2.3 Mohr Circle of Stress

To make practical use of the stress matrix in Eq. (2.13), we must be able to find the stress components in directions different from the reference directions. One possibility was demonstrated by the principal stress matrix in Eq. (2.23). The Mohr circle of stress is a second, simple graphical method of transforming the stress tensor. As the discussion of stress is algebraically simpler in 2D than in 3D, the Mohr circle is introduced in 2D. Many problems in rock mechanics are essentially 2D as stresses do not vary along one Cartesian coordinate perpendicular to the free surface (cf. stresses around boreholes, Chap. 7). Hence, it is worthwhile to study the properties of 2D stress tensors.

Consider the arbitrary plane *ds* (unit length assumed in the *z*-direction) in a deformable body whose normal makes an angle α with the orientation of maximum principal stress σ_1 (Fig. 2.6). We look for the normal stress (σ) and shear stress component (τ) acting on the surface element *ds* as a function of principal stresses and tilt angle (σ_1, σ_2, α). Again, two equilibrium conditions must be fulfilled. For infinitesimal volumes, the balance of moments leads to the condition that pairs of shear stresses must be equal (symmetry of stress tensor). The balance of forces prevents the prism from translation and rotation. Note that in the 2D description of the Mohr circle, we have to operate with prisms (Fig. 2.6, triangles with unit length in the third direction), since stress is defined as force per unit area.

The balance of forces (force = stress times area) at the small prism surface *ds* (area = length (*ds*) times unit length (1)) reads

$$\begin{aligned} \sigma_2 \sin \alpha ds - \sigma \sin \alpha ds + \tau \cos \alpha ds &= 0 \\ \sigma_1 \cos \alpha ds - \sigma \cos \alpha ds - \tau \sin \alpha ds &= 0 \end{aligned}, \tag{2.29}$$

Using trigonometric identities

$$\begin{aligned} \cos 2\alpha &= 2\cos^2\alpha - 1 = 1 - 2\sin^2\alpha \\ \sin 2\alpha &= 2 \sin \alpha \cos \alpha \end{aligned}, \tag{2.30}$$

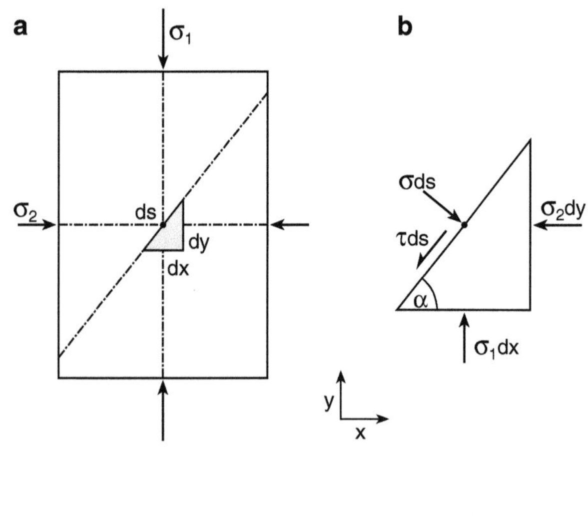

Fig. 2.6 In a deformable body loaded by minimum and maximum principal stress **a** a small prism is shown **b** where the balance of forces is computed

the normal stress σ and shear stress τ on the prism surface ds follow as a function of the two principal stresses

$$\begin{aligned} \sigma &= \frac{\sigma_1 + \sigma_2}{2} + \frac{\sigma_1 - \sigma_2}{2}\cos 2\alpha \\ \tau &= \frac{\sigma_1 - \sigma_2}{2}\sin 2\alpha \end{aligned}, \tag{2.31}$$

These are the equations of a circle in the (σ, τ) plane with its centre at the point $(\sigma=(\sigma_1+\sigma_2)/2, \tau=0)$ and with radius $(\sigma_1-\sigma_2)/2$. Equation (2.31) defines the *Mohr circle of stress*. This representation of stress was first proposed in 1882 by the German engineer Otto Mohr (1835–1918). It relates the principal stresses applied to a deformable body to the normal and shear stresses of an arbitrary oriented plane inside the body (Fig. 2.6). Equations (2.29)–(2.31) express the crucial difference between forces and stresses, which is the key to understanding the concept of stress. The resolution of a normal force requires, e.g., $\cos\alpha$, the resolution of a normal stress component however requires $\cos^2\alpha$. One $\cos\alpha$ is due to the resolution of normal force and one $\cos\alpha$ is due to the resolution of the slicing plane on which the force is acting. Due to trigonometric identities (Eq. (2.30)), the double resolution of stresses is hidden in the term $\cos 2\alpha$ of the Mohr circle equation (Eq. (2.31)).

Each point P on the Mohr circle (Fig. 2.7a) states values of normal and shear stress on the arbitrary plane within the body. For a given set of principal stresses (reference frame σ_1, σ_2), we can compute a second set of normal and shear stresses (frame σ, τ) at arbitrary (angle α) oriented surfaces. The values of normal and shear stress versus angle is seen in Fig. 2.7b for $\sigma_1=1$ MPa and $\sigma_2=0.5$ MPa. The transformation $[\sigma_1, \sigma_2]\rightarrow[\sigma, \tau]$ of stress components in 2D is analogous to the eigenvalue problem of the stress matrix in 3D visualized in Fig. 2.4. The points where the Mohr circle intersects the σ-axis represent principal planes. The associated σ-values are the principal stresses, σ_1 and σ_2. The Mohr circle shows that the principal stresses are the maximum and minimum values of normal stresses in a body. The points

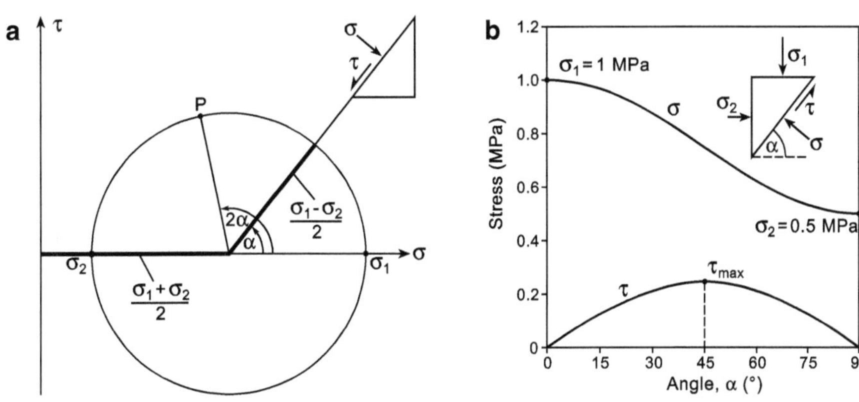

Fig. 2.7 Mohr circle of stress **a** in *stress space*, and **b** in *physical space*. The Mohr circle relates principal stresses (σ_1, σ_2) to normal and shear stresses on an arbitrary tilted plane (σ, τ, α) inside a deformable body

representing principal planes lie opposite the diameter in stress space. In physical space (Fig. 2.6, small prism), the planes are perpendicular since twice the rotation takes place on the Mohr circle. The *maximum shear stress* τ_{max} occurs when $2\alpha = 90°$. Thus the plane of maximum shear stress is oriented at 45° to the principal planes. Note that in rock mechanics notation, the positive τ-axis is upside down and positive shear stresses plot below the σ-axis. Also in times of fast personal computers, the Mohr circle of stress does not lose its significance for displaying stress, in particular with respect to the presentation of rock failure criteria (Chap. 3).

The Mohr circle is a way of plotting the normal and shear components for traction vectors associated with all possible planes through point *P*. Note the difference between the stress tensor at a point (Sect. 2.1) and the traction vector acting on some given plane through that point (Sect. 2.3). In *Mohr space* (= *stress space*), the normal and shear stress components of the stress vector with respect to the given plane are displayed. In *physical space*, the fracture plane within a rock mass after failure is typically inspected (Chap. 3). The orientation of the fracture plane is also governed by the internal friction of the material and therefore differs from the stress space.

Note-Box The Mohr circle of stress is obtained from balancing forces on a small prism within a deformable body under applied principal stresses. The Mohr circle relates normal and shear stress acting on an arbitrary slicing surface element of the prism to the applied principal stresses. The Mohr circle is a graphical method of transforming the stress tensor, and one way to visualize the stress field of a deformable body.

2.4 Visualizing Stress

In order to completely specify the state of stress in 2D (3D) with three (six) pieces of information, it is necessary to know the values of the stress components at each point on the body or, alternatively, to know the two (three) principal stresses and one (three) principal stress direction(s). Although it is difficult to display all of these data, there are a number of graphical methods that are useful in giving a partial picture of the stress field. As we know from Chap. 1, a stress field describes the way that the state of stress varies through space in a body. Since stress is not a straightforward descriptive quantity (Sect. 1.1, fictitious term; Sect. 2.1, abstract concept), we require appropriate techniques to display stress components and stress orientations.

One way to visualize stress magnitudes in 2D is *Lame's stress ellipse* combining Cartesian components of the stress vector (σ_x, σ_y) with components of principal stresses (σ_1, σ_2). Balancing forces according to the small prism presented in Fig. 2.8a it follows that

$$\left(\frac{\sigma_x}{\sigma_1}\right)^2 + \left(\frac{\sigma_y}{\sigma_2}\right)^2 = 1. \tag{2.32}$$

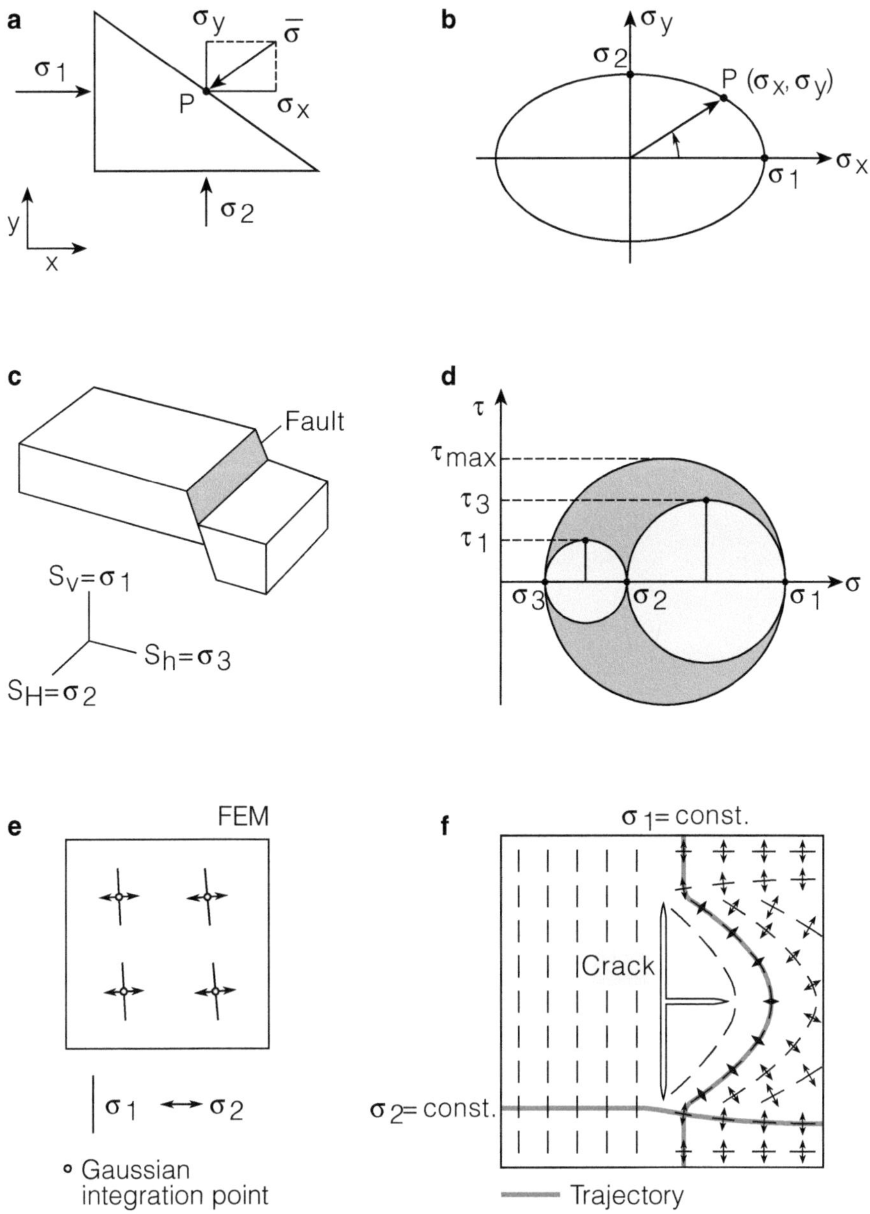

Fig. 2.8 Visualizing stress in physical space (*left*) and stress space (*right*). The relationship in 2D between a Cartesian frame and principal stresses (**a**) is shown by a stress ellipse (**b**). The relationship in 3D between a Mohr frame (normal and shear stress) and principal stresses of a normal fault in the Earth's crust (**c**) is shown by a Mohr circle of stress (**d**). Principal stress directions at points (**e**) are shown as stress trajectories (**f**)

Each vector from the origin to a point on the ellipse (Fig. 2.8b) represents a traction vector that acts on some plane passing through the point at which the principal stresses are (σ_1, σ_2). However, although Lame's stress ellipse shows the various traction vectors that act on different planes, it does not indicate the plane on which the given traction acts. It can be shown that in 3D the locus of all points traced out by the stress vector for all possible orientations of the plane on which it is acting defines an stress ellipsoid, the equation of which is given, e.g., in Jaeger and Cook (1979). The *stress ellipsoid* is one way to provide a 3D description of the state of stress at a single point in a deformable body. Note that the concept of stress can also be associated with that of a continuum and therefore is of value only at a scale at which the continuum concept is valid. The minimum volume for which an equivalent volume can be defined is termed the representative elementary volume (Chap. 10, REV). Generally, the continuum concept is of interest when the volume under investigation is at least two orders of magnitude larger than that of the REV (Hudson et al. 2003). In conclusion, a finite REV is needed for the physical definition of the concept of stress, while a point (zero volume) in Eq. (2.1) is needed for the mathematical definition of the concept of stress.

Our second way of visualizing stress components is the Mohr circle but, instead of 2D (Sect. 2.3), we now consider a 3D state of stress determined by three principal stresses and their orientations. Using Cauchy's second law, the stress vector components can be written

$$\begin{aligned} \sigma^2+\tau^2 &= \sigma_i\sigma_i = \sigma_1{}^2 n_1{}^2+\sigma_2{}^2 n_2{}^2+\sigma_3{}^2 n_3{}^2 \\ \sigma &= \sigma_i n_i = \sigma_1 n_1{}^2+\sigma_2 n_2{}^2+\sigma_3 n_3{}^2 \end{aligned} . \tag{2.33}$$

With the help of $n_i n_i=1$, e.g., the identity

$$\left(\sigma-\frac{\sigma_2-\sigma_3}{2}\right)^2+\tau^2=-\sigma(\sigma_2+\sigma_3)+\left(\frac{\sigma_2+\sigma_3}{2}\right)^2+(\sigma^2+\tau^2), \tag{2.34}$$

we obtain the expression

$$\left(\sigma-\frac{\sigma_2+\sigma_3}{2}\right)^2+\tau^2=n_1{}^2(\sigma_1-\sigma_2)(\sigma_1-\sigma_3)+\left(\frac{\sigma_2-\sigma_3}{2}\right)^2. \tag{2.35}$$

Mathematically, this is the equation of a circle with its centre at ($\sigma=(\sigma_2+\sigma_3)/2, \tau=0$) and a radius which depends on n_1. Since $0\le n_1{}^2\le 1$, the minimum centre point distance of stress points is $(\sigma_2-\sigma_3)/2=\tau_1$ for $n_1=0$, whereas the maximum distance is $\sigma_1+(\sigma_2-\sigma_3)/2$ for $n_1=\pm 1$. Analogue findings on two additional equations following from (2.35) by cyclic permutation of indices lead to the *Mohr circles of stress in 3D*. Arranging principal stresses with respect to magnitude ($\sigma_1\ge\sigma_2\ge\sigma_3$), we obtain the stress points within the shaded area of Fig. 2.8d. The circles with radii τ_i correspond to slicing planes with the surface normal perpendicular to one of the three principal axes.

Our third way of visualizing stress is restricted to the direction of principal stresses. Like in *Finite Element* (FE) codes where principal stress directions are displayed at Gaussian integration points within a finite element (Fig. 2.8e, dots), we can define stress trajectories. A line whose tangent at every point is in the direction of a principal stress component is called a *stress trajectory*. In the 2D rock model with a T-shaped crack (Fig. 2.8f), two stress trajectories are shown, one for the direction of the largest and the other for the direction of the least principal stress. As principal stresses are always at right angles to each other, stress trajectories form an orthogonal set of lines (Fig. 2.8f, σ_1 perpendicular σ_2). In Exercise 2.2, the reader can complete a dense network of stress trajectories for the model sketched in Fig. 2.8f using stresses at all Gaussian integration points within the full finite element model.

Analogue to the lines of electric (or magnetic) flux where electric (or magnetic) fields are visualized, stress trajectories are used to visualize the elastic stress field in a deformable body. An early picture of electric flux lines is shown in Fig. 2.9, where the Russian Jakob von Narkievicz-Jodko electrified a human hand in 1895 and saved the picture on a photographic plate. A bunch of electric lines was captured in silver gelatine to give an idea of the complex internal electric field of a human hand. You can imagine that also in the case of a complex rock body with or

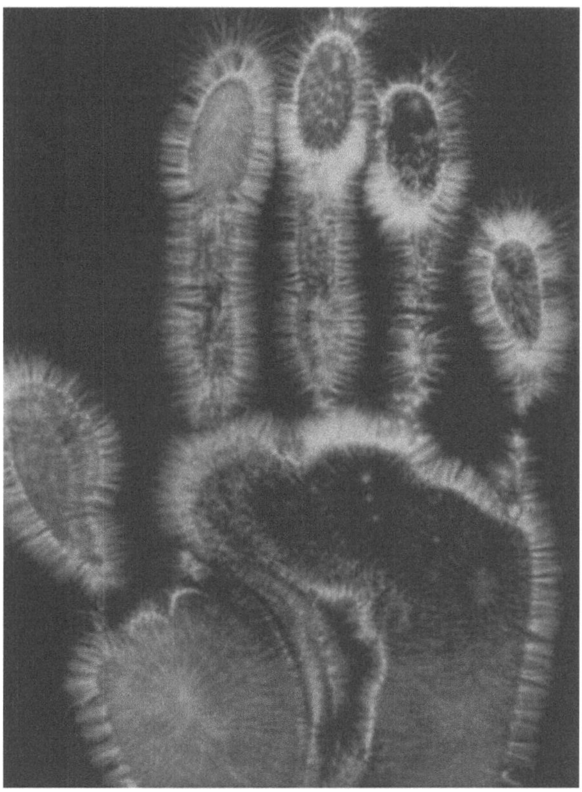

Fig. 2.9 In 1895, Jakob von Narkievicz-Jodko electrified a human hand in order to capture the electric flux line network in silver gelatine

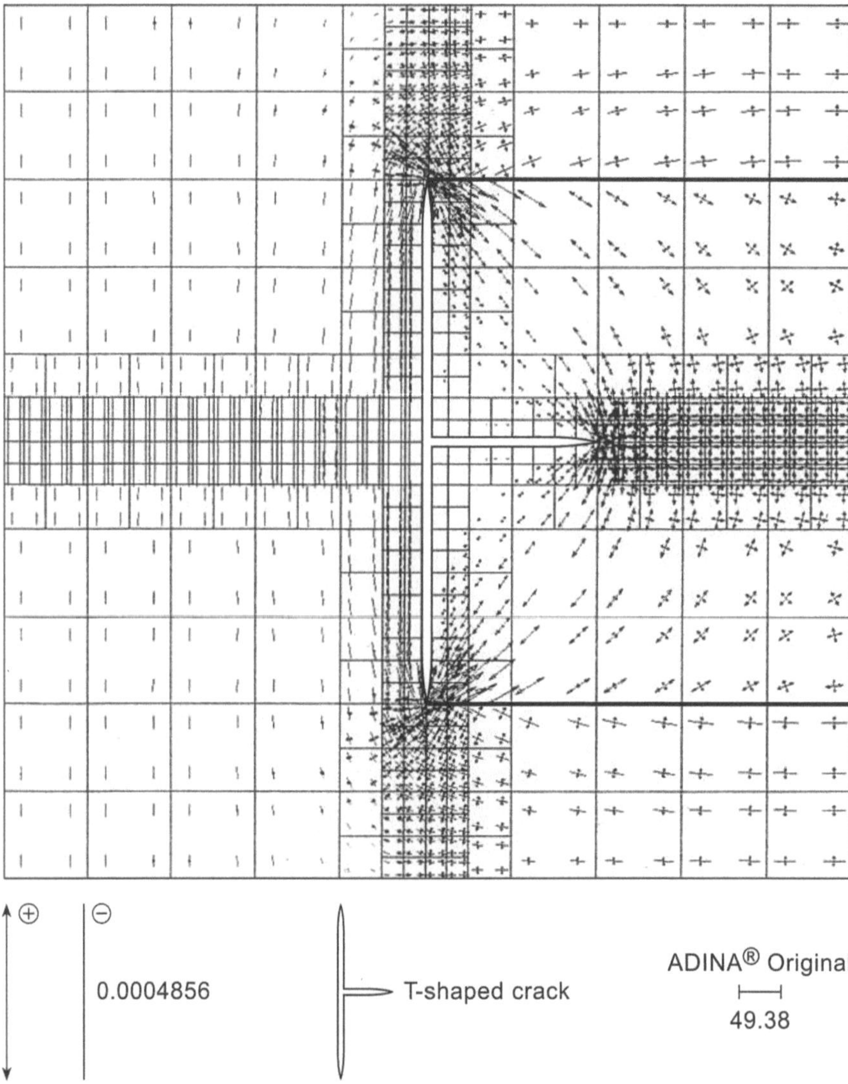

Fig. 2.10 Finite element mesh of a T-shaped crack formed during cooling of a 2D periodic rock model showing directions and magnitudes of σ_1 and σ_2 at every single Gaussian integration point within each of the 436 quadratic finite elements (Zang 1991)

without excavation surfaces, many data points of principal stresses are needed to draw a realistic network of stress trajectories characterizing the elastic stress field (see Exercise 2.2).

Among the experimental techniques to visualize stress are *isochromatics* (constant maximum shear stress, see Sect. 6.4), *isopachs* (curves along which the mean normal stress is constant, e.g. Durelli et al. (1958) electrical conducting paper), *isoclines* (curve on which principal axes make a constant angle with a given fixed reference direction, photoelasticity), and fault slip lines (curves on which the shear stress is maximum, structural geology).

Exercise 2.2 Stress Trajectories. Consider the finite element mesh with a T-shaped crack in Fig. 2.10. Put a transparency foil on top of the figure.

(a) Draw a set of solid lines onto the foil following the σ_1-orientation in the rock model using the stress orientations given at Gaussian integration points. Add a second set of lines with a different color, which follows the σ_2-orientation.
(b) What may be the reason for the rotation of the stress field in the rock model near the two marked bold solid lines?

Note that at crack tips, stress magnitudes increase (Fig. 2.10, mesh refinement) and stress orientations may change quickly over short distances. The reason for this will be expanded in Chap. 3, where crack-tip stress singularities are treated. As is true for large rock mass excavation surfaces, also small cracks in rock serve as free surfaces and therefore are principal normal stress planes.

Note-Box The way that the state of stress varies through space (stress field) can be visualized, e.g., by firstly a stress ellipse (ellipsoid), secondly a Mohr stress circle, and thirdly stress trajectories. Along a stress trajectory (line), the direction of principal stress is tangential in every point. In the first two methods, the state of stress is described at a point of a deformable body. Lame's stress ellipse defines the locus of all points traced out by the stress vector for all possible orientations of the plane on which it is acting. The normal and shear stress components of the stress vector with respect to the given plane are visualized in the Mohr circle.

Chapter 3
Rock Fracture Criteria

When stresses in the Earth's crust exceed the crustal strength, the rock fractures and fails (Brace 1964; Hoek 1968; Obert 1972). A fracture criterion describes the conditions for which failure occurs in a material. In principle, one distinguishes between phenomenological and mechanistic failure theories. *Phenomenological theories* (Coulomb, Coulomb-Mohr, Mohr or Hoek-Brown) quantify the spatial orientation of fracture planes with respect to the stress field producing the discontinuities. *Mechanistic theories* (Griffith, McClintock & Walsh, fracture mechanics models) start from the premise that fracture initiates from existing flaws acting as stress concentrators through which the brittle fracture process in rock is controlled. Both theories are used to determine the stress state in the Earth's crust and to evaluate the stresses for some of the stress measurement techniques.

3.1 Phenomenological Theories

In the late eighteenth century, C.A. Coulomb (1736–1806) postulated that a shear fracture occurs in isotropic material, if the maximum shear stress is reached. According to Eq. (2.28) our first rock fracture criterion (*maximum shear stress criterion*) reads

$$\tau_{\max} = \frac{\sigma_1 - \sigma_3}{2}. \tag{3.1}$$

The sign of the shear stress only affects the direction of sliding after failure, so usually the absolute value of τ appears in the failure criterion (see Eq. (2.28)). We ignore the absolute sign of shear stress in (3.1) which is convenient for mathematical manipulations, but take it into account in visualizing the failure criterion (Fig. 3.1a). Equation (3.1) defines a horizontal straight line in (σ,τ) stress space that intercepts the τ-axis at $\tau_{\max}$ (Fig. 3.1a). If principal stresses are such that the circle touches the failure line, the rock will fail in shear (Fig. 3.1a, P^+ and P^-). Circles that extend above the horizontal line are not represented in nature, since failure occurs

A. Zang, O. Stephansson, *Stress Field of the Earth's Crust*,
DOI 10.1007/978-1-4020-8444-7_3,

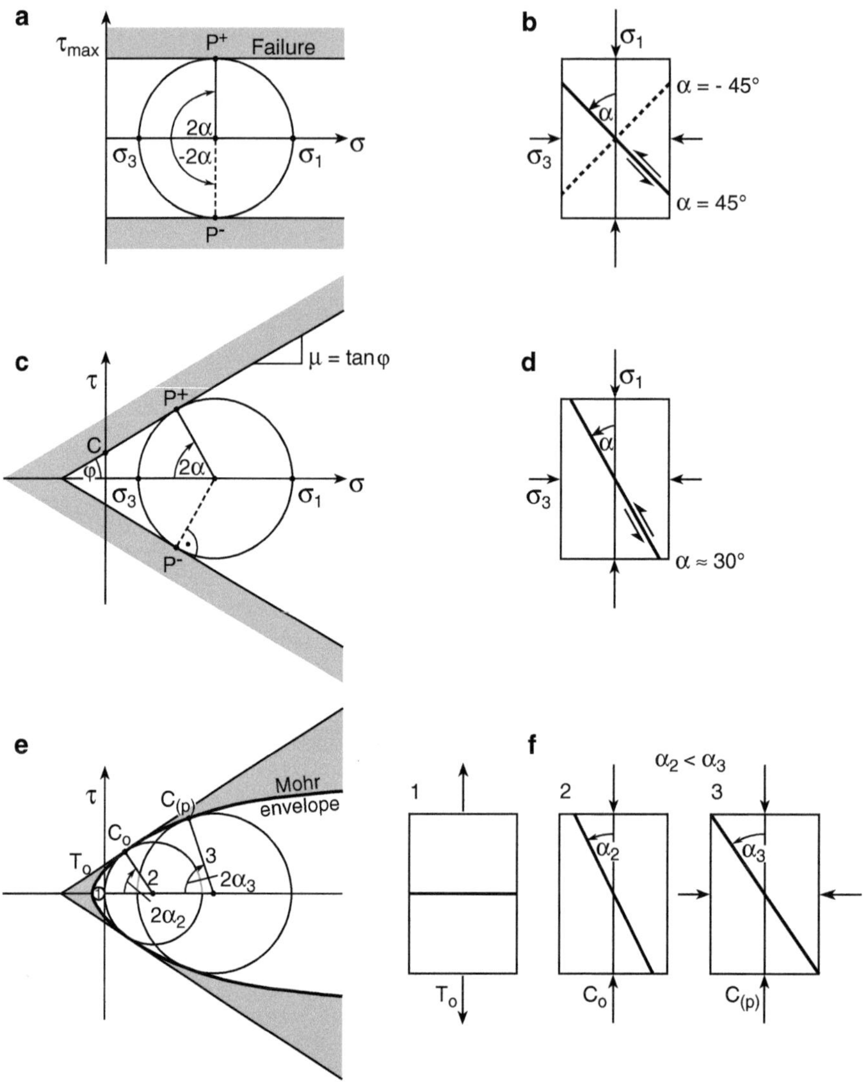

Fig. 3.1 Rock failure envelopes in Mohr diagrams (*left*, stress space) and orientation of rupture planes in laboratory fracture tests (*right*, physical space). In **a**, **b** the maximum shear stress criterion (3.1), in **c**, **d** the Coulomb fracture criterion (3.2) and in **e**, **f** the generalized Mohr criterion (3.4) is presented with failure envelopes. The shaded area shown is the failure regime of rock

as soon as the Mohr circle first touches the failure line. Points $P^{\pm}$ at which the circle is tangent to the failure line represent the stress state on the plane of failure. Hence, the angle by which the failure plane is oriented to the σ_1 direction is given by $\alpha = \pm 45°$. The two possible planes of shear failure resulting from the absolute sign of shear stress in Eq. (2.28) are referred to as *conjugate failure planes*. Note that in physical space (Fig. 3.1b) only one failure plane will develop, whereby local defects in the rock material play a key role as stress concentrator and initiator of the failure plane (Sect. 3.3). The horizontal failure line in Fig. 3.1a is also called a *critical Mohr envelope*, since it represents the envelope of all Mohr circles at the state of failure. The Mohr envelope separates stress conditions where rock is intact (Fig. 3.1a, empty region) from stress conditions where rock failed (Fig. 3.1a, light grey shaded region).

The next simplest and most widely used failure criterion is that of Coulomb-Mohr, where the shear strength of rock is made up of two parts, a normal stress-dependent frictional part $\mu\sigma$ and a constant term C. Our second rock fracture criterion (*Coulomb criterion*) reads

$$\tau = \mu\sigma + C \tag{3.2}$$

and is illustrated in Fig. 3.1c, d. The parameter μ is known as the *coefficient of internal friction*, as it applies along the hypothetical slicing plane (Chap. 2) that is internal to the rock before failure occurs. Equation (3.2) defines a straight line in (σ,τ)- space that intercepts the τ-axis at C, and has slope μ. In original form, Coulomb wrote $\tau = C + \mu\sigma$ instead. The angle φ that this line makes with the σ-axis is given by $\varphi = \tan^{-1}\mu$, and is known as the *angle of internal friction*. The parameter C is called *cohesion* since sliding at zero normal stress occurs at finite values of shear stress (abscissa in the Mohr diagram at $\sigma = 0$). Again, shear failure of rock will occur if principal stress circles touch the failure line (Fig. 3.1c, $P^{\pm}$). The angle by which the failure plane is oriented to the maximum principal stress direction is

$$\alpha_{1-4} = \pm\left(\frac{\pi}{4} \mp \frac{\varphi}{2}\right). \tag{3.3}$$

Equation (3.3) covers four fracture angles in total. As implied by Mohr's hypothesis, the orientation of the fracture surface is inclined to σ_1 at a fracture angle, α, equal to $\alpha_1 = 45° - (\varphi/2)$ or $\alpha_3 = -45° + (\varphi/2)$ (conjugate), where tan φ is the slope of the failure envelope in Mohr space (Fig. 3.1c). This pair of fracture angles (α_1, α_3) is valid for the compressive state of stress in triaxial testing, i.e. $\sigma_1 > \sigma_3$ (Scholz 1990). Since the failure envelope is a continuous function that crosses the transition from a compressive to a tensile stress state (Fig. 3.1c, e), a second pair of fracture angles exists with $\alpha_2 = 45° + (\varphi/2)$ or $\alpha_4 = -45° - (\varphi/2)$ (conjugate) which is valid for the tensile stress state, i.e. $\sigma_1 < \sigma_3$ (Paterson and Wong 2005). The progressive change in fracture orientation with confining pressure for the transition from tensile to compressive stress regimes was investigated "par excellence" by Ramsey and Chester (2004) in triaxial tests on dog-bone-shaped samples of Lorano Bianco Carrara marble.

Depending on the selected *friction coefficient* of rock (e.g., Table 3.1a, $0.5 < \mu < 1$) failure plane angles α range between 22° and 32°, with a most likely value at $\alpha \approx 30°$ corresponding to $\mu \approx 0.58$ (Fig. 3.1d). We have two final remarks about the Coulomb failure criterion. First, the orientation of rupture planes according to Eq. (3.3) is used in *structural geology* and tectonics to distinguish between different types of crustal faulting (Chap. 5). Second, the failure angle α in Eq. (3.3) is given solely by

Table 3.1 (**a**) Coefficient of internal friction, μ and cohesion, C of rocks (after Landolt-Börnstein 1982, Chapter by Fritz Rummel, pp. 198–199), and (**b**) dry and wet friction coefficients of some minerals (after Horn and Deere 1962)

(**a**)

Rock type	Coefficient of internal friction, μ	Cohesion, C (MPa)	Pressure range p, (MPa)
Limestone, Solnhofen	0.53	105	< 100
Marble, Carrara	0.62	28	6 to 60
Sandstone, Ruhr	0.88	37	< 100
Granite, Westerly	1.03	70	80 to 150

(**b**)

Mineral	Quartz	Calcite	Biotite	Talc	Muscovite	Serpentinite
μ(dry)	0.11	0.14	0.31	0.36	0.43	0.52
μ(wet)	0.42	0.68	0.13	0.16	0.23	0.29

the friction angle φ which is a constant in the Coulomb failure criterion. Thus, the orientation of the failure plane is independent of the confining stress. Laboratory experiments, however, show that the failure angle decreases with increasing confining pressure. Therefore, a modification of the Coulomb criterion is necessary and will be demonstrated in the following paragraph.

Mohr (1900) suggested a more general, non-linear failure relation of the form (our third fracture criterion, the *generalized Mohr fracture criterion*)

$$\begin{aligned} \tau &= f(\sigma) \\ F(\sigma_1, \sigma_3) &= 0 \end{aligned}, \tag{3.4}$$

which is displayed in stress space in Fig. 3.1e. In principle, this curve can be determined experimentally as the envelope of all of the Mohr circles that correspond to the states of stress that cause failure. The generalized Mohr criterion is usually concave downward (Fig. 3.1e) and the angle of failure α will decrease with increasing confining stress. Experimentally, the generalized Mohr criterion can be determined by plotting the Mohr circles for the stresses at failure, as found in a series of laboratory tests conducted under different confining stresses. The failure curve will then be given by the envelope of these circles. The Mohr envelope can be determined from three types of laboratory rock tests (Fig. 3.1f). In the uniaxial tensile test (Fig. 3.1f, 1) *uniaxial tensile strength*, T_0 is determined which falls on the negative σ-axis (Fig. 3.1e, $\sigma=-T_0$; tension negative as in rock mechanics convention). In the uniaxial compression test (Fig. 3.1f, 2) *uniaxial* (unconfined) *compressive strength* C_0 is determined. This value is shown on the critical Mohr circle through the origin (σ_1, $\sigma_3=0$) in stress space (Fig. 3.1e, 2) that touches the generalized Mohr fracture criterion envelope. In a suite of triaxial compression tests (Fig. 3.1f, 3) the *confined compressive strength* of rock, $C(p)$ is determined for various confining pressures. Note that each triaxial compression test requires one test specimen, since the critical Mohr circle touches the failure envelope only in the case where the specimen fails. The nonlinear fracture criteria of type (3.1)–(3.4) are all empirical in nature. One such failure law is capable of fitting data for many different rock types (Hoek and Brown (1980) criterion). The interested reader can find a compilation of such empirical rock failure criteria in the two monographs by Andreev (1995) and Sheorey (1997).

As an example, Fig. 3.2 shows a right cylindrical core of Aue granite before and after testing in triaxial compression. In Fig. 3.2a, the servo-controlled 4600 kN loading frame of MTS (*Material Test Systems Inc.*) with a stiffness of 11 GN/m
13.1, 3.2 (Fig. 3.2a, 1), a pressure vessel (Fig. 3.2a, 2) and a lower end plug with the rock sample wrapped in rubber are shown (Fig. 3.2a, b, 3). The rubber prevents the penetration of hydraulic oil into the specimen when the confining pressure of 40 MPa is applied. The sample bristling with acoustic emission (AE) sensors before testing is shown in Fig. 3.2b. The sensors are made of piezoceramics which convert the change in mechanical stress (e.g., crack opening) into the change of electrical signals. One application of AE in rock mechanics is discussed in Sect. 8.2. The specimen after failure at a vertical load of 900 kN is seen in Fig. 3.2c. At a typical laboratory strain rate of 10^{-5} s^{-1} (corresponding to 0.02 mm min^{-1}), the Aue granite core fails at a differential stress of 410 MPa (σ_1=450, σ_3=p_C=40 MPa). The shear rupture plane is oriented 30° from the maximum principal stress (acting vertical) at the top of the core and oriented about 25° from the σ_1-direction at the bottom of the core (Fig. 3.2d). Using arrays of AE sensors, we are able to map shear rupture propagation, even in non-transparent (opaque) rock material (e.g., Lockner et al. 1991; Zang et al. 1998).

In general, rock failure depends on all three principal stresses, also the intermediate principal stress σ_2 which has thus far been neglected. Imagine that a rock specimen is taken to failure by increasing the three principal stresses in some manner. The point of failure is plotted in the principal stress space, i.e. the space spanned by the three principal axes. If this procedure is repeated for different stress paths, all failure points measured form a surface (2D) in the stress space (3D) called the *failure surface*. The failure surface can be described by the equation

$$F(\sigma_1, \sigma_2, \sigma_3) = 0, \tag{3.5}$$

which forms our *fourth fracture criterion*. In this simplified "picture" which is a generalization of Eq. (3.4), the assumption is that the rock is intact at stress states inside the failure surface, while it fails for any stress state outside the failure surface (Chap. 8, Fig. 8.12). Since it is difficult to draw a surface in 3D, failure surfaces are often presented by cross-sections instead. The most common is the cross-section in a π-plane (Fig. 3.3). These planes are normal to the hydrostatic stress axis, i.e. the axis where σ_1=σ_2=σ_3. Among others, the cross-section of the Coulomb fracture criterion is shown in Fig. 3.3. It is seen to be an irregular hexagon with sharp corners. The *modified Lade criterion* is a simple criterion which appears to account for the influence of the intermediate principal stress on shear strength in a realistic way. The *Drucker-Prager criterion* is circular in the π-plane, while the Coulomb

Fig. 3.2 Experimental setup for a triaxial compression test in rock. **a** Servo-controlled 4.6 MN loading frame of MTS Systems Inc. with pressure vessel, **b** core sample Ag56r in rubber tube jacket with 10 piezoceramic sensors before testing, **c** Aue granite core with shear rupture plane after testing with 40 MPa confining pressure, and **d** close-up of shear rupture plane for determining the angle of fracture (modified after Zang et al. 2002). Experiments were carried out by S. Stanchits, C.F. Wagner and A. Zang

and the modified Lade criteria have a three-fold symmetry. As suggested by Nadai (1950), the driving force for failure is the second invariant of the deviatoric stress (Chap. 2, J_2). Nadai's assumption can be written in terms of the octahedral shear stress or the mean normal stress

$$\tau_{OCT}^2 = \frac{2}{3} J_2 = f(\sigma_m), \tag{3.6}$$

where f is some increasing function of the mean normal stress σ_m. Our fifth J_2-*fracture criterion* takes into account all three principal stresses. The deviatoric stress invariants have a straightforward geometrical interpretation in the principal stress space (Fig. 3.3). For example, $\sqrt{3J_2}$ is the equation of a circle centred on σ_m, with the normal pointing along the hydrostatic axis. Our π-plane is the plane normal to the (1, 1, 1) direction in the principal stress space, and is also called the *octahedral plane* or deviatoric plane. The normal stress and the shear stress on this plane are $\sigma_{OCT}=\sigma_m$ and $\tau_{OCT} = \sqrt{(2/3)J_2}$.

Experiments conducted under true-triaxial conditions (e.g. Haimson and Chang 2002; Haimson 2006; Mogi 2007) have shown that the intermediate principal stress σ_2 has a significant impact on the strength of several rock types (e.g. Colmenares and Zoback 2002). Typically, it is found that rocks are stronger when $\sigma_1>\sigma_2>\sigma_3$ than for the situations where $\sigma_1=\sigma_2$ or $\sigma_2=\sigma_3$. It is possible that any increase of σ_2 above σ_3 may cause additional strengthening. In their probabilistic failure model, Fjaer and Ruisten (2002) showed that the rock appears stronger – statistically – for the low stress symmetry case $\sigma_1>\sigma_2>\sigma_3$ than for the case of higher stress symmetry $\sigma_1>\sigma_2=\sigma_3$ or $\sigma_1=\sigma_2>\sigma_3$. The explanation of the role of the intermediate principal stress, however, lies in the factors involved in the crack propagation rather than the crack initiation (Sect. 3.3).

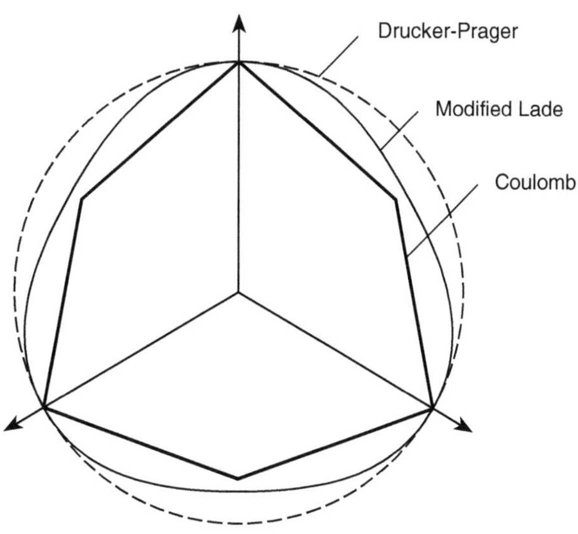

Fig. 3.3 Cross-sections for some failure criteria in the octahedral plane. The arrows represent projections of the principal axes onto the plane. The friction angle is 22.5°. Criteria are scaled so that they have equal intercepts (modified after Fjaer et al. 2008)

Exercise 3.1 *Coulomb fracture criterion.* In a laboratory triaxial compression test with σ_3=40 MPa confining pressure, an Aue granite (Erzgebirge, Germany) core fails at 410 MPa differential stresses ($\sigma_1-\sigma_3$=410 MPa). The shear rupture plane observed is tilted 30° from the direction of largest principal stress σ_1.

(a) Draw a Mohr circle for the laboratory test conditions of Aue granite.
(b) Display the Coulomb criterion with a tangent touching the Mohr circle and graphically determine the two characteristic parameters of the fracture criterion cohesion (zero-offset) and friction coefficient (slope).
(c) Answer question (b) again, if the fracture angle observed is 25°.

3.2 Mechanistic Failure Theories

The theoretical ultimate strength of solids is inferred from forces breaking atomic bonds across a lattice plane (Marder and Fineberg 1996). In 1920, Griffith recognized that real materials contain imperfections that result in failure at much lower stresses than the theoretical strength. The *Griffith criterion of fracture* is based on an energy balance for crack propagation. If a glass rod loaded in tension contains a crack of length *2c* perpendicular to loading direction (Fig. 3.4a), our sixth fracture criterion (the *uniaxial Griffith criterion*) reads

$$\begin{aligned} \sigma_c &= T_0 = \sqrt{\frac{2\gamma E}{\pi c}} \propto \sqrt{\frac{\gamma}{c}}, \\ c_c &= \frac{2\gamma E}{\pi \sigma^2} \propto \frac{\gamma}{\sigma^2} \end{aligned} \tag{3.7}$$

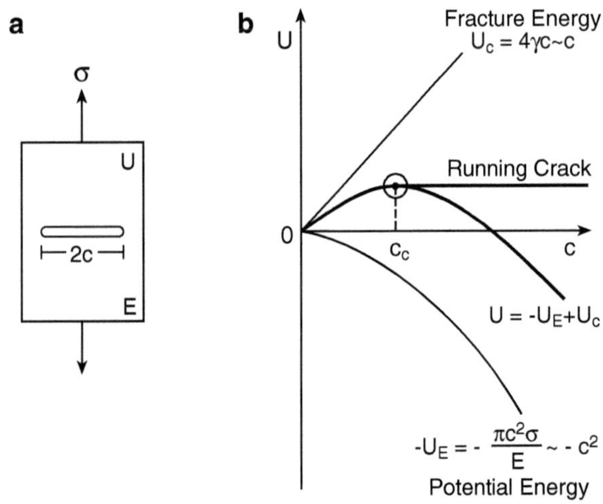

Fig. 3.4 Model of Griffith for crack propagation in a glass rod (**a**), and the balance of energies for the process (**b**)

Table 3.2 Specific fracture surface energies of rocks and quartz (after Landolt-Börnstein 1982, Chapter by Fritz Rummel, pp. 189–191)

Rock/Mineral	Specific fracture surface energy, γ[Jm^{-2}]	Type of test
Granite, Barre VT	60	direct pull
Quartzite, Chilhowee	50	3-point bending, notched specimen
Granite, Westerly RI	30	indentation test
Sandstone, Witten	28	3-point bending, notched specimen
Limestone, Solnhofen	5	3-point bending, notched specimen
Marble, Carrara	3	double torsion
Quartz, single crystal	0.5–1	cleavage type test

where σ_c denotes the *critical stress* and c_c the *critical crack length* for crack extension. E is Young's modulus of material and γ is the *specific fracture surface energy* required to create the unit area of a new crack surface. Data of γ result from laboratory tests (cleavage tests of quartz single crystals, three-point bend test of notched rock beams, see Table 3.2). The difference in effective fracture surface energy of rocks and minerals (factor 100) led Friedman et al. (1972) to speculate about the width of the fracture zone in sandstones due to the release of residual strains in the quartz aggregates during rupture growth.

In the following, we quote a paragraph from the article "How things break" by Marder and Fineberg (1996) related to the Griffith criterion and highlighting three important energies for a crack of length c which grows at rate $v=dc/dt$ in a plate. "The *potential energy* of the system decreases as the crack extends, and since the size of the region where this happens scales as c^2, the potential energy released scales as $-c^2$ (Fig. 3.4b, negative parabola); it also scales as the square of the applied stress (see Eq. (3.7)). The *fracture energy* makes the crack move forward. It requires breaking bonds, creating new surfaces and generating heat; it scales as specific fracture surface energy times crack length γc (Fig. 3.4b, positive straight line). The total *kinetic energy* due to the motion of crack scales as $(cv)^2$, since the amount of mass that moves as the crack opens scales as c^2. For the *quasi-static crack growth* problem of Griffith (very slowly growing crack), only the first two terms contribute to the energy balance (Fig. 3.4b, curve representing the sum of potential and fracture energy, $U=-U_E+U_C$). At the critical crack length (*Griffith point*, Eq. (3.7)) and Fig. 3.4b circled dot), the potential energy overwhelms the fracture energy and, from here on, more energy is released than consumed by crack extension. Now extension is rapid and spontaneous (Fig. 3.4b, running crack). Since (1) the sum of fracture and potential energy decreases as $(c-c_c)^2$ for $c>c_c$ and (2) the energy is conserved by converting potential energy to kinetic energy, we find that the crack speed is (Freund 1998)

$$v = \frac{dc}{dt} = v_{\max}\left(1 - \frac{c_c}{c}\right). \tag{3.8}$$

Therefore, the critical stress to fracture a body with a crack of initial size c scales as $c^{-1/2}$ (see Eq. (3.7)). The velocity v_{max} is the *Rayleigh wave speed*." A Rayleigh wave is a surface wave which propagates in media with a free surface parallel to this surface. Its amplitude decreases with distance from the surface. The crack speed is controlled by the Rayleigh wave speed, since any crack ligament can be regarded as a free surface (Chap. 2). For practical strength values computed from Young's modulus (*E/500*, Marder and Fineberg 1996), a critical crack length of about 1 μm can be computed. Such a pre-existing crack, predicted by Griffith theoretically and not visible in light microscopy, is called a *Griffith crack.*

In 1924, Griffith developed a 2D fracture criterion assuming that macroscopic failure can be identified with the initiation of cracking from the largest, most critically oriented Griffith crack. He analyzed the stresses around an elliptical crack in a biaxial stress field and found the most critical orientations that yielded the greatest tensile stress concentrations. Comparing these results with that for a crack in uniaxial tension, he obtained the *biaxial Griffith criterion* (our seventh fracture criterion):

$$\begin{matrix} (\sigma_1 - \sigma_3)^2 - 8T_0(\sigma_1 + \sigma_3) = 0 \\ \sigma_3 = -T_0 \end{matrix} \quad \text{if} \quad \begin{matrix} \sigma_1 > -3\sigma_3 \\ \sigma_1 < -3\sigma_3 \end{matrix}. \tag{3.9}$$

Compressive stresses are taken as positive. The condition $\sigma_1 > -3\sigma_3$ defines the state of predominantly compressive character. In Mohr space, this fracture criterion reads

$$\tau^2 = 4T_0(\sigma + T_0), \tag{3.10}$$

indicating that the corresponding Mohr failure envelope is a parabola (Fig. 3.1e). For the tensile fracture portion of this failure envelope, the most critically oriented crack is normal to the least principal stress. For the shear portion, it is inclined at an angle α from the largest principal stress direction given by

$$\cos 2\alpha = \frac{\sigma_1 - \sigma_3}{2(\sigma_1 + \sigma_3)}. \tag{3.11}$$

The criterion combines tensile and shear failure in a single criterion predicting a ratio of uniaxial compressive strength to tensile strength of eight ($C_0 = 8T_0$). The criterion is based on a micro-mechanistic failure theory, but analogous to the phenomenological failure theories discussed, the criterion neglects the effect of intermediate principal stress on rock strength. The angle α in Eq. (3.11) is measured from the major axis of the elliptical crack to the σ_1-direction. Since the maximum tensile stress along the crack ligaments does not occur exactly at the end of the major axis (see Sect. 3.3.1, wing crack), the initial crack tends to propagate out of its plane.

McClintock and Walsh (1962) considered friction along closed crack ligaments and reformulated the biaxial Griffith criterion. With the help of the *crack closing pressure* P_{cc}, they obtained the so-called *modified Griffith criterion* (our eighth fracture criterion)

$$\left(\sqrt{1+\mu^2} - \mu\right)(\sigma_1 - \sigma_3) = 4T_0\sqrt{1 + \frac{P_{cc}}{T_0}} + 2\mu(\sigma_3 - P_{cc}), \tag{3.12}$$

where μ is now identified with the friction coefficient of closed crack ligaments. The term P_{cc} is the critical macroscopic compressive stress perpendicular to the crack that is needed for closing the crack. The Mohr envelope in the normal and shear stress space is

$$\tau = 2T_0\sqrt{1 + \frac{P_{cc}}{T_0}} + 2\mu(\sigma - P_{cc}). \tag{3.13}$$

Mechanistic fracture theories (Griffith 1924; McClintock and Walsh 1962) both in principal stress space and Mohr stress space are compared in Scholz (1990). Criteria following Eqs. (3.1) to (3.4) account for the first-order strength properties of rock and cannot be distinguished on the basis of experimental data. Whereas the Coulomb criterion is strictly empirical, the Griffith and modified Griffith criterion predict macroscopic failure based on a description of the micromechanics. The approach Eq. (3.7) is incorrect when investigating rock failure under polyaxial compression. Note that the identification of rock failure with the extension of a single, isolated crack is a gross oversimplification of the complex process involved in the zone next to a propagating shear rupture (Paterson and Wong 2005).

3.3 Fracture Mechanics

Understanding rock failure under ambient crustal conditions requires (1) starter crack models resulting from the heterogeneous nature of rock which "work" in compression (Kemeny and Cook 1987, see Sect. 3.3.1), (2) the investigation of mixed-mode crack propagation (Whittaker et al. 1992, see Sect. 3.3.2), and (3) the computation of the interaction of the stress fields from many individual cracks (see Sect. 3.3.3). In linear fracture mechanics (LFM), mathematical cracks are analyzed in terms of stress intensity factor, energy release rate and *J*-integral (Liebowitz 1968; Lawn and Wilshaw 1975; Atkinson 1987; Gross and Seelig 2006). In nonlinear fracture mechanics (NFM), cohesive crack models indicate a region of intense micro-cracking ahead of the tip of a propagating macro-fracture.

The historical roots of fracture mechanics (FM) go back to Leonardo da Vinci (1452–1519) who drew a setup to test the strength of metal wires. One end of the wire is suspended while the other end is attached to a basket that is slowly filled with sand. When the wire breaks, the sand collected in the basket can be weighed and the strength determined. Another important ingredient in the history of rock fracture is the introduction of the coefficient of friction by Amoton (1699). Based on this, Coulomb (1776) developed his fracture criterion first applied to soil mechanics and later to rock mechanics (Sect. 3.1). The next milestone in FM was in 1913, when Inglis quantified stresses in a disk with a hole to explain the difference between practical and theoretical strength of materials (Orowan 1949). As we know from Sect. 3.2, significant modification of the Inglis imperfection model was made in 1920 by

Griffith (Sect. 3.2) explaining tensile cracking in glass rods with an energy balance concept. Irwin (1958) modified the Griffith concept for quasi-static crack by replacing the stress singularity ahead of the crack tip (LFM approach) by a plastic zone characterized by the attenuation of stress magnitudes to realistic values (NFM approach). From here on, the fracture mechanics of rocks splits into three parts crack initiation, crack propagation, as well as crack interaction and coalescence.

3.3.1 Crack Initiation

Under overall compressive loading as expected in the deeper Earth's crust, crack initiation in rock can start (1) at sliding contacts of pre-existing cracks (wing cracks), (2) at pores (pore cracks) and (3) at inclusions or other stress concentrators like contacts of neighboring mineral grains (Hertzian cracks). The "birth" of the "wing crack model" was in 1963 when Brace and Bombolakis showed that planar cracks inclined with respect to compressive principal stresses do propagate by producing small daughter tension cracks (*wing cracks*) at the tips of the sliding mother crack (Fig. 3.5a). Nemat-Nasser and Horii (1982) and Horii and Nemat-Nasser (1985) observed wing cracks in transparent materials like epoxy resin (PMMA). Ashby and Hallam (1986) formulated the conditions for wing-crack initiation (our ninth rock fracture criterion, the *wing-crack fracture criterion*)

$$\sigma_1 = A\sigma_3 - B, \tag{3.14}$$

where A, B are material properties and (σ_1, σ_3) are the 2D principal stress components (Fig. 3.5a). The material constants depend on the properties of the mother crack (friction coefficient, μ and crack half length, a) and the critical stress intensity factor K_{IC} for tensile cracking.

$$A = \frac{\sqrt{1+\mu^2}+\mu}{\sqrt{1+\mu^2}-\mu}; B = \frac{\sqrt{3}}{\sqrt{1+\mu^2}-\mu}\frac{K_{IC}}{\sqrt{\pi a}}. \tag{3.15}$$

The driving force for wing-crack nucleation is K_{IC} (see Sect. 3.3.2). Assuming randomly oriented pre-existing sliding cracks, the wing cracks should first nucleate from those sliding cracks oriented at

$$\psi = \frac{1}{2}\tan^{-1}\left(\frac{1}{\mu}\right) \tag{3.16}$$

In Fig. 3.5b, the stress on the ligament of an elliptical mother crack is shown (after Lajtai 1971). Using a mother crack which is tilted by ψ=30° from the maximum principal stress direction and having an aspect ratio of 0.1, the first wing crack will develop close to the upper tip of the mother crack (Fig. 3.5b, *wing 1*), where

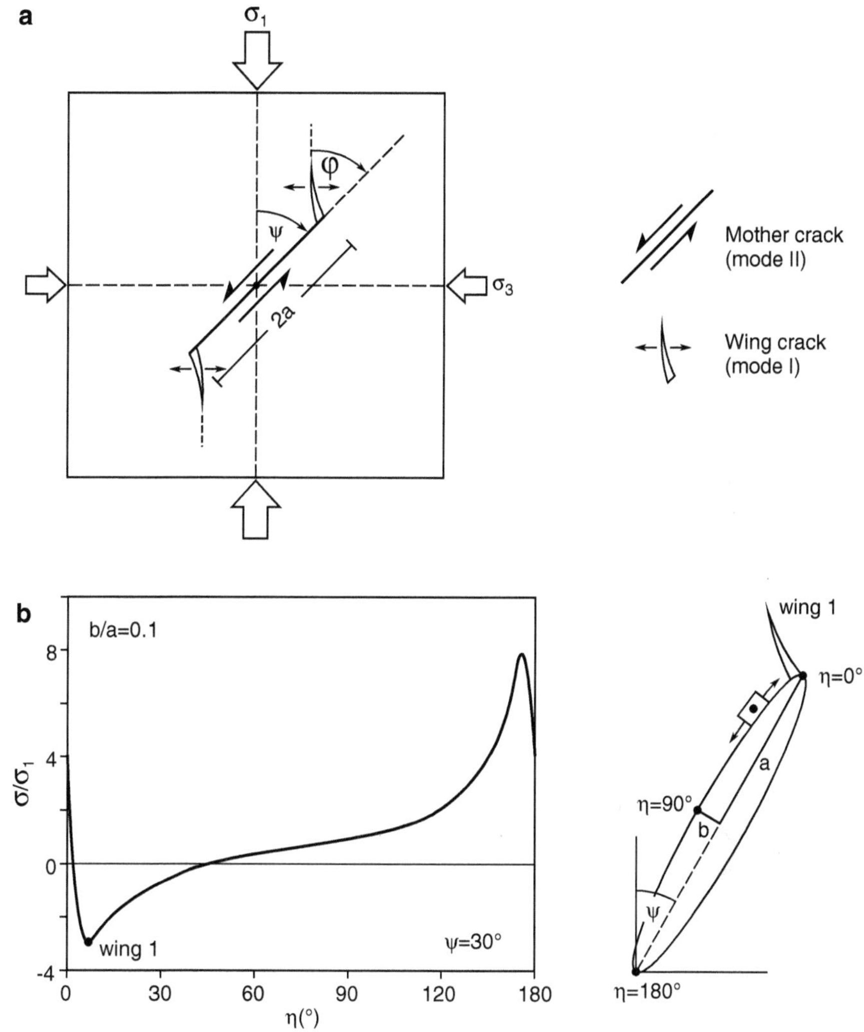

Fig. 3.5 Model of wing cracks (Ashby and Hallam 1986) developing at the tips of a sliding shear crack (**a**), and normalized stress along the periphery of an elliptical, tilted mother shear crack with respect to maximum principal stress direction (**b**)

maximum tensile stress occurs. The second wing crack will develop at the lower tip of the mother crack (Fig. 3.5b, π-periodicity), accordingly. Note that in order to formulate the wing crack criterion (Eqs. (3.13) and (3.14)), a *mixed mode fracture criterion* is needed since daughter crack (mode I, tensile) and mother crack (mode II, in plane shear) propagate in different ways (Whittaker et al. 1992). The definition of different modes of fractures in rock mechanics is given in Sect. 3.3.2. Assuming that the wing cracks initiate normal to the orientation of maximum tensile stress (Erdogan and Sih 1963), the initiation angle of the wing crack and the sliding crack is $\varphi{\sim}70°$ (Fig. 3.5a). Varying the mixed mode fracture criterion, the wing initiation angle is $\varphi{\sim}75°$ computed from the maximum energy release-rate criterion, and $\varphi{\sim}82°$ from the minimum energy-density criterion (Dumstorff et al. 2006). Assuming the maximum tensile stress criterion, this angle is relatively constant over a range of mother-crack orientations from 20 to 50° with respect to the largest principal stress direction for $0 \leq \mu \leq 0.6$. For rocks with friction coefficient $\mu{\approx}0.6$ at closed mother cracks, the wing crack criterion simplifies (Ashby and Sammis 1990):

$$\sigma_1 = 3.12\sigma_3 - 3.06\frac{K_{IC}}{\sqrt{\pi a}}. \tag{3.17}$$

Crack initiation can be detected experimentally by the onset of acoustic emissions (Sect. 8.2), by the nonlinearity of the stress-strain curve or by *dilatancy* (volume expansion under compression) of the rock specimen. Brace et al. (1966) and Holcomb and Costin (1986a, b) showed crack initiation data based on each of the three techniques and verified formula (3.14) on triaxially compressed Westerly granite samples. They showed that the $\sigma_1(\sigma_3)$-relationship of experimental data generally approximately follow linear trends, with slopes corresponding to values of μ varying from 0.55 to 0.64. They computed a representative mother-crack length of $2a{\approx}1$ mm from the intercept in (3.14) using published data of the fracture toughness $K_{IC}=1$ MPa $\sqrt{m}$. Although developed for compact rock, the wing-crack model has also been used to analyze brittle behaviour in porous rock (Pestman and van Munster (1996), see Chap. 8).

For porous rock, the pore-crack model is a physically more adequate description of crack initiation. The behavior of pore emanated cracks is qualitatively similar to that of a wing crack induced by sliding on an inclined crack. *Stable* propagation of pore cracks initially results in dilatancy. As the cracks propagate further, they begin to interact and if their mean spacing is sufficiently small, crack propagation becomes *unstable* leading to crack coalescence and catastrophic rock failure. When porous, brittle solids are loaded in compression, small cracks grow from the pores. Based on compression tests of glass and brittle plastic plates with cylindrical holes, Sammis and Ashby (1986) developed the *pore crack model*. The cracks grow from the pore surface towards the maximum principal stress direction (Fig. 3.6a) until the stress intensity factor K_I becomes equal to the fracture toughness K_{IC} of the material. Then the stable crack length, $L=l/a$, is defined by our tenth fracture criterion (the *pore-crack fracture criterion*)

$$\begin{aligned} \frac{\sigma_1\sqrt{\pi a}}{K_{IC}} &= -F(\lambda, L) \\ F(\lambda, L) &= L^{-1/2}\left\{\frac{1.1(1-2.1\lambda)}{(1+L)^{3.3}} - \lambda\right\}^{-1}, \\ F(0, L) &= \left\{\frac{(1+L)^{3.3}}{1.1\sqrt{L}}\right\} \end{aligned} \tag{3.18}$$

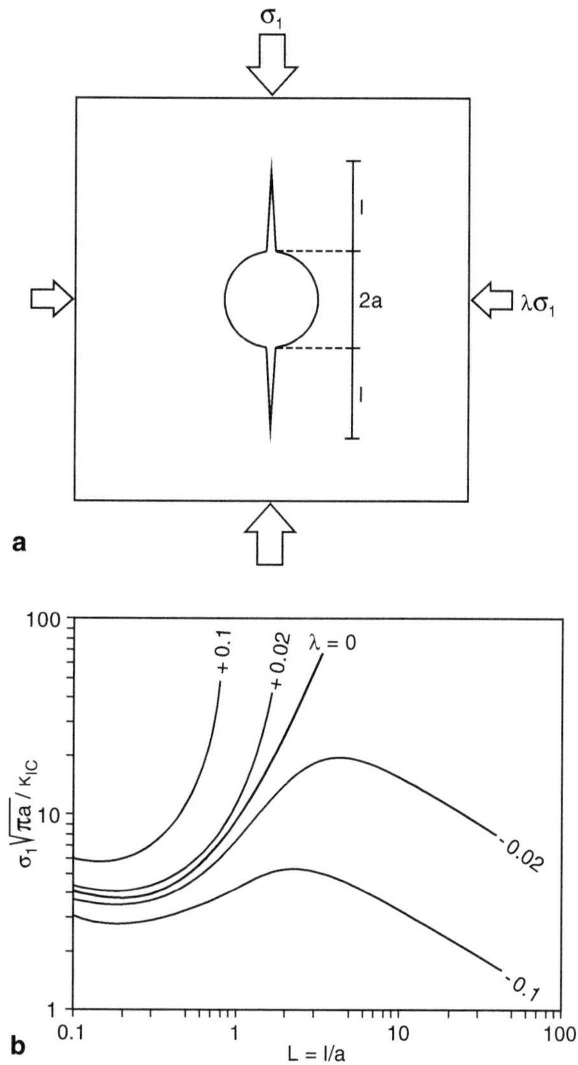

Fig. 3.6 Pore-crack model (Sammis and Ashby 1986) under biaxial compressive stresses (**a**), and normalized stress values versus the crack extension $L = l/a$ in an infinite plate, compression is positive (**b**)

where σ_1 is the magnitude of the largest compressive principal stress and λ denotes the ratio of the two principal stress magnitudes ($\lambda = \sigma_3/\sigma_1$). The result of the crack extension L is plotted as a function of normalized stress, $\sigma_1\sqrt{\pi a}/K_{IC}$ in Fig. 3.6b. For $\lambda \geq 0$ including the case of simple compression ($\lambda = 0$), the crack growth is stable (after an initial pop-in to $L = 0.2$). Cracks ultimately become unstable for $\lambda < 0$. During *unstable crack growth* the crack propagates catastrophically without any stop (arrest) until the sample fails. At elevated confining pressures ($\lambda \geq 1/3$), crack growth is suppressed since circumferential stresses at the pore surface vanish. In addition, Sammis and Ashby (1986) assured their theoretical findings with experimental results on crack growth in glass and plexiglass plates using fracture toughness values of glass 0.6 and PMMA 1.0 MPa $\sqrt{m}$.

Weakly cohesive rock, however, involve more complex micro-mechanisms like sliding, rotation and crushing of mineral grains (Wong 1990, Wu et al. 2000) not covered by the pore-crack model. Therefore, Zhang et al. (1990) developed a Hertzian fracture model to analyze the onset of grain crushing and pore collapse in sandstones. In this study, porous rock is approximated by a randomly packed assembly of spherical grains. In the vicinity of the grain contact region, the two particles are subjected to local normal force F, which is related to effective pressure (Chap. 5), porosity Φ and grain radius R (Fig. 3.7). The local stress field is governed by the magnitude of F, the radius R and the *elastic moduli* of the grains (Young's modulus, E and Poisson ratio, ν). Using the assumptions that (1) the maximum circumferential stress is attained at the perimeter of the contact area and (2) the stress intensity factor K (Sect. 3.3.2) of a crack located at the contact region is that of an edge crack, the critical effective pressure for the onset of grain crushing (our eleventh fracture criterion, the *Hertzian fracture criterion*) is

$$P_{cr} = 2.2\frac{(1-\nu^2)^2}{(1-2\nu)^3}\frac{K_{IC}^3}{E^2}(\alpha\Phi R)^{-3/2}, \tag{3.19}$$

where $\alpha = c/R$ is the ratio of the initial crack length to the grain radius. The ratio α was estimated from experimental data giving values of $\alpha = 3.6 \times 10^{-5}$ for quartz grains and $\alpha = 1.7 \times 10^{-4}$ for feldspar grains, respectively (Wong et al. 1992). Mechanical data on sandstone suggest that low-porosity rock and rock with smaller grain size require higher critical pressures (Wong et al. 1997).

All rigorous theoretical investigations of the deformation of aggregates of elastic grains are based on the classical solution to the problem of normal (Hertz 1882) and shear (Mindlin 1949) interaction of elastic spheres (Johnson 1985). The influence of cohesive bonding on the elastic deformation of porous rock was considered by Digby (1981). The *Hertzian contact* had to be modified by cementing agents preventing the grains from fracturing (Dvorkin et al. 1991). Effective properties of cemented granular material (Fig. 3.7b) are calculated analytically (Dvorkin et al. 1994) and by using numerical techniques (Zang and Wong 1995). In the latter approach, finite-element computations for models with various cement fractions and various cement stiffness show that high tensile stresses develop near triple junctions where grains, cementing material and pore space meet (Fig. 3.7b, dots). These points are favorable localities for the onset of grain breakage and cracking in cemented granular materials. Since tensile stresses are inhibited in the cement layer, it may be shielded from extensile cracking. "Weak" cement (low stiffness) fails through shear fracturing and, in this way, absorbs most of the elastic energy stored in the system preventing "strong" grains from failure.

In Fig. 3.8, one example of Hertzian fracture and one example of cement fracture is shown for a Flechtingen sandstone (Bebertal, Germany) deformed in the laboratory. The optical micrograph in Fig. 3.8a shows Hertzian fractures in

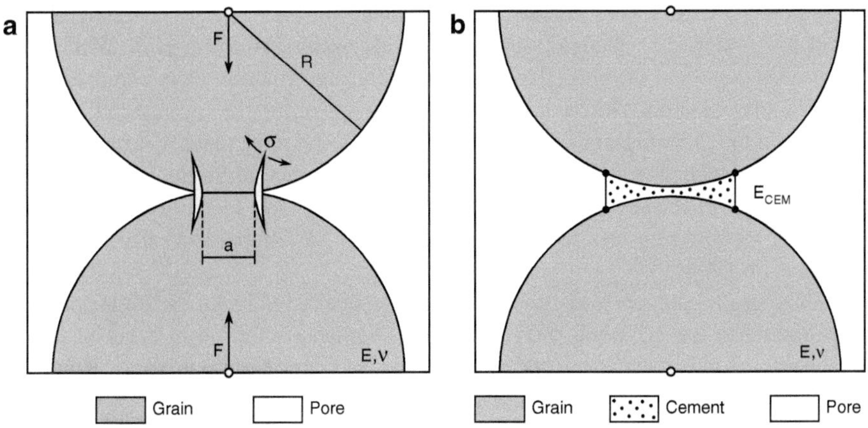

Fig. 3.7 Hertzian fracture model (Zhang et al. 1990) for granular materials (**a**), and favourable points for cracking in cemented granular material (Zang and Wong 1995) at triple junctions (*dots*) where grain, cement and pore space meet (**b**)

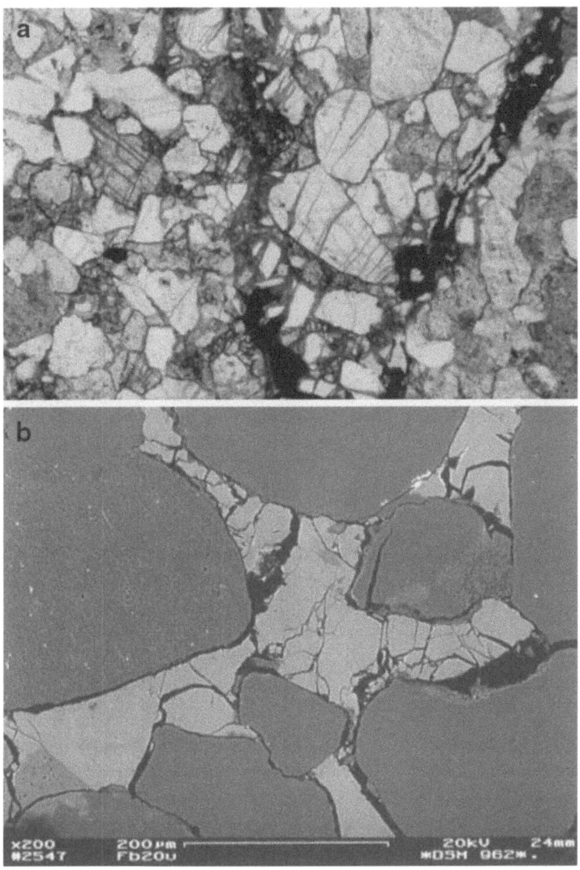

Fig. 3.8 Microscopic fracture inspection in Flechtingen sandstone after triaxial testing. **a** Optical micrograph of Hertzian fractures in neighbouring quartz grains in dry sample Fb36 after asymmetric, triaxial testing and **b** SEM photograph showing microcracks in calcite cement of wet core Fb20 after a symmetric, triaxial test (Zang et al. 1996c)

neighbouring quartz grains deformed within the process zone of a shear rupture (Sect. 3.4) induced by triaxial (asymmetric) indenter tests (Zang et al. 2002). For crack visualisation purposes, the rock cores after deformation were saturated with epoxy under vacuum. The backscattered scanning electron microscope (SEM) photograph in Fig. 3.8b highlights cement fracturing in the same material after standard triaxial (symmetric) testing. Quartz grains remain intact while the calcite cement absorbs most of the shear rupture energy by cracking. Wet Flechtingen sandstones (Fig. 3.8b) result in a less localized fracture zone as compared to dry specimens (Zang et al. 1996c).

When grain contacts undergo complete slip, distinct element methods are more appropriate to use (Cundall 1971, Jing and Stephansson 2007) in order to model kinematics, force distribution and development of brittle failure in granular materials. One commercial software package based on discrete elements is the Particle Flow Code (PFC, Itasca Ltd.), see Chap. 8.

3.3.2 Crack Propagation

Once initiated, the wing crack grows out of the initial plane of the sliding crack. Using very brittle epoxy material (FTPE), Yamaguchi et al. (2008) mapped out the propagation path of wing cracks at different pressure levels starting from a single sliding crack (Fig. 3.9, *M1*). At a critical pressure level of our ninth fracture criterion (3.14)–(3.15), a first pair of wing cracks initiates (Fig. 3.9a, *W1*) and grows in *mode I* until the sliding mother crack is energetically more favourable to grow (Fig. 3.9b, *M2*). At a second critical pressure level, a second pair of wing cracks initiates (Fig. 3.9c, *W2*) and subsequently the *mode II* mother crack is re-activated. In conclusion, the simple wing crack growth turns out to be a complex process of energy partition between mode I and mode II cracking, a so-called *mixed mode cracking* phenomenon.

In fracture mechanics (FM), mathematical cracks are assumed and the key parameter that controls the nucleation and propagation of cracks is the critical stress-

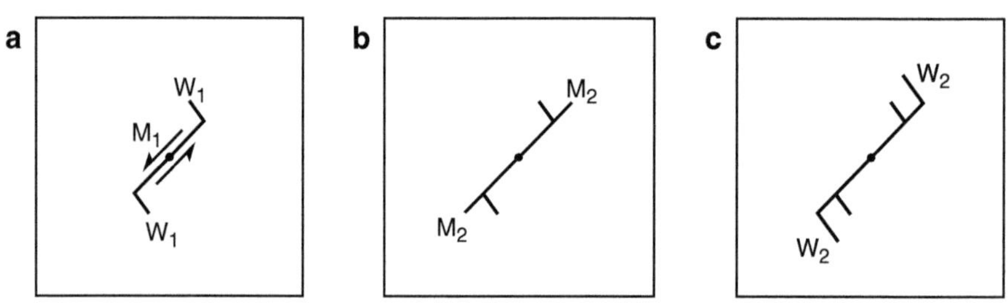

Fig. 3.9 According to Yamaguchi et al. (2008) in experimental data on brittle epoxy (FTPE), a single sliding crack results in multiple pairs of wing cracks when the energy jumps back and forth in the mixed mode crack propagation problem. In **a**, the first pair of wing cracks is emitted from the sliding mother crack. In **b**, the mother crack is reactivated in its own plane. In **c**, the second pair of wing cracks develops from the extended mother crack. Stress increases from left **a** to right **c**

intensity factor. A *mathematical crack* is defined as an ideal flat, perfectly sharp double surface of zero thickness (Paris and Sih 1965). In 2D, the crack consists of a crack tip (Fig. 3.10a, dot) and two crack ligaments (Fig. 3.10a, straight lines). Depending on the crack-opening displacement with respect to the crack propagation direction, we can separate *tensile cracks* (Fig. 3.10a, tensile *mode I*), *in-plane shear cracks* (Fig. 3.10b, *sliding mode II*) and *anti-plane shear cracks* (Fig. 3.10c, *tearing mode III*). In general, the 3D crack tip stress field is a superposition of the near-field stresses resulting from all three crack opening modes. In *linear fracture mechanics* (*LFM*), the behaviour of a cracked body (Fig. 3.10a–c) is assumed to be linear elastic. In *nonlinear fracture mechanics* (*NFM*), inelastic processes occur within a process zone propagating ahead of the moving crack tip (Fig. 3.10d–f). The *fracture process zone* is the region next to the crack tip (2D) or crack front (3D) in which the micro-physically complex process of breaking atomic bonds, creating new crack surfaces and generating heat takes place (Fig. 3.10d–f, process zone).

No one has published the actual geometry of the process zone of an anti-plane shear rupture (Fig. 3.10f, tearing mode). An attempt to experimentally investigate the geometry of a mode III fracture process zone in Solnhofen limestone was initiated by Cox and Scholz (1988a, b). Unfortunately, the laboratory results were complicated by the effect of two closely spaced, interacting mode III cracks. A second attempt was made by Sabato Ceraldi et al. (2006) for mode III fractures in Berea sandstone. It still seems to be a discrepancy between laboratory work focusing

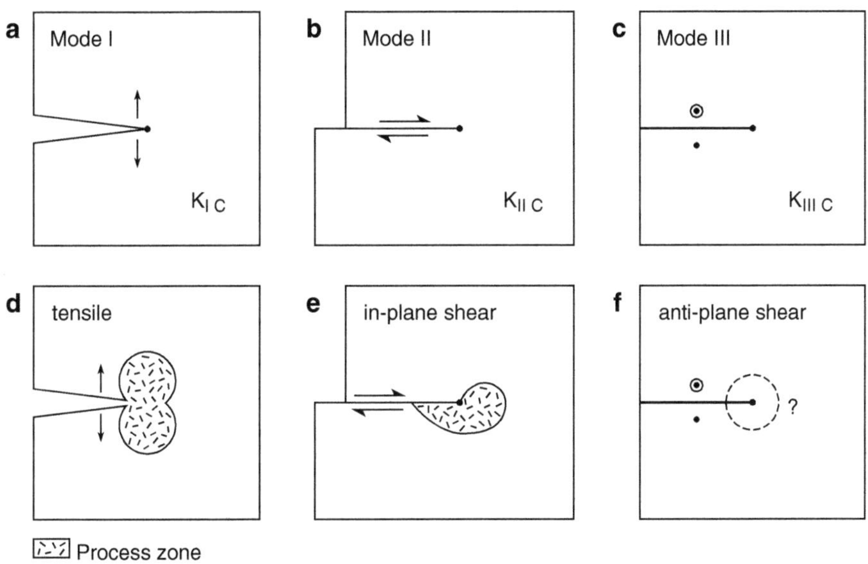

Fig. 3.10 Idealized pictures of mode I, II and III fracture propagation in LFM (*top*) and NFM (*bottom*). An idealized, sharp cut in elastic, homogeneous, isotropic material represents **a** mode I tensile crack, **b** mode II in-plane shear crack and **c** mode III anti-plane shear crack growth in LFM. The idealized shape of the process zone of a **d** tensile fracture, **e** sliding mode shear fracture and **f** tearing mode shear fracture is contoured by approximate solutions in NFM

mostly on mode II fractures and source models of mode III ruptures used in earthquake mechanics (Udias 1999).

In linear fracture mechanics, our twelfth fracture criterion (the *linear fracture mechanics criterion*) is based on the critical stress intensity factor, the fracture toughness of rock

$$K_i = K_{iC} \text{ with } i = I, II, III. \tag{3.20}$$

Fracture (the onset of crack propagation) occurs if the stress intensity factor of a certain fracture mode K_i reaches a critical value, K_{ic} (the so-called *fracture toughness* which is a material parameter only). The *stress intensity factor* characterizes the stress singularity at the crack tip and depends on (1) the geometry of the cracked solid and (2) the external loads. In Fig. 3.11 (curve LFM), the stress component varies inversely as the square root of the distance from the crack tip. This drop in stress magnitude is mathematically called *stress singularity*. In linear fracture mechanics (LFM), all crack-tip singularities are of type $r^{-1/2}$, where r is the distance from the sharp crack tip. Since real rocks cannot sustain infinite stresses, inelastic processes close to the crack tip (microcracks, dislocation loops) alleviate the singular stress values to a finite value T_0, the tensile strength of the material (Fig. 3.11, curve NFM). Due to stress redistribution near the crack tip, the energy indicated by regime A is dislocated to energy indicated by regime B (Fig. 3.11). For quasi-static crack propagation (e.g. Griffith cracks), energy terms calculated from stress-strain curves are supposed to be equal (Fig. 3.11, $A = B$).

Fracture toughness values from laboratory experiments of some rock types are listed in Table 3.3. Note that (1) fracture toughness values of tensile cracks are lower compared to those of shear cracks, and (2) that K_{IC} tests are standardized

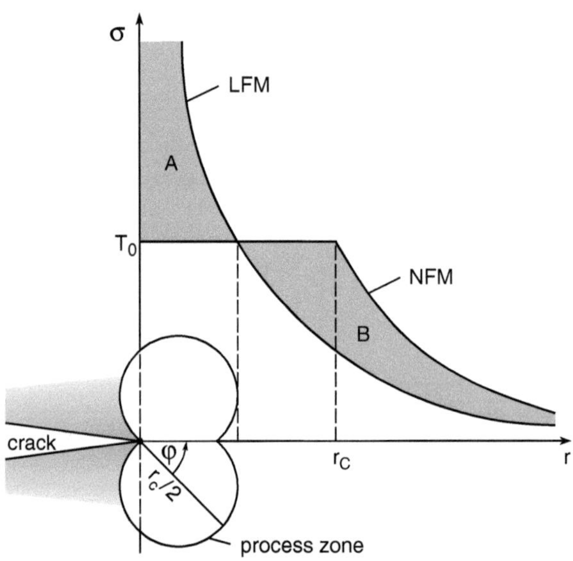

Fig. 3.11 Stress singularity at the tip of a linear FM crack compared to the finite-stress value (tensile strength) in the process-zone model of nonlinear FM (after Schmidt 1980)

Table 3.3 Fracture toughness values (MPa $\sqrt{m}$) of rocks. After Scholz (1990) © Cambridge University Press and Backers (2005) © GFZ German Research Centre for Geosciences

Rock	K_{IC}	K_{IIC}^{0}	K_{IIC}^{30} #	K_{IIIC}
Granite				
– Aue, Germany	1.6	4.1	10.0	–
– Mizunami, Japan	2.4	4.9	13.2	–
– Westerly (RI), USA	1.7[s]	–	–	2.4[s]
Marble, Carrara, Italy	2.4	3.1	6.3	–
Limestone				
– Rüdersdorf , Germany	1.1	2.3	5.5	–
– Solnhofen*, Germany	1.0[s]	–	–	1.3[s]

$K_{IIC}^{0}, K_{IIC}^{30}$ mode *II* fracture toughness from Punch Through Shear tests (Backers 2005) at zero and 30 MPa confining pressure
[S]fracture toughness data from Scholz (1990)
*perpendicular bedding

(e.g., three-point bending test on rock cores with a Chevron notch) whereas K_{IIC} and K_{IIIC} tests are non-standardized techniques (Backers et al. 2002, Backers 2005). Values of the critical stress intensity factor depend on the type of material (grain size, mechanical behaviour), and the conditions of pressure and temperature (see Table 3.3, K_{IIC}^{0} and K_{IIC}^{30}).

If some critical energy density in the material is exceeded, the fracture will extend. The mechanical energy associated with the work being done by displacements at the far field must be balanced by the energy associated with the two fracture surfaces being formed plus the energy stored in the elastic field around the fracture. Energy can also be dissipated as heat or in other irreversible processes. The *critical energy release rate* is G_c and is twice the energy released per unit area Q_f of the two fracture surfaces formed, $G_c=2Q_f$.

The critical strain-energy release rate must be reached to extend the fracture and therefore the following relationships exist between the energy release rate and the fracture toughness for the three fracture modes of the material when plane strain conditions are imposed (Irwin 1958)

$$G_{IC} = \frac{K_{IC}^2}{E^*}; \; G_{IIC} = \frac{K_{IIC}^2}{E^*}; \; G_{IIIC} = \frac{K_{IIIC}^2}{2\mu^*}, \tag{3.21}$$

where $E^*=1(1-\nu^2)^{-1}$, $\mu^*=E(2(1+\nu))^{-1}$ and E, ν are the elastic constants. The original strain energy release rate criterion (*G-criterion*) of Griffith (1920) states that when the strain-energy release rate in the direction of maximum G-value reaches the critical value G_c, the fracture tip will propagate in that direction. The G-criterion does not distinguish between the different modes of failure. Since both mode I and mode II fracturing are common in rocks, Shen and Stephansson (1993a) introduced the so-called *F-criterion* (our thirteenth fracture criterion), which states that in an arbitrary direction (θ) at a fracture tip, the F-value is

$$F(\theta) = \frac{G_I(\theta)}{G_{IC}} + \frac{G_{II}(\theta)}{G_{IIC}}, \tag{3.22}$$

where $G_I(\theta)$, $G_{II}(\theta)$ *and* G_{IC}, G_{IIC} are the strain-energy release rates at the fracture tip and critical energy release rates for mode I and mode II, respectively. When the maximum F-value of Eq. (3.22) reaches one, fracture propagation will occur. The sum of the normalized G-values in the F-criterion is used to determine the failure load and the direction of failure.

3.3.3 Crack Interaction and Coalescence

Crack interaction arises from the overlap of the local stress fields between two neighbouring cracks, which will be a function of the crack spacing. Extreme cases of crack interaction are those in which the cracks are, respectively, collinear and parallel (Kemeny and Cook 1991). *Crack coalescence* means a very strong crack interaction where initially separated cracks are growing and come together, often by mixed tensile and shear modes (mixed-mode cracking).

Three approaches can be used to analyze the mechanics of crack interaction (1) continuum damage mechanics (Kachanov 1986), (2) stochastic damage mechanics (Gross and Seelig 2006), and (3) fracture mechanics simulation using boundary element methods (Shen and Stephansson 1993b; Vásárhelyi and Bobet 2000; Shen et al. 2006). A common feature of damage models is their prediction that the failure envelope is linear. For example, Ashby and Sammis (1990) connect the empirical coefficient of friction in Eq. (3.2) to the friction coefficient of the sliding crack model in Eq. (3.15), and the macroscopic parameter of cohesion in Eq. (3.2) to the microphysical parameter $K_{IC}\,(\pi a)^{-0.5}$ in Eq. (3.17).

3.4 Nonlinear Fracture Mechanics

The term process zone describes the region at the crack tip that involves non-linear, irreversible deformation in the form of microcracking or crystal plasticity. In situations where this process zone is large compared to sample dimensions, the linear fracture-mechanics approach breaks down. While the fracture process zone is small for metals, it cannot be neglected in concrete, rock and ceramics. Hoagland et al. (1973) postulated a finite zone of distributed crack damage ahead of a moving fracture tip. Evans et al. (1977) calculated the geometry of this zone ahead of a tensile fracture (Figs. 3.10d, 3.11). Several theories have been formulated to account for the effect of inelastic deformation at the tip of a moving crack. Following Dyskin (1997), process-zone models based on nonlinear FM can be separated into cohesive zone-crack models (Dugdale 1960; Barenblatt 1962), postpeak behaviour models (Goodier 1968), bridging models (Hoagland and Embury 1979; Rossmanith 1983) and non-singular stress models (Wieghardt 1907; Neuber 1937, 1984). For shear cracks, Rice (1980) developed a slip weakening instability model and calculated the fracture energy from the work done in shear during the stress relieve.

Independently, Schmidt (1980) and Ouchterlony (1980) described a process-zone model for rock assuming a zone of microcrack damage ahead of a tensile crack. Similar to the published plastic zone for crack growth in metals (Irwin 1958), they developed a 2D process zone model appropriate for brittle rocks (Fig. 3.11). Microcracking ahead of the fracture occurs if the stresses at the crack tip reach the local tensile strength value, T_0 of the material. They computed the region where the local stress at the crack tip exceeds the tensile strength of material ($\sigma \geq T_0$) and obtained a *nonlinear FM criterion* (our fourteenth fracture criterion)

$$r_c(\varphi) = \frac{1}{\pi}\left(\frac{K_{IC}}{T_0}\right)^2 \left\{\cos\left(\frac{\varphi}{2}\right)\left[1+\sin\left(\frac{\varphi}{2}\right)\right]\right\}^2. \tag{3.23}$$

Note that the zone where stress redistribution occurs at the crack tip (Fig. 3.11 $0 \leq r \leq r_c$) is double the region ($0 \leq r \leq r_c/2$) as quantified by Eq. (3.23). The onset of unstable cracking is characterized by the critical radius, $r_c(\varphi)$ which reaches its maximum value at $\varphi = 60°$ (Fig. 3.11). Müller (1987) experimentally found that for different rocks, r_c varies from 1 mm (rock salt), 3 mm (marble, Carrara), 5 mm (sandstone, Buntsandstein) to 22 mm (granite, Falkenberg). If one considers friction along partially opened or closed crack ligaments (Fig. 3.12, c_B= breakdown crack length) the process zone is smaller compared to the case of a frictionless crack.

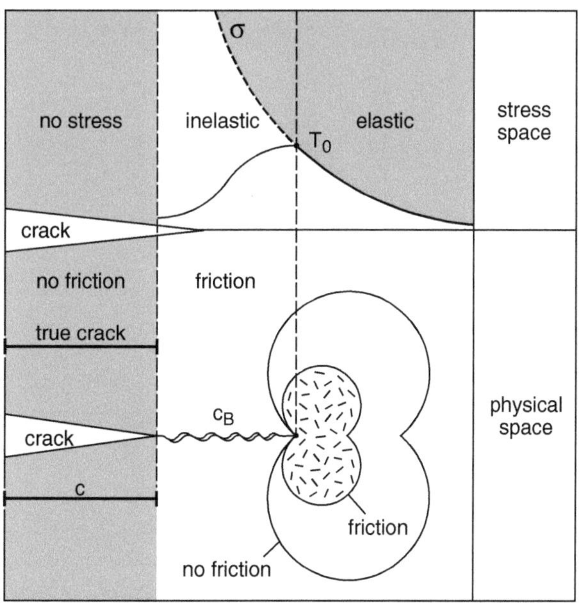

Fig. 3.12 Comparison of stresses (*top*, stress space) and size of process zone (*bottom*, physical space) for NFM cracks with and without friction along closed crack ligaments

Leaving the sharp tips of cracks in linear FM, blunted cracks in nonlinear FM would result in higher fracture toughness values for the same material. In addition, the fracture toughness depends on state variables (e.g., pressure (Table 3.3), temperature). Equation (3.23) can be easily modified for a confining pressure term p_c added to T_0 in the denominator ($K_{IC}/(T_0+p_c)$), resulting in a smaller process zone as compared to the unconfined case. Note that the process zone of tensile cracks (due to the symmetry of the crack-opening mode) is symmetric (Fig. 3.10d) while the process zone of a shear crack has an asymmetric shape (Fig. 3.10e, after Rossmanith 1995).

According to today's knowledge of FM, the zone of microcracking ahead of the tip of a moving rupture plays a key role in understanding the failure of rock in the Earth's crust. Note that the width of the process zone (and also the fracture toughness) scales with the length of the rupture in rock mass (Vermilye and Scholz 1998; Scholz 2002). In Fig. 3.13 we compiled scaling data of the fracture process zone from laboratory and natural faults over six orders of magnitude. In this double logarithmic plot of process zone width versus fault length, the scatter of single data points is as large as one order of magnitude. The regression slope of 1/62 from natural faults (Fig. 3.13, solid line from Vermilye and Scholz 1998) and the regression slope of 1/50 from laboratory faults (Zang et al. 2000), however, give a rule of thumb for this fracture-scaling relationship.

The above considerations are valid for the quasi-static crack growth situation. For high-speed cracks like earthquake ruptures propagating at the speed of km/s the crack tip stresses must be calculated using the equation of motion in place of the equation of equilibrium. This analysis is called crack kinetics (Freund 1998) and

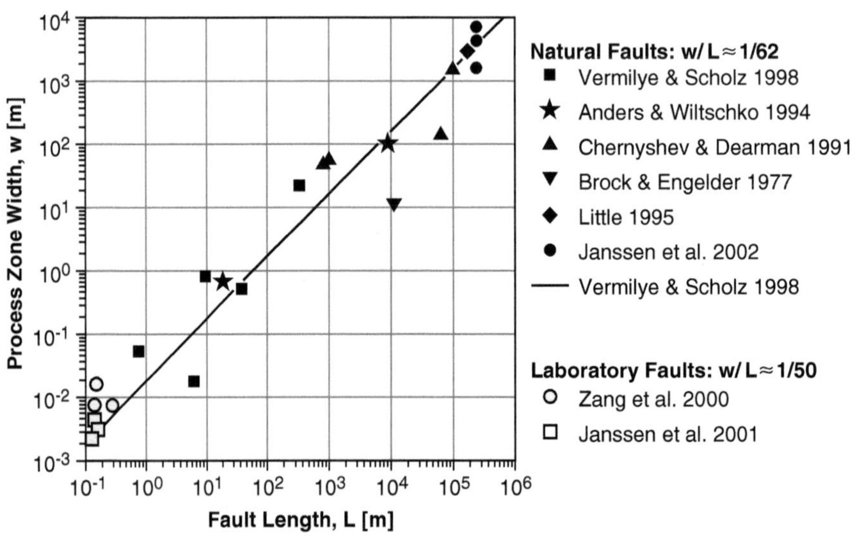

Fig. 3.13 Fault-scaling relationship in rock. The width of the fracture process zone determined on natural and laboratory faults is plotted versus the length of fault

leads to the definition of dynamic stress intensity factors. Crack dynamics apart from subcritical crack growth is neglected in this chapter. Further, damage models, probabilistic and statistical rock fracture (Alava et al. 2006) and size effects in brittle rock fracture (Bazant and Planas 1998) are neglected. Armed with our 14 rock fracture criteria summarized in Table 3.4, we are prepared to enter the rock stress terminology and develop simple models of Earth stresses varying with depth.

Table 3.4 Summary of rock fracture criteria

#	Author	Fracture Formula	Name of criterion and applicability
	Phenomenological Theory		
1	C.A. Coulomb (1736–1806)	$\tau_{\max} = \frac{\sigma_1 - \sigma_3}{2}$	*Maximum shear stress criterion* for isotropic homogeneous rock
2	C.A. Coulomb	$\tau = \mu\sigma + C$	*Coulomb criterion* for compact rock without appreciable open cracks
3	O. Mohr (1835–1918)	$F(\sigma_1, \sigma_3) = 0$	*Generalized Mohr criterion* for rock with appreciable open cracks and concave failure envelope
3A	Hoek and Brown (1980)	$\tau = A(\sigma_N + B)^C$	*Hoek-Brown criterion* is an empirical law of type (3) fitting many different types of rock with three parameters.
4	Handin et al. (1967)	$F(\sigma_1, \sigma_2, \sigma_3) = 0$	*Failure surface* of rock under three principal stress components
5	A. Nadai (1950)	$\tau_{OCT}^2 = \frac{2}{3}J_2$	*Nadai's fracture criterion* uses the second deviatoric stress invariant and is applicable to true-triaxial rock tests
	Mechanistic Theory		
6	A.A. Griffith (1920)	$\sigma_c = T_0 \sim \sqrt{\frac{\gamma}{c}}$	*Griffith criterion* for quasi-static single tensile crack growth based on specific surface fracture energy, γ
7	A.A. Griffith (1924)	$(\sigma_1 - \sigma_3)^2 - 8T_0(\sigma_1 + \sigma_3) = 0$ $\sigma_3 = -T_0$	*Biaxial Griffith criterion* for failure of rock in compression and tension with tensile strength, T_0
8	McClintock and Walsh (1962)	$\tau = 2T_0\sqrt{1 + \frac{P_{cc}}{T_0}} + 2\mu(\sigma - P_{cc})$	*Modified Griffith criterion* as applicable to rock with friction, μ along closed crack ligaments (closure pressure, P_{cc})

Table 3.4 (continued)

#	Author	Fracture Formula	Name of criterion and applicability
	Linear Fracture Mechanics based on stress intensity factor K		
9	Ashby and Hallam (1986)	$\sigma_1 = 3.12\sigma_3 - 3.06\frac{K_{IC}}{\sqrt{\pi a}}$	*Wing crack criterion* for compact rock under biaxial compression
10	Sammis and Ashby (1986)	$\frac{\sigma_1\sqrt{\pi a}}{K_{IC}} = -F(\lambda, L)$	*Pore crack criterion* for porous rock with initial crack length, *a* under moderate biaxial stress $\lambda=\sigma_3/\sigma_1<1/3$
11	Zhang et al. (1990)	$P_{cr} \sim \frac{K_{IC}^3}{E^2}(\alpha\Phi R)^{-3/2}$	*Hertzian fracture* for weakly cohesive rock accounts for grain (radius *R*) crushing and pore collapse (porosity Φ)
12	LFM	$K_i=K_{iC}$ with $i=I, II, III$	LFM approach of critical *K* for mode I (tensile), mode II (sliding-) and mode III (tearing shear) fractures
13	Shen and Stephansson (1993a)	$F(\theta) = \frac{G_I(\theta)}{G_{IC}} + \frac{G_{II}(\theta)}{G_{IIC}}$	*F-criterion* for mode I and mode II fracture propagation of rock material
	Non-Linear Fracture Mechanics based on rock process zone		
14	F. Ouchterlony (1980) and R.A. Schmidt (1980)	$r_c \sim \left(\frac{K_{IC}}{T_0}\right)^2$ for mode *I*	NFM approach reduces the stress singularity from (12) to finite strength values, T_0 by assuming a process zone with radius, r_c ahead of fracture

Note-Box A brittle fracture criterion describes the conditions for which failure occurs at a given stress field in rock. Phenomenological theories quantify the spatial orientation of fracture planes with respect to the stress field producing the rupture or form empirical power-law strength criteria. Mechanistic theories start from the premise that fracture initiates from existing flaws acting as stress singularity (linear fracture mechanics) through which the brittle fracture process is controlled. Non-linear fracture mechanics idealizes brittle rock fracture as a cloud of small daughter cracks propagating ahead of the growing mother fracture limiting stresses at the crack tip to a finite-strength value of the material. In fracture mechanics, we distinguish between three modes of fractures and the mixture of these modes. Brittle rock-fracture criteria are summarized in Table 3.4.

Chapter 4
Rock Stress Terminology

There is no internationally agreed terminology for words describing the state of stress in a rock mass (Hudson et al. 2003). In Fig. 4.1 we present a terminology for the classification of rock stress modified from a diagram published by Amadei and Stephansson (1997). The diagram is a structured compilation of terms used earlier (Bielenstein and Barron 1971; Lindner and Halpern 1978; Hyett et al. 1986; Cornet 1993) and in the discussion of descriptive stress terms by Harrison and Hudson (2000). For didactic reasons, we modified two aspects of the diagram. First, each type of rock stress is symbolized by a pictogram illuminating a typical cause of the stress component shown. Second, the term structural stresses (Jaeger and Cook 1979) is added to account for the influence of material properties (anisotropy, heterogeneity) on the in-situ state of rock stress.

Rock stresses at the first hierarchical level can be classified into four major groups (Fig. 4.1, A–D). One pair of stresses involves homogeneous rock mass (Fig. 4.1, AB), the other pair inhomogeneous or anisotropic rock material (Fig. 4.1, CD) at the
M4 scale of the stress component involved. *In-situ stress*, also called virgin, primary, pre-mining, natural or primitive stress, exists in the rock mass prior to any man-made or artificial disturbance (Fig. 4.1, A). Man-made stress, also called induced, secondary or mining stress, is the natural stress state as perturbed by engineering (excavating, explosion, drilling, pumping). We call this stress state *perturbed in-situ stress*. As an example we show stresses around a circular cavity (Fig. 4.1, B). Orientations of principal stresses in the vicinity of the hole are shown at selected points at the left and as stress trajectories at the right. For anisotropic or heterogeneous rock material, the term *structural stress* (Fig. 4.1, C) or *perturbed structural stress* (Fig. 4.1, D) has to be used instead (Sect. 4.4).

On the second hierarchical level, in-situ stress (or any other stress component from the first-level stresses B, C or D) is subdivided into four types of stresses originating from different forces (Fig. 4.1, A1–A4). The Earth's self-gravity force is responsible for the *gravitational stress* component (Fig. 4.1, A1) which can be obtained from the weight of the overlying rock mass (Sect. 4.1). Gravitational stress increases with depth in the Earth's crust and also accounts for stress resulting from the Earth's topography near the surface. *Tectonic stress* (Fig. 4.1, A2) is caused by

A. Zang, O. Stephansson, *Stress Field of the Earth's Crust*,
DOI 10.1007/978-1-4020-8444-7_4, © Springer Science+Business Media B.V. 2010

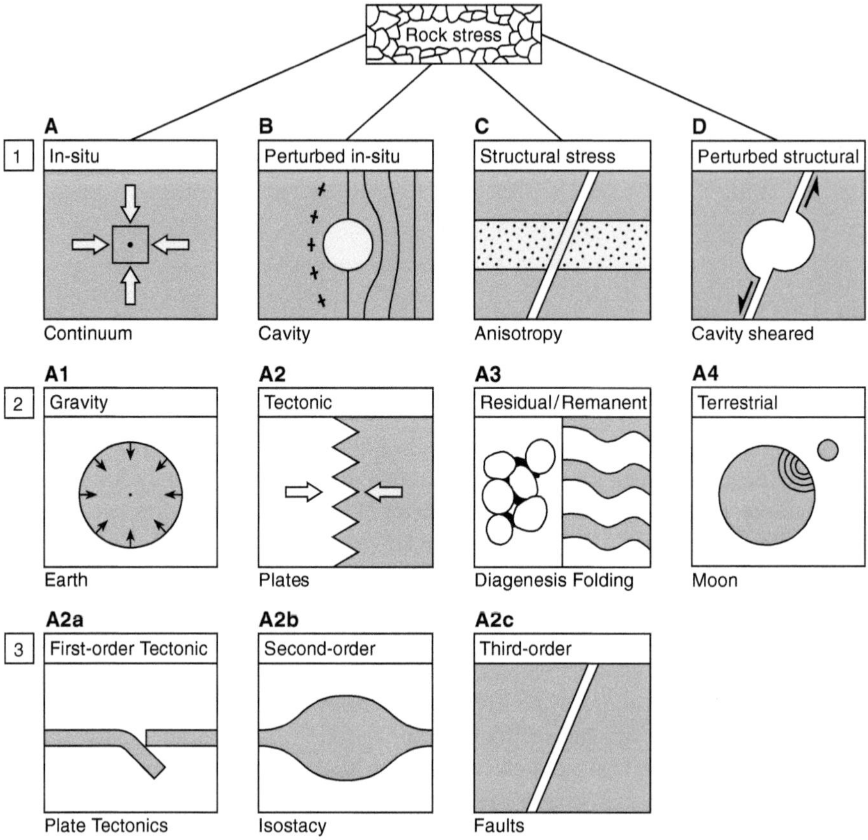

Fig. 4.1 Rock stress terminology at three hierarchical levels. Level 1 separates homogeneous (*AB*) from heterogeneous (*CD*), and solid (*AC*) from excavated rock mass (*BD*). Level 2 separates in-situ stresses (*A*) according to their origin forces like gravity (*A1*), tectonics (*A2*), residual (*A3*), and tidal (*A4*). Level 3 separates tectonic stresses according to their coherent domain into first order (*plate scale*), second order (*mountain range*), and third order stresses (*fault scale*)

a pervasive force field imposed by active tectonics (neo-tectonics) or past tectonic events (remnant stresses) and is discussed in Sects. 4.2 and 5.3.

On the third hierarchical level, active tectonic stresses due to present-day straining of the Earth's crust, are further subdivided into *global tectonic stress* due to the relative displacement of tectonic plates (Fig. 4.1 A2a) and *local tectonic stress* from bending plates or isostacy (Fig. 4.1, A2b). This subdivision of tectonic stress was proposed by Zoback et al. (1989) when analyzing stress patterns from the *World Stress Map* (WSM) data base. Following Zoback (1992), the terms global and local tectonic stress are replaced by *first order* (plate scale) and *second order* (mountain scale) *stresses*. Heidbach et al. (2007) down-scaled tectonic stress to the *third order* (fault scale) as discussed in Chap. 11. The different order tectonic stresses are scaled according to their *coherent domain* in the region in which a stress component is supposed to be uniform (in magnitude and/or orientation). Fault-scale stresses may also be regarded as structural stress (Fig. 4.1, C). If the fault is located close to an underground excavation, the term perturbed structural stress is used (Fig. 4.1, D).

Remnant tectonic stress due to palaeo-tectonic events in our stress pictogram (Fig. 4.1, A3 right) is regarded as a kind of large-scale residual stress (Fig. 4.1, A3 left) with a higher hierarchical level compared to other tectonic stress components. The stress field of the Earth's crust as discussed in this book mainly refers to the tectonic state of stress (Fig. 4.1, A2) as the sum of different order tectonic stresses (A2a, A2b, A2c) plus a possible remnant component (A3 right). In places where mining induced stress is of interest (Fig. 4.1, B), the rock stress terminology on the next lower hierarchical level would read B1, B2, B3, and B4. For example, B2 means the tectonic stress components as perturbed by the excavation.

Residual stresses in rock are defined as a system of stresses that exist in or approach equilibrium in the interior of a rock when neither normal nor shear stresses are being transmitted through its surface (Voight 1966a). The on-line encyclopedia *Wikipedia* defines residual stresses as stresses that remain in a body even after the originating mechanism has been removed. In Fig. 4.1, residual stresses are illustrated at a grain scale for a cemented grain aggregate (A3 left) and at a structural geology scale for folding a layered rock mass upon mountain-building (A3 right). At both scales, due to the physico-chemical effects involved, we refer to residual stresses. These stresses are discussed in Sect. 4.3. *Terrestrial stresses* are induced by diurnal and seasonal variations of temperatures, Moon pull, and Coriolis force (Fig. 4.1, A4). They only account for in-situ stress at a very shallow depth (Scheidegger 1982; Swolfs and Walsh 1990; Berest et al. 1992).

Of the four second-level contributors to the in-situ stress tensor (Fig. 4.1, A1–A4), apart from gravity, only the tectonic stress component can reach a large scale, regional significance, since its evolution is connected to genesis, motion and subduction of lithospheric plates. The determination of the in-situ state of stress in a rock mass is therefore equivalent to the determination of the tectonic stress plus the gravitational stress (Fig. 4.1, A1 and A2). Other stress components (Fig. 4.1, A3 and A4) have large standard deviations in orientation and magnitude. In the following, we evaluate on the gravity stress field (Sect. 4.1, A1), tectonic stress (Sect. 4.2, A2) and residual stress (Sect. 4.3, A3). Structural stresses (Fig. 4.1, CD) are discussed in Sect. 4.4. Simple Earth-stress models, as discussed in Chap. 5, are based on gravity and tectonic stress alone. In places where structural stresses are of interest (Fig. 4.1, CD), the rock stress terminology can be changed accordingly. For example, D2a represents the plate tectonic stress component as perturbed in the vicinity of an underground opening.

The stress field which exists in a rock mass is always a result of many processes and mechanisms. It is of particular importance in rock engineering, since the stress field defines the boundary conditions for underground modelling and design in rock mechanics' work. Because the exact geological history of a rock mass will never be known precisely (Sect. 4.3), because the constitutive equations describing the mechanical behaviour of the rocks remain approximate (Sect. 5.5) and because the detailed structure of the rock mass in the deeper crust cannot be measured exactly (Sect. 4.4), it is impossible to determine the in-situ stress field by computation, although numerical studies can assist in interpretation. One way to solve this problem is to determine the local stress tensor no matter which mechanisms may have caused them (Cornet 1993). A second way is to determine the best estimated rock-stress model as proposed by Stephansson (2003), see Sect. 10.4.

4.1 Gravity Stress

According to Chap. 1, the rock-stress component caused by the Earth's gravity field (Fig. 4.1, A1) is called vertical stress S_v and is governed by the weight of an overburden ρgz (Sect. 2.1). If the rock density is variable $\rho(z)$, the vertical stress is computed by

$$S_V = g \int_0^z \rho(z)dz. \tag{4.1}$$

The simple assumption that principal stresses are vertical and horizontal is no longer valid when the surface of the Earth's crust shows significant topography. As a zero-order approximation, the unloaded surface of a homogeneous, isotropic and semi-infinite rock mass representing the Earth's surface is seen in Fig. 4.2. Only gravity affects the rock mass (Fig. 4.2, acceleration vector of gravity, g). The surface topography is shown schematically as valley (Fig. 4.2a, V) and mountain (Fig. 4.2a, M). The deviation of stress trajectories from a vertical section is displayed in Fig. 4.2b. As we know from Chap. 2, principal stress trajectories cut free surfaces (principal stress planes) at right angles. Since the crustal cavity (Fig. 4.2, negative rock volume) and the crustal additive (Fig. 4.2, positive rock volume) can be regarded as half-inclusions, the evaluation of structural stresses in terms of inclusion models is applicable (Sect. 4.4).

In general, it is difficult to determine analytically the in-situ stress field in a rock mass with arbitrary surface irregularities using the theory of elasticity (Ling 1947; Savage 1994) or a perturbation method (Gao 1991). Therefore, numerical methods like *finite elements* (FEM) and *boundary element methods* (BEM), are the only alternative. In crucial areas (e.g., the potential site for radioactive waste at Yucca mountain, Nevada), where stresses have to be determined as precisely as possible, the effect of topography needs to be considered even with analytical models (Swolfs and Savage 1985). The variation in vertical stress as experienced through stress measurements (Part II of this book) is discussed in Sect. 10.2. To give an idea of the magnitude of gravitational stress, we use a typical gradient of 27 MPa km^{-1} (Chap. 5) and compute a number of 405 MPa (~4 kbar) for the vertical stress magnitude at a seismogenic depth of 15 km within the Earth's crust.

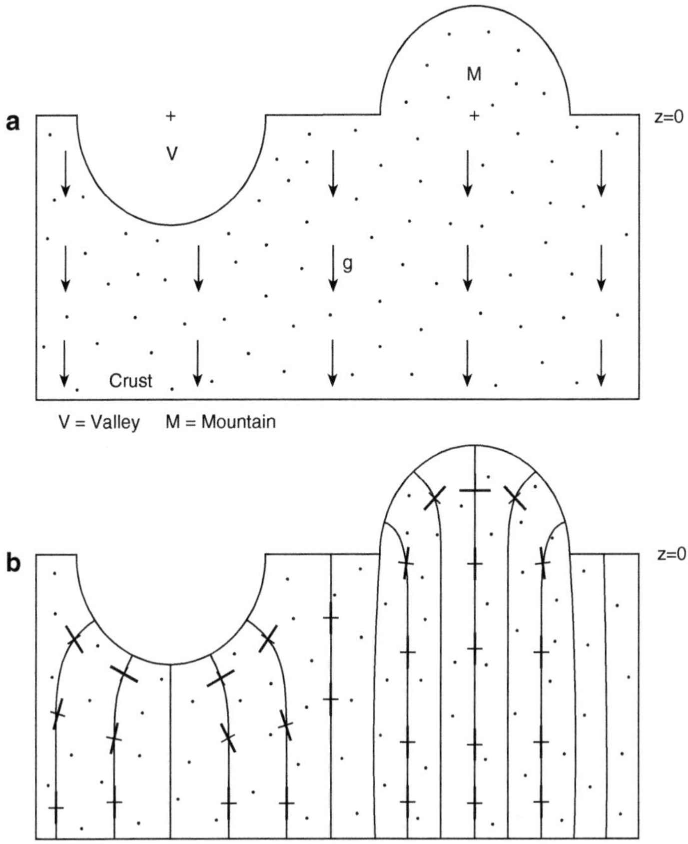

Fig. 4.2 Rock stress due to gravity demonstrated **a** in a semi-infinite body with valley V and mountain M under acceleration g, and **b** corresponding stress trajectories in the excavated and added half-inclusion model

4.2 Tectonic Stress

The Earth is not an inert body because movements in the crust occur continuously during the build-up of stress caused by plate tectonics and discontinuously during the partial relief of stress caused by earthquake ruptures. As we know, coherent domains of stress can exist at different scales. Global patterns of tectonic stress (Fig. 4.1, A2a) as analyzed by Zoback et al. (1989) involve two types of forces. Broad-scale tectonic forces act on lithospheric plates, while local scale forces are caused, for example, by bending of the lithosphere at plate boundaries. First-order tectonic stress domains are, per definition, uniform over large areas of continents.

As an example, the contemporary intraplate stress field of the Australian continent is shown in Fig. 4.3. Measured stress data, as obtained from the WSM data set, are plotted in Fig. 4.3a for several stress provinces defined by Hillis and Reynolds

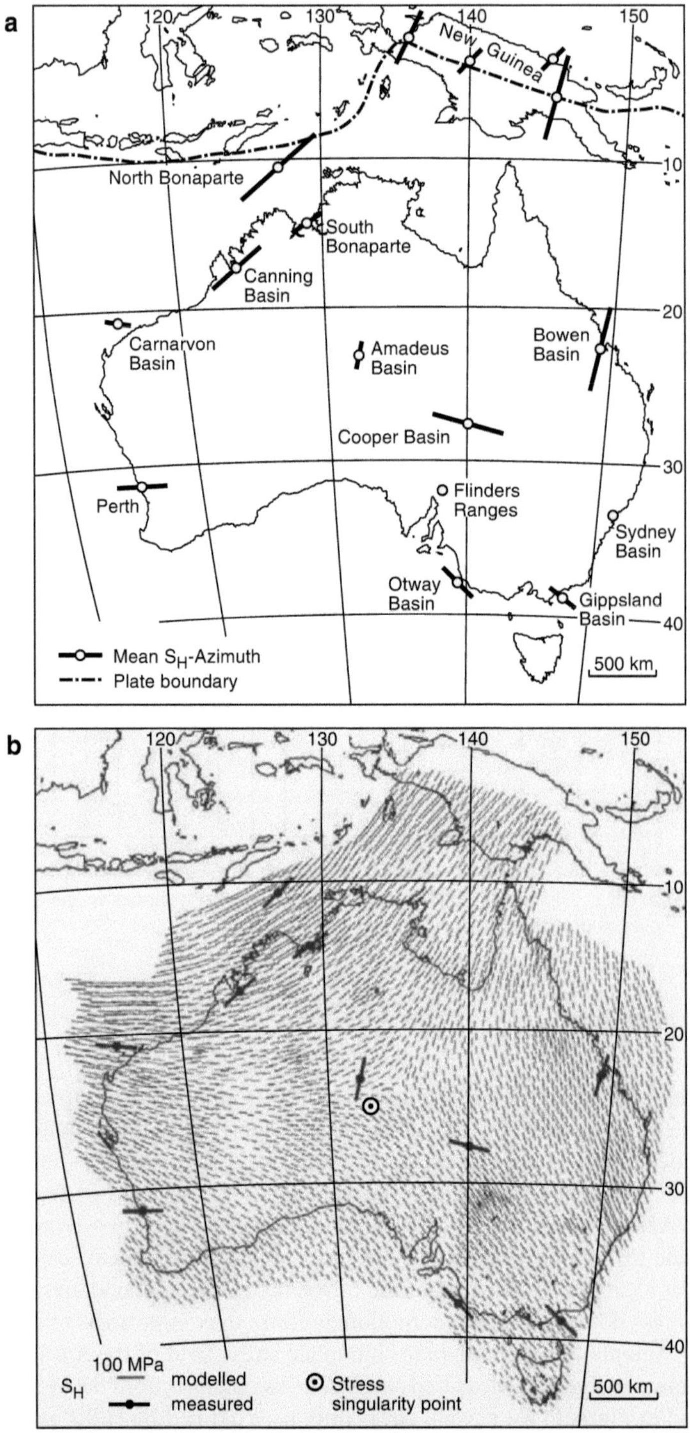
a
120
130
140
150
New Guinea
10
20
30
40
North Bonaparte
South Bonaparte
Canning Basin
Carnarvon Basin
Amadeus Basin
Bowen Basin
Cooper Basin
Perth
Flinders Ranges
Sydney Basin
Otway Basin
Gippsland Basin
Mean S_H-Azimuth
Plate boundary
500 km
b
100 MPa
S_H
modelled
measured
Stress singularity point

(2000) to constrain the predicted stress field from numerical modelling (Reynolds et al. 2002). The orientation of the bar in each stress province indicates the mean value of the S_H-azimuth from WSM data. The length of each bar is proportional to the weight used in calculating the misfit value compared to the numerically predicted stress field of Reynolds et al. (2002). They carried out a 2D planar finite-element analysis with 2527 triangular elements, which provides a resolution of $2° \times 2°$ in latitude and longitude. Modelled boundary forces acting on the Indo-Australian plate include ridge push, boundary traction and buoyancy forces (see Chap. 11). In Fig. 4.3b, predicted stress orientations from numerical modelling are shown for the plane stress model of Dyksterhuis and Müller (2004) using 24,400 finite elements providing resolution of $0.2° \times 0.2°$. Their result, obtained with a homogeneous material model, shows large-scale stress domains of the present-day stress field in continental Australia. According to this, Australia splits into three major stress domains with more or less uniform orientation of the maximum horizontal compressive stress (Fig. 4.3: NW Shelf with NE–SW trending of S_H-azimuth, SW Shelf with E–W trending S_H, and E territory with NNE–SSW trending of S_H). This triptych of first-order stresses is in agreement with the existence of a *stress singularity point* (Fig. 4.3b, circled dot) as computed from stress trajectories by the method of Mukhamediev et al. (2006), see Sect. 10.2. This model not only explains the observed overall rotation of the Australian stress field (S_H-axis in WSM data rotates clockwise at an angle of 180° moving anti-clockwise around the continental margin of Australia), but also is able to explain the more local 90° stress rotation from NS in the Amadeus Basin to EW in the Cooper-Ergomangan Basins (Fig. 4.3a). This stress rotation is due to the large curvature of stress trajectories in the vicinity of the singularity point (Fig. 4.3b). As we see in Chap. 11, attempts to fit global stress data with plate motion models, like a multi-continent stress puzzle, also work quite well.

On the contrary, second-order tectonic stresses (Fig. 4.1, A2b) can vary drastically over short distances. According to results presented in Sect. 4.4, this distance is related to about three times the diameter of the heterogeneity. For example, the variation of stress trajectories in the vicinity of a volcano (heterogeneity) is shown in Fig. 4.4. Stress trajectories are displayed as the frozen-in dyke pattern from Spanish Peaks, Colorado (Johnson 1961). In the two schematic sketches (Fig. 4.4a, b), the influence of an isotropic and biaxial stress field on the alignment of volcanic dykes is displayed. For isotropic tectonic stress (Fig. 4.4a, $S_H = S_h$), a radial pattern of dykes and eruptive fissures is expected, whereas in a biaxial tectonic stress field (Fig. 4.4b, $S_H > S_h$), the dykes and fissures (lines in the plan view cross-section) are deviated towards the direction of S_H. From Fig. 4.4c showing the mapped dyke swarm in the vicinity of Spanish Peaks, it is evident that at the time the dykes were intruded, a

Fig. 4.3 Australian intraplate stress field. **a** Stress provinces based on WSM stress data to constrain the predicted numerical stress field. Direction of each bar indicates the mean value of the maximum horizontal stress azimuth. Length of each bar is proportional to the weighted misfit value between a measured and modelled stress field (after Reynolds et al. 2002). **b** Computed present-day maximum horizontal stress azimuth for the Australian plate assuming homogeneous elastic material properties (after Dyksterhuis & Müller 2004). In **b** also the stress singularity point at 133°E, 25°S is displayed (*circled dot*) as determined by Mukhamediev et al. (2006)

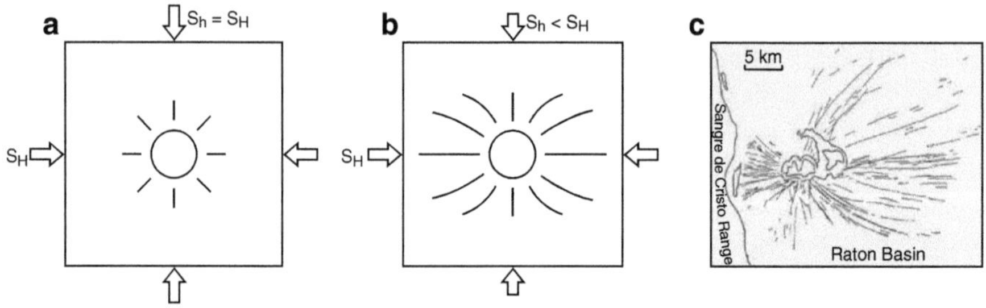

Fig. 4.4 Local tectonic stress near the caldera of a volcano as inferred from stress trajectories for **a** isotropic stress and **b** biaxial stress. In **c** dike swarm and eruptive fissures mapped near Spanish Peaks, Colorado indicate frozen-in stress trajectories (modified after Johnson 1961)

biaxial in-situ stress field prevailed at the volcano. Using the diameter of the volcanos caldera (10 km) as the scale of the heterogeneity, stress perturbations should drop off for distances larger than 30 km. Today, linear arrays of monogenetic cinder cones (volcanoes) are used to infer paleo-S_H-azimuths (Paulsen and Wilson 2009).

Third-order tectonic stresses (Fig. 4.1, A2c or C) are discussed in Sect. 4.4. How structural stress rotates close to a fault discontinuity is investigated in Sects. 4.4.1 and 10.2 when the influence of mountain ranges on horizontal stresses is interpreted. In Sect. 11.2, stress decoupling in salt pillows are addressed as examples of third-order tectonic stresses. Stress-depth variations are presented in Chap. 5 and in Sect. 10.1

4.3 Residual Stress

In 1964, Emery stated that "any rock must have in it more or less conserved elastic strain energy". All definitions of residual stress in rock (Tullis 1977; Holzhausen and Johnson 1979; Hoskin and Russell 1981) are based on that of Voight (1966a), namely that residual stresses in rocks are self-equilibrating stresses that are locked-in while the boundary of the rock sample is stress-free. Zang and Berckhemer (1989) applied two helpful classifications of residual stresses (RS) to rocks. First, in material sciences, RS are classified in terms of their *coherent domain* (analogous to different-order tectonic stress domains), i.e. the sphere of influence in which residual stresses are assumed to be distributed homogeneously (Macherauch and Hauk 1987). *RS of the first kind* have a coherent domain of a few grains. *RS of the second kind* are homogeneous within a single mineral grain, and *RS of the third kind* have a coherent domain at subgrain dimensions, e.g. lattice defects at atomic distances. Second, in geosciences, a rock physicist classifies RS according to their physical origin. Rocks are polycrystalline materials formed (*igneous rocks*), re-crystallized (*metamorphic rocks*) or cemented (*sedimentary rocks*) at elevated pressure and temperature. Therefore, genetic causes of RS are (1) crystallization of grains out of a melt, (2) plastic deformation, (3) phase transition, (4) re-crystallization and (5) diagenetic consolidation of mineral aggregates. In addition, (6) thermo-elastic stresses can be generated in rock due to tectonic uplift or sink of rock mass within the Earth's crust (Bruner 1984; Fredrich and Wong 1986; Carlson 1990; Zang 1993a, b; Ito et al. 1997b). Mismatch in mineral properties is the reason for the development of type (6) RS (Heinze and Goetze 1974; Tvergaard and Hutchinson 1988).

An example of type (5) RS is the development and visualization of RS in cemented sandstone given experimentally by Friedman (1972) and numerically by Zang and Stöckl (1991). The photographs of Gallagher's (1971) Ph.D. thesis in Friedman's (1972) article consist of vertically loaded, naturally shaped grains of transparent material of type CR-39 (Fig. 4.5a) and the same model cemented with epoxy and the load removed (Fig. 4.5b). Isochromatic fringes in the cured epoxy cement indicate that some of the previous load stored in the vertically stressed grain framework is transferred into the cementing agent when the external load is removed. Stress redistribution in the unloading process (approaching self-equilibrium in the RS system) depends on the mismatch in Young's modulus of the grains (Eg = 2.4 GPa) and the cured cement (Em = 2.7 GPa) as discussed in Sect. 3.3.1. As we see in Sect. 6.4 (photoelasticity), the number of isochromatic fringes is proportional to the difference in principal stresses.

In their numerical model of calcite-cemented sandstone, Zang and Stöckl (1991) simulate the burial diagenesis in two scenarios. In the first scenario, a porous quartz

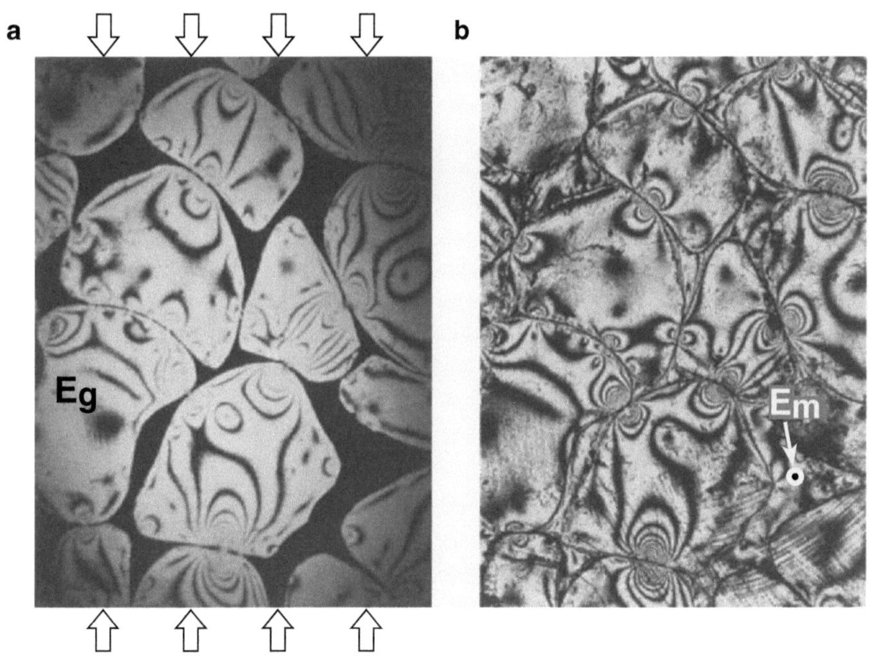

Fig. 4.5 Photographs of **a** uncemented, vertically loaded, naturally shaped grains of CR-39, and **b** the same model cemented with CIBA epoxy and load removed studied in terms of isochromatic fringes from photoelasticity (after Friedman 1972)

grain aggregate is consolidated and unstressed cement is built into the pore space in one step similar to Friedman's (1972) grain-cement model. In the second scenario, diagenesis is modelled by filling the pore space in eight steps at increasing stress (Fig. 4.6a, b). Thereafter, when the primary stress field is removed, stress-relief microcracking is simulated in the RS system, highlighting tensile cracks in the cementing agent (Fig. 4.6c). Finally, crack closure is modelled (Fig. 4.6d) showing that the crack closure pressure (Sect. 3.2) is equal in magnitude to the pre-existing in-situ pressure at cementation.

In a rock containing residual stress, equilibrium is achieved by counterbalancing elements which are in compression (grains, Fig. 4.5a) with elements that are in tension (cement, Fig. 4.5b). The *equilibrium volume* is the smallest unit of rock in which all forces balance (Varnes and Lee 1972). This locking-domain of a self-equilibrating unit in rock may vary in scale from a single mineral grain to the size of a granite pluton (Engelder 1993). Identifying the size of the equilibrium volume is difficult and without knowledge of this volume, the interpretation of RS data is always speculative (Sect. 10.3). According to Bielenstein and Barron (1971), a major drawback to the proposal that residual stresses exist on a regional scale is that large volumes of rock must exist in tension and tensile stresses are seldom measured because of non-existing. Also, Martin et al. (1990) pointed out that the same problem holds for bedded-sedimentary rocks.

The detection of RS in rock is usually done by X-ray diffraction (Friedman 1967) or by double (secondary) overcoring (Nichols 1975; Tullis 1977). Various measuring techniques for residual stresses in industry material are reviewed by Lu (1996). In Chap. 6, the physical basis of the different techniques is described as applicable to geomaterial. Note, however, that the magnitude of RS measured increases as the volume of rock decreases (Swolfs et al. 1974; Hyett et al. 1986; Engelder 1993). This trend is attributed to the fact that as the rock volume increases, discontinuities are more likely to be found (Sect. 10.3), and each discontinuity is unable to transfer RS in tension. Careful selection of rock material and a combination of different techniques is required to obtain reliable stress data from double-coring experiments (Zang et al. 1996b).

In general, it is difficult to distinguish between active and remnant tectonic stress. The significance of RS compared to other in-situ stress components was emphasized by Lang et al. (1986) for granitic rock from the Underground Research Laboratory (URL) in Pinawa, Canada. According to their work, the contribution of RS to the total in-situ stress at depth is less than 3%. In addition, since stress is a point property (Chap. 2), there is high potential for scatter also in heterogeneous rock mass (Hudson and Cooling 1988).

4.4 Structural Stress

The term structural stress was mentioned by Jaeger and Cook (1979, p. 382) but sunk into oblivion in later studies of stress terminologies (Engelder 1993). *Structural stresses* are caused (1) by anisotropy and (2) by heterogeneity of rock mass.

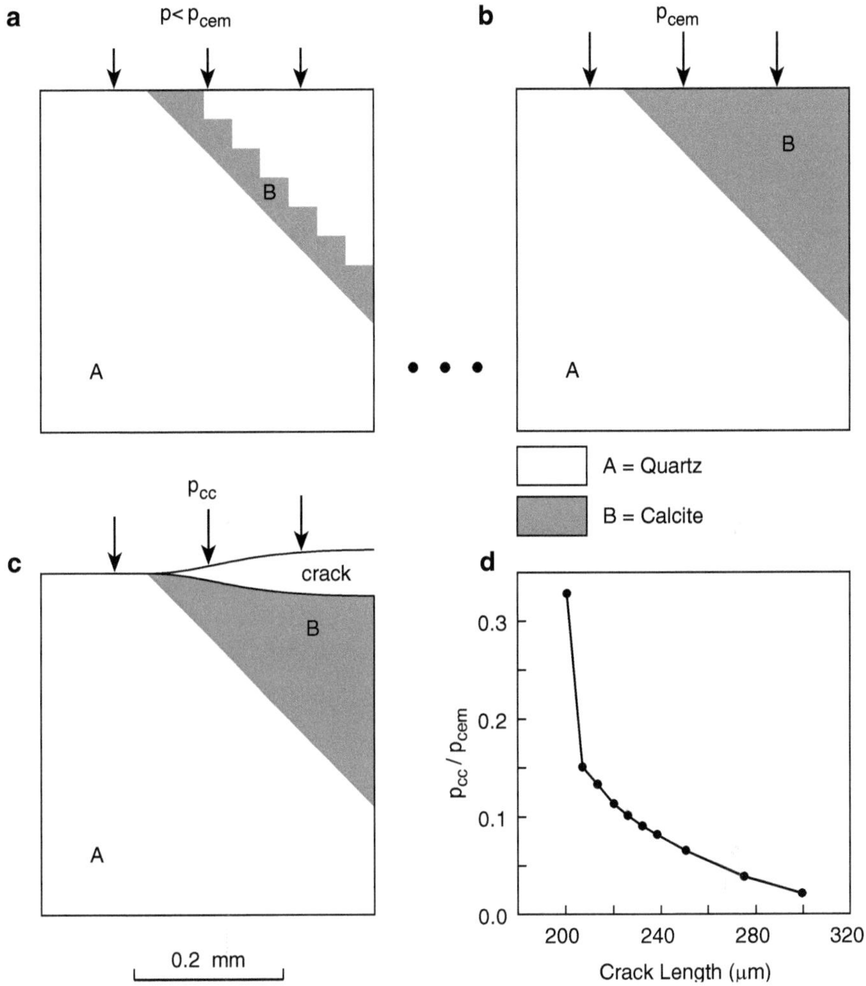

Fig. 4.6 Numerical model of stress changes in calcite-cemented sandstone. Burial diagenesis starts at small pressure (**a**), and is completed in eight steps at cementation pressure p_{cem} (**b**). Upon stress relief during in-situ coring, the calcite-cemented pore space is cracked (**c**). By reloading the cracked system (**d**), the crack closure pressure, p_{cc} is computed (modified after Zang and Stöckl 1991)

To demonstrate the effect of structural stress, in Fig. 4.7 homogeneous, idealized anisotropic and heterogeneous rock material is shown with and without externally applied loads. The biaxial compressive stress field ($S_1 > S_2$) is traced into the homogeneous material (Fig. 4.7a and d) without any distortions. Principal stress orientations in selected body points (Fig. 4.7d, crosses in a regular grid indicating *near field stress*) are oriented parallel to the applied S_1 and S_2 stress components (Fig. 4.7d, open arrows indicating *far-field stress*). Stress trajectories form a chess-board pattern (Fig. 4.7d, light grey lines), and far-field and near-field stresses are identical both in magnitude and in orientation.

In case of anisotropic material (Fig. 4.7b and e), the biaxial stress field applied is perturbed by the presence of the planes of anisotropy (Fig. 4.7b, tilted set of parallel light grey lines). Principal stress orientations in the near field are rotated towards the orientation of the rock anisotropy (Fig. 4.7e, crosses and chessboard pattern of trajectories). Stress trajectories form an orthogonal network of lines which is aligned parallel and perpendicular to the anisotropy planes. The change in stress magnitude and orientation depends on whether the anisotropy is caused by non-zero material contrast (rock fabrics like bedding, layering or foliation) or zero material contrast (fissuring or jointing). One example of non-zero contrast (bedding planes with different elastic properties of neighbouring layers) and one example of zero contrast (open joints) are treated later in Sect. 4.4.1.

In case of heterogeneous material (Fig. 4.7c and f), besides orientation, also the magnitude of principal stresses is perturbed in the close vicinity of the defect. From the stress trajectories in Fig. 4.7f, it is evident that near-field stresses around inclusions (Fig. 4.7c, filled circles) do not represent the direction or the magnitude of the externally applied far-field stresses. Note that near-field stresses around inclusions either of soft material (hole) or rigid material (stiff inclusion) within the host rock, can be treated analytically with thermo-elastic stress incompatibilities (Sect. 4.4.2). As a rule, near-field stresses decrease rapidly with distance from the defect (r^{-2} in 2D, r^{-3} in 3D). Therefore, in real Earth's crustal settings far-field stresses can be treated as undisturbed at distances of about three times the diameter of the defect.

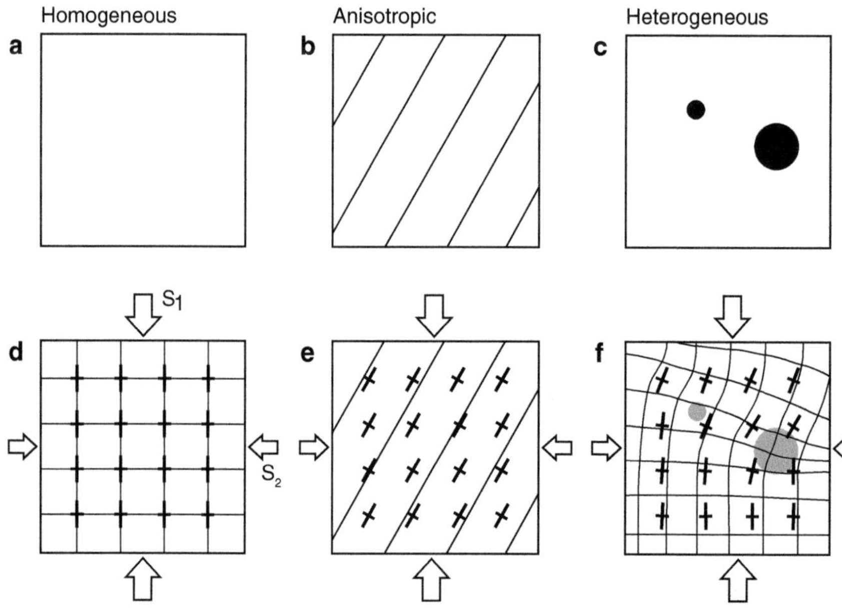

Fig. 4.7 Rock material properties **a** homogeneous, **b** anisotropic, and **c** heterogeneous effect principal stress orientation. Homogeneity results in a chessboard pattern of stress trajectories (**d**), anisotropy produces a pattern tilted towards preferred anisotropy (**e**), and rock heterogeneity produces complex near-field stresses around inclusions (**f**) when biaxial remote stress is applied

4.4.1 Anisotropy

In *anisotropic rock*, the rock properties vary with direction. *Hooke's law* for a general anisotropic, linear elastic material states that stress σ_{ij} is linearly proportional to strain ε_{ij} as expressed by (Mavko et al. 1998)

$$\sigma_{ij} = c_{ijkl} \times \varepsilon_{kl}. \tag{4.2}$$

In this equation, summation is implied over the repeated subscripts k and l. The elastic stiffness tensor c_{ijkl} is a fourth-rank tensor and has a total of 81 components from which, due to symmetry of the second-rank tensors of stress and strain, only 36 coefficients are independent $(c_{ijkl} = c_{jikl} = c_{ijlk} = c_{jilk})$. The existence of a unique strain-energy potential $(c_{ijkl} = c_{klij})$ further reduces the number of independent constants to 21. This is the maximum number of independent elastic coefficients (*aelotropy*) that any medium can have (Table 4.1, triclinic system). The number of elastic coefficients is reduced as the symmetry of the system is increased. The end member is the linear elastic *isotropic rock* system described by only two elastic constants. *Transverse isotropy* (Table 4.1, hexagonal case) is often used to describe the symmetry of rocks with one dominant system of layers (e.g., bedding, layering, foliation). In that case, five elastic constants are used in a reference frame attached to the rock layers. Similarly, nine elastic constants are used for an *orthotropic* material, e.g. a rock mass with three sets of orthogonal fractures having different properties (Crampin and Kirkwood 1981).

Another approach to layered rock systems is the anisotropic bimaterial model shown in Fig. 4.8. The rock is comprised of a repetitious sequence of stiff (μ_1) and soft (μ_2) layers with thickness t_1 and t_2, respectively (Fig. 4.8a). As the layered model is loaded vertically (Fig. 4.8b), the soft material bulges while the ends of the stiff material bars remain nearly straight. Soft bars will store higher strains while stiff bars attract higher stresses.

Table 4.1 Quantifying anisotropic, elastic rock behaviour by number and relationships between coefficients of the fourth-rank stiffness tensor

Symmetry of system	Number of coefficients	Relationships between c_{ijkl}	Also called
Triclinic	21	Maximum anisotropy	Aelotropy
Monoclinic	13		
Orthorhombic	9	$c_{11}, c_{12}, c_{13}, c_{22}, c_{23}, c_{33}, c_{44}, c_{55}, c_{66}$	Orthotropic
Trigonal	6		
Tetragonal	6		
Hexagonal	5	$c_{11} = c_{22} = A, c_{33} = C, c_{44} = c_{55} = L,$ $c_{66} = N, c_{12} = c_{11} - 2c_{66}, c_{13} = c_{23} = F$	Transvers isotropic
Cubic	3		
Isotropic	2 (e.g. K, μ)	$c_{11} = c_{22} = c_{33} = K + 4/3\mu, c_{44} = c_{55} =$ $c_{66} = \mu, c_{23} = c_{13} = c_{12} = c_{11} - 2c_{44}$	

For this rock model, Holzhausen and Johnson (1979) calculated residual stresses based on the idea of Friedman's (1972) locked-in stresses in the grain-cement model. After loading (Fig. 4.8b), the interfaces between pairs of soft and stiff layers are bonded together and the external stress field (Fig. 4.8b, p_y) is removed. For residual stress in the stiff and soft layers, they find (horizontal load neglected)

$$\begin{aligned} \sigma_{xx}^{soft} &= -\sigma_{xx}^{stiff}\left(\frac{t_1}{t_2}\right) \\ \sigma_{xx}^{stiff} &= -p_y\left(1-\frac{\mu_1}{\mu_2}\right)\left\{1+\left(\frac{t_1}{t_2}\right)\left(\frac{\mu_1}{\mu_2}\right)\right\}^{-1}, \end{aligned} \tag{4.3}$$

where t_1/t_2 is the ratio of the layer thicknesses, p_y is the applied load and μ_1/μ_2 is the ratio of shear moduli, respectively. In the case of equally thick layers ($t_1 = t_2$), Eq. (4.3) simplifies

$$\begin{aligned} \sigma_{xx}^{soft} &= -\sigma_{xx}^{stiff} \\ \sigma_{xx}^{stiff} &= -p_y\left(\frac{1-x}{1+x}\right), \end{aligned} \tag{4.4}$$

where $x = \mu_1/\mu_2$ is the ratio of the shear modulus. As is evident from (4.4), horizontal residual stresses arise within the layers which are equal in magnitude but opposite in sign. The stress magnitude can be computed from the applied load and the shear modulus of neighbouring rock layers. Holzhausen and Johnson (1979) illustrate the correctness of their analytical result by investigating bimaterials (gelatine (soft) and rubber (stiff) layer) with photoelasticity (Chap. 6). The fringe pattern shows that stresses are uniform within the layers (i.e., Eq. (4.4) is valid), but become complex a short distance from the free end of the rods, as suggested by Saint Venant's principle (i.e., Eq. (4.4) not valid).

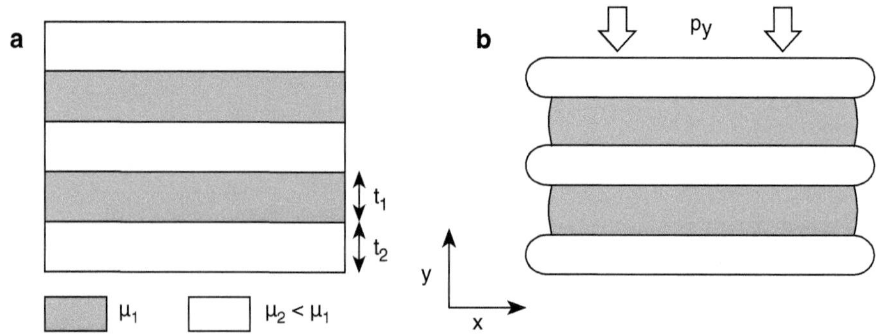

Fig. 4.8 Anisotropic rock model comprised of a repetitious sequence of soft and stiff layers **a** without and **b** with vertical applied load

The rock model discussed in Fig. 4.8 serves to understand stresses in anisotropic rock bodies at different scales (bedding or foliation at grain scale (Gottschalk et al. 1990; Shea and Kronenberg 1993), layering of rock at a metre scale (Teufel 1986) and stratification of rock mass at 10–100 m scale (Haimson and Rummel 1982). Depending on the relative stiffness between rock layers, in-situ stresses can vary substantially from one layer to another (Plumb 1994). Amadei et al. (1988) proposed an analytical solution for stresses in horizontal strata without lateral displacement. Like in the model of Fig. 4.8, the jump in the magnitude of horizontal stress across stratum contacts is due to the contrasts in elastic properties of neighbouring layers. Another approximation to a multi-layered stratum is the equivalent anisotropic continuum approach (Salomon 1968).

A second system of anisotropic rock is a cracked, fissured or jointed rock mass presented in Fig. 4.9 as often found in the upper Earth's crust. Young's modulus inside the open discontinuity is assumed to be zero (Fig. 4.9a). In 1976, Goodman noted the highly non-linear nature of closure-stress curves in jointed rocks and attributed this irreversible process to the inelastic crushing and splitting at asperities of contacts that could not be recovered during unloading. Bandis et al. (1983) show that the following hyperbolic type of equation fits well their experimental data for mated surfaces

$$\sigma_N = \frac{d}{\alpha - \beta d}, \tag{4.5}$$

where σ_N is the normal effective stress (Sect. 5.4) and d is the joint closure (Fig. 4.9b). The constants α, β approach $d_m = \alpha/\beta$ for $\sigma_N \rightarrow \infty$, where d_m is the maximum value of joint closure (asymptotic value d_m in Fig. 4.9c, data from Bandis et al. 1983). Hysteresis in joint closure curves (Fig. 4.9c, arrows) and the difference in absolute values from one to the next loading cycle (Fig. 4.9c: $d_m^{(i)}$ with $i = 1,2,3$) are both explained by irreversible processes like grain crushing and asperity shearing at the grain scale of the joint ligaments (Fig. 4.9b, close-up of irregular joint surfaces). If asperities are partially broken in early cycles, surfaces become smoothed and joint closure occurs at much lower values (Fig. 4.9c, third cycle) compared to the virgin loading curve (Fig. 4.9c, first cycle). The key parameter in treating hydro-mechanical behaviour of a highly anisotropic, jointed rock mass is the specific stiffness of the joint. In his "Jaeger Memorial Dedication Lecture", Cook (1992) defines the *specific joint stiffness* as

$$\kappa = \frac{\partial \sigma_N}{\partial d} \quad \text{with} \quad \frac{1}{\kappa} = \alpha\left(1 - \frac{\beta}{\alpha} d\right)^2, \tag{4.6}$$

where

$$\kappa_0 = \frac{1}{\alpha} \quad \text{and} \quad \kappa = \kappa_0\left(1 - \frac{d}{d_m}\right)^{-2} \tag{4.7}$$

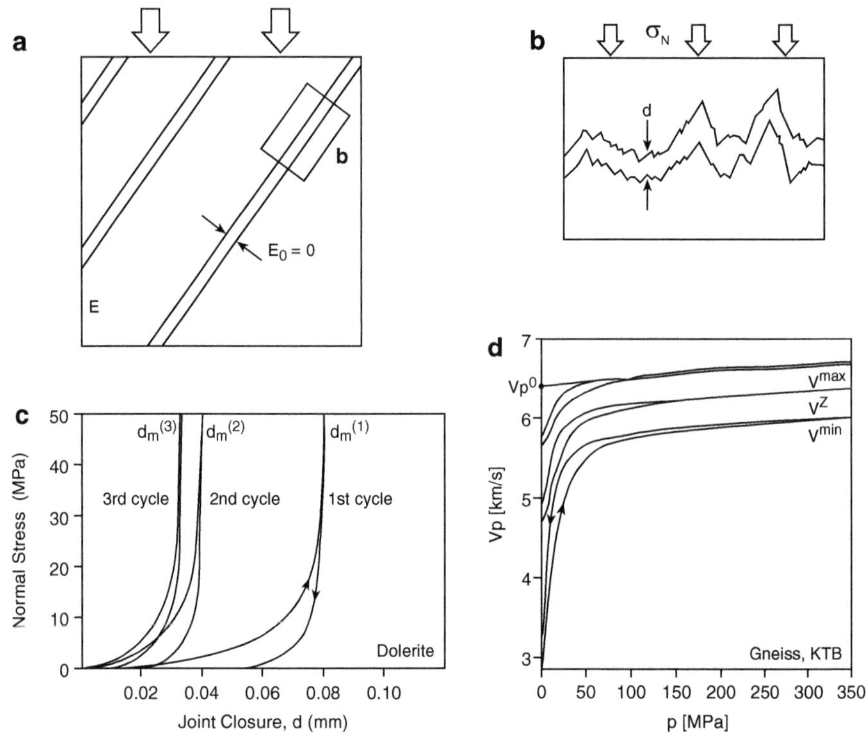

Fig. 4.9 Jointed rock mass under applied vertical load (**a**) and close-up of rough ligaments of a single joint under effective normal load (**b**). Measurements of joint closure in dolerite (**c**) after Bandis et al. (1983) and the effect of crack closure on ultrasonic *P*-wave velocity in gneiss from the KTB deep borehole (**d**) after Zang et al. (1996a)

is the specific joint stiffness at zero stress κ_0 and the specific stiffness at any stress κ, respectively. The stiffness relates the deformation of a fracture to the stress. Stiff fractures displace less, because they tend to have smaller apertures and larger regions of contact between ligaments than compliant fractures. Equations (4.6) and (4.7) establish an empirical relation between joint stiffness and closure as shown in Fig. 4.9b, c. Other empirical relationships include the "bed of nails model" by Carlson and Gangi (1985) and a semi-logarithmic constitutive equation between joint closure and normal stress (Brown and Scholz 1985, 1986). Single fractures under normal stress with emphasis on fluid flow were investigated in the laboratory and numerically by Pyrak-Nolte and Morris (2000). Theoretically, a fracture is a displacement discontinuity, i.e., stresses across the fracture are continuous but displacements are not. Small physical modifications of the fracture can cause large changes in the mechanical, seismic and hydraulic response of a fracture system (Pyrak-Nolte 2007). Saturation of fractures with fluid, combined with high stress, makes the fracture mechanically and seismically invisible. When shearing of rock joints is considered, the distribution of the contact area and void space determine the fracture dilatation and hydraulics (Lanaro and Stephansson 2003). A recent review on constitutive models of rock fractures can be found in Jing and Stephansson (2007).

The sensitivity of cracks to stress can also be demonstrated by measured wave velocities under increasing confining pressure in the laboratory (Birch 1960, 1961; Simmons 1964). In Fig. 4.9d, the curve with the arrow directed upward indicates pressurizing the sample (crack closure), whereas the curve with the arrow directed downward stands for the unloading process (crack opening). The difference in absolute values of *P*-wave velocity for the three curves depends on the direction the ultrasonic waves are transmitted through the anisotropic (foliated) gneiss core (Sect. 6.5). The slope of the $v_p(p)$-curves at high confining pressure (Fig. 4.9d, $p > 100$ MPa) reflects the behaviour of the crack-free material (Kern 1978; Kern and Richter 1979). While d_m in Fig. 4.8c is the asymptotic value in the joint closure curve, the *P*-wave velocity of the crack-free material, v_P^o is the intercept of the asymptotic slope at high pressure with the ordinate in Fig. 4.9d at $p = 0$. Velocity variations in rocks of anisotropic symmetry are studied both theoretically (Crampin 1977; Hudson 1981, 1986; Sayers 1988, 1990; Hudson 1991) and by ultrasonic laboratory techniques (Kern and Wenk 1990; Sayers and van Munster 1991; Popp and Kern 1994). A good monograph on this topic is Barton (2007).

A synoptic symbiosis for the relation between the anisotropic rock model I (Fig. 4.8, layered rock) and the anisotropic rock model II (Fig. 4.9, jointed rock) is shown in Fig. 4.10. In this schematic sketch, the direction of the maximum principal stress is shown for different scenarios of stiff and soft rock layers. For the open discontinuity (Fig. 4.10a, $E = 0$), the major principal stress direction is deviated parallel to the orientation of the discontinuity. For the homogeneous material (Fig. 4.10b), the layer is mechanically transparent and no change in the principal stress occurs in magnitude or in orientation. For a rock layer filled with very stiff material (Fig. 4.10c, $E \to \infty$), the major principal stress direction is deviated perpendicular to the discontinuity. For most rock engineering circumstances, the stress distribution will be between those shown for the case of Figs. 4.10a, b (Hudson and Harrison 2000).

The effect of stress rotation in a rock mass transected by a single fault (case Fig. 4.10a) is illustrated in Fig. 4.11. For this purpose, Su and Stephansson (1999) carried out a 2D stress modelling with the distinct element code UDEC. The two blocks of rock separated by a fault are subjected to a biaxial tectonic stress field (Fig. 4.11a). Elastic and isotropic properties of intact rock are assumed and the Coulomb fracture criterion of fault strength is applied (Chap. 3). The modelling was conducted in three stages: (1) the boundary stress ratio ($\lambda = \sigma_1/\sigma_2$) was considered, (2) the angle α between the fault and the regional maximum principal stress was

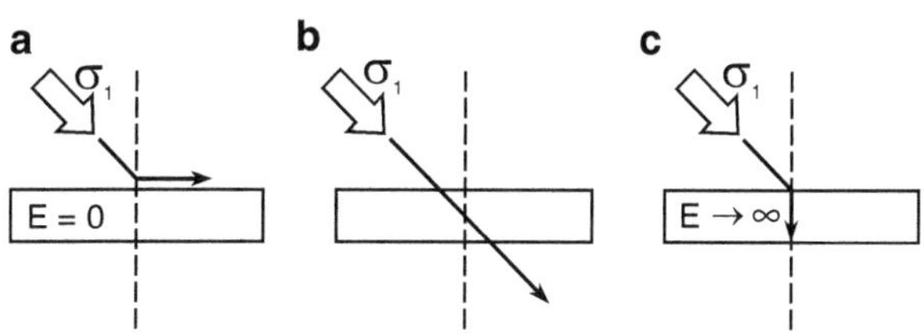

Fig. 4.10 Effect of stiffness of rock layer filling material on the stress state. In **a** the rock layer has a modulus of zero. In **b** the layer has the same modulus as the surrounding rock. In **c** the rock layer is very rigid and has a modulus of infinity (after Hudson and Harrison 2000)

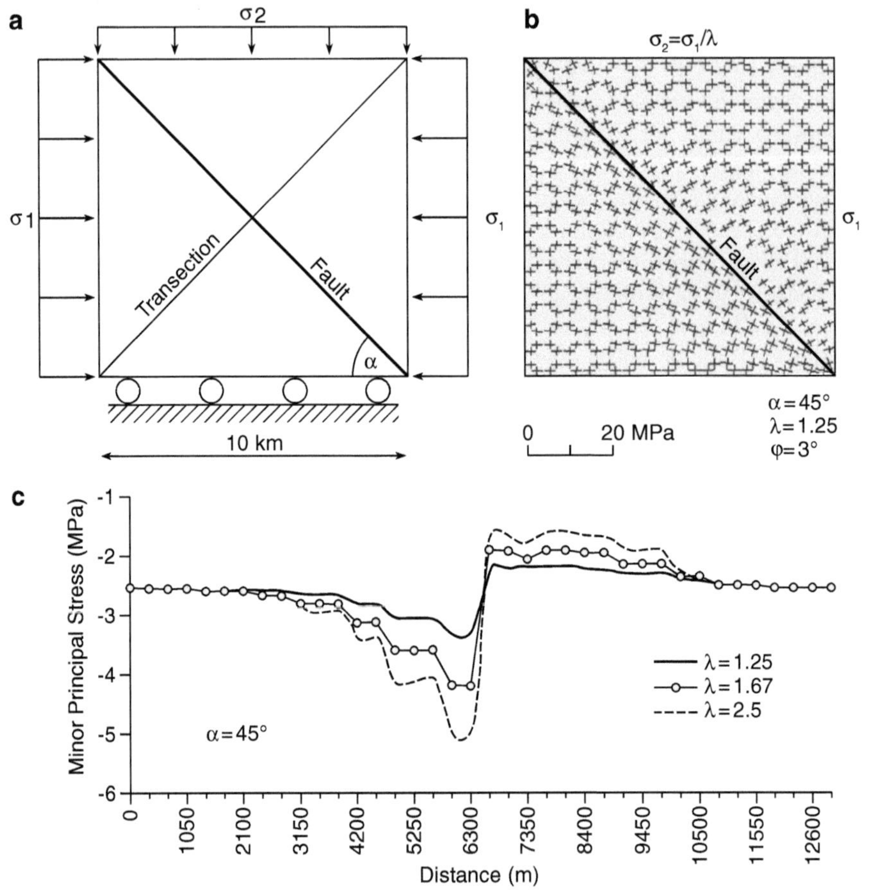

Fig. 4.11 Distinct element-stress modelling of faulted rock in 2D. **a** Block geometry and boundary condition, **b** reorientation of stresses near the fault with $\lambda = 1.25$ and friction angle $\varphi = 3°$ and **c** variation of the minimum principal stress across a fault ($\alpha = 45°$) for different stress ratios λ (modified from Su and Stephansson 1999)

considered, and (3) a series of sensitivity analyses were carried out to determine the effect of varying the rock mechanics input parameters. Reorientation of stresses near the fault with $\lambda = 1.25$ and a low friction angle $\varphi = 3°$ is presented in Fig. 4.11b. Notice the reorientation of principal stresses close to the fault. The range of reorientation of principal stresses diminishes with the increase of the friction angle of the fault. Figure 4.11c illustrates the variation of the minor principal stress along the transection across a fault with $\alpha = 45°$ and for a different stress ratio λ. A natural example for the stress rotation in the vicinity of Chelungpu fault, Taiwan has recently been published by Lin et al. (2007). In this study, the current stress state and the rotation of principal stresses in the vicinity of the fault induced by the 1999 Chi-Chi earthquake has been investigated. Near the fault, (1) the S_H-azimuth changes by 90° from the far-field tectonic stress (N 120° E) and (2) S_H-magnitudes dropped from 60 MPa to 20 MPa (Wu and Zoback 2008).

4.4.2 Heterogeneity

Rock mass is rarely uniform, in particular in the continental crust. Heterogeneities at different scales, e.g. elastic mismatch at grain scale, inclusion at outcrop scale, ore body and geologic structure at regional scale and subduction at the continental scale, all cause perturbed in-situ stresses. Depending on the distance from the heterogeneity we distinguish between *near-field stresses*, the local stress perturbation in the vicinity of the heterogeneity and *far-field stresses*, the global stress system applied at infinity (Fig. 4.7d–f). Using the terms near-field and far-field stress is necessary only if structural stresses in a body become important. Structural stresses are also called local or regional, depending on the size of their coherent domain (Cornet 1993). Perturbed stress fields caused by heterogeneities can be understood by applying inclusion models.

A very simple geometry of heterogeneity is that of a circular (2D) or spherical inclusion (3D) in Fig. 4.12. The inclusion K with radius R is embedded either in an infinite matrix M (Fig. 4.12a) or in a finite matrix annulus that is coaxial with the circular inclusion (Fig. 4.12b). If the composite material in Fig. 4.12 undergoes pressure (Δp) or temperature changes (ΔT) within the Earth's crust, stress and strain incompatibilities arise due to the mismatch in the thermo-elastic properties of the inclusion and the matrix material. To quantify stresses (or strains) around the inclusion, we have to refer to the thermo-elastic material law of an isotropic, homogeneous body (Fung 1965; Timoshenko and Goodier 1970)

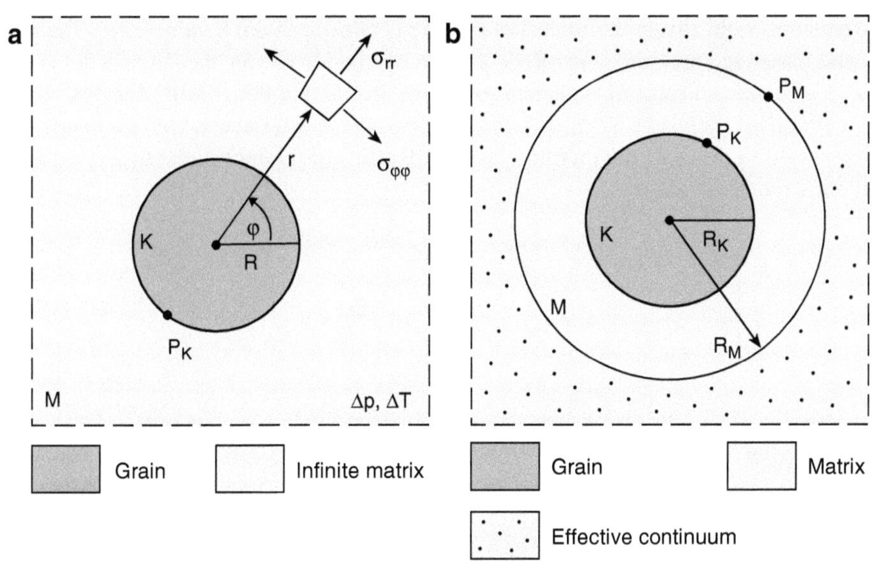

Fig. 4.12 Simple inclusion models and stress components. In **a** the inclusion K is circular within an infinite matrix M. In **b** the bimaterial consists of a circular inclusion K within a coaxial matrix annulus M surrounded by a material with effective properties of K and M

$$\begin{aligned}\varepsilon_{ij} &= \frac{1}{E}\left\{(1+\nu)\sigma_{ij} - \nu\sigma_{kk}\delta_{ij}\right\} + \alpha\Delta T\delta_{ij} \\ \sigma_{ij} &= \frac{E}{1-\nu}\left\{\varepsilon_{ij} + \frac{\nu}{1-2\nu}\varepsilon_{kk}\delta_{ij}\right\} - \frac{\alpha E}{1-2\nu}\Delta T\delta_{ij}\end{aligned}, \tag{4.8}$$

where E, ν are the elastic constants, α is the linear thermal expansion coefficient and δ_{ij} is the Kronecker unit tensor. The state of equilibrium in an elastic solid is characterized by the fact that forces are balanced by internal stresses in each body point (Landau and Lifschitz 1970)

$$2(1-\nu)grad \circ div(\bar{u}) - (1-2\nu)rot \circ rot(\bar{u}) = 0, \tag{4.9}$$

where $\bar{u}$ denotes the displacement vector. Equation (4.9) can be simplified for spherical inclusions since the displacement field is free of shear components $(rot(\bar{u}) = 0)$. Solving the differential equation for the displacement field, we find the stress field around the inclusion in spherical coordinates

$$\begin{aligned}\sigma_{rr}(r \le R) &= \sigma_{\varphi\varphi}(r \le R) = p_K \\ \sigma_{rr}(r > R) &= \Delta p - (\Delta p - p_K)\left(\frac{R}{r}\right)^3 \\ \sigma_{\varphi\varphi}(r > R) &= \Delta p + \frac{\Delta p - p_K}{2}\left(\frac{R}{r}\right)^3\end{aligned}. \tag{4.10}$$

The stress field inside the inclusion ($r \le R$) is constant with a value of p_K, the so-called interface pressure at $r = R$. Outside the inclusion ($r > R$), the stresses decay as r^{-3} as expected for a 3D spherical problem. In 2D, the decay with distance from a cylindrical inclusion is r^{-2}. The radial stress at the interface $\sigma_{rr}(r = R) = p_K$ is expressed in terms of the thermo-elastic properties of the inclusion (E_K,ν_K,α_K) and the matrix (E_M,ν_M,α_M)

$$\begin{aligned}p_K &= \frac{3(1-\nu_M)}{z}\Delta p + \frac{2(\alpha_M - \alpha_K)E_M}{z}\Delta T \\ z &= 1 + \nu_M + 2(1-2\nu_K)\frac{E_M}{E_K}\end{aligned}, \tag{4.11}$$

where Δp, ΔT are the changes in state variables applied to the bimaterial. The solution for the spherical inclusion in a continuum under external pressure Δp, was given by Goodier (1933) and the solution for the same problem after cooling ΔT, was given by Selsing (1961).

In Fig. 4.12b, the model of Fig. 4.12a is modified by a coaxial matrix annulus before the inclusion is embedded in an elastic continuum having the effective properties (averaged values) of the two materials. The reason for solving this *self-con-*

sistent approach is that a realistic contrast in modulus results at the grain boundary (inclusion/matrix interface) since the effective medium is able to simulate the average properties of a multi-inclusion model with various concentrations of components K and M (Budiansky 1965; Hill 1965). In the self-consistent approach (Fig. 4.12b), two interface stresses have to be computed (p_K = stress at inclusion/matrix interface, p_M = stress at matrix/continuum interface). The solution for the first interface stress at $r = R_K$ is

$$\begin{aligned} \sigma_{rr}(r = R_K) &= p_K \\ \sigma_{\varphi\varphi}(r = R_K) &= \frac{1}{c_M}\left\{\frac{3}{2}p_M - \left(c_K + \frac{1}{2}\right)p_K\right\}, \end{aligned} \tag{4.12}$$

and the solution for the second interface stress at $r = R_M$ is

$$\begin{aligned} \sigma_{rr}(r = R_M) &= p_M \\ \sigma_{\varphi\varphi}(r = R_M) &= \frac{1}{c_M}\left\{\left(1 + \frac{c_K}{2}\right)p_M - \frac{3c_K}{2}p_K\right\}, \end{aligned} \tag{4.13}$$

where $c_K = (R_K/R_M)^3$ is the concentration of the inclusion and $c_M = 1-c_K$ is that of the matrix. Again, p_K and p_M can be computed as a function of Δp and ΔT

$$\begin{aligned} p_K(\Delta p, \Delta T) &= \frac{z_2 z_5 \Delta p - \left[(\alpha_M - \alpha)z_2 - (\alpha_M - \alpha_K)z_4\right]\Delta T}{z_1 z_4 - z_2 z_3} \\ p_M(\Delta p, \Delta T) &= \frac{z_1 z_5 \Delta p - \left[(\alpha_M - \alpha_K)z_3 - (\alpha_M - \alpha)z_1\right]\Delta T}{z_1 z_4 - z_2 z_3}, \end{aligned} \tag{4.14}$$

whereby the constants $(z_1, z_2, \ldots, z_5)$ are expressed as

$$\begin{aligned} z_1 &= \frac{1 - 2\nu_K}{E_K} + \frac{(1 - 2\nu_K)c_K}{E_M c_M} + \frac{1 + \nu_M}{2E_M c_M} \\ z_2 &= \frac{3(1 - \nu_M)}{2E_M c_M}; \quad z_3 = z_2 c_K \\ z_4 &= \frac{1 - 2\nu_M}{E_M c_M} + \frac{(1 + \nu_M)c_K}{2E_M c_M} + \frac{1 + \nu}{2E} \\ z_5 &= \frac{3(1 - \nu)}{2E} \end{aligned} \tag{4.15}$$

The effective elastic properties of the continuum (E, ν) necessary to compute z_4 and z_5 are calculated from the Hashin (1962) algorithm in Watt et al. (1976), while the thermal expansion coefficient is averaged according to $\alpha = 0.5(\alpha_K - \alpha_M)$. Using the set of Eqs. (4.12)–(4.15), the following stress relationships in a heterogeneous rock system can be investigated (Zang 1991)

$$
\begin{aligned}
p_{K,M} &= p_{K,M}(\Delta p, \Delta T, c_K) \\
p_{K,M} &= p_{K,M}\left(\frac{E_K}{E_M}, \frac{\nu_K}{\nu_M}, \frac{\alpha_K}{\alpha_M}\right).
\end{aligned}
\tag{4.16}
$$

In Fig. 4.13, interface stresses $\sigma_{rr}(r = R_K)$ and $\sigma_{\varphi\varphi}(r = R_K)$ are plotted for three different scenarios simulating (1) unloading of rock mass by $\Delta p = 100$ MPa, (2) cooling of rock mass by $\Delta T = -100°C$, and (3) the combined effect of $\Delta p = 300$ MPa with $\Delta T = -200°C$. The latter scenario simulates stresses in rock which is cut from the in-situ stress field at a depth of about 10 km within the Earth's crust. For all computations, the ratio of modulus varied according to $0 \le E_K/E_M \le 2.5$ with $E_M = 100$ GPa, $\nu_K = \nu_M = 0.25$ and $c_K = 0.1$ fixed. In scenario (1), radial stress $\sigma_{rr}(r = R_K)$ increases from zero to 132 MPa at $E_K/E_M = 2.5$, whereas circumferen-

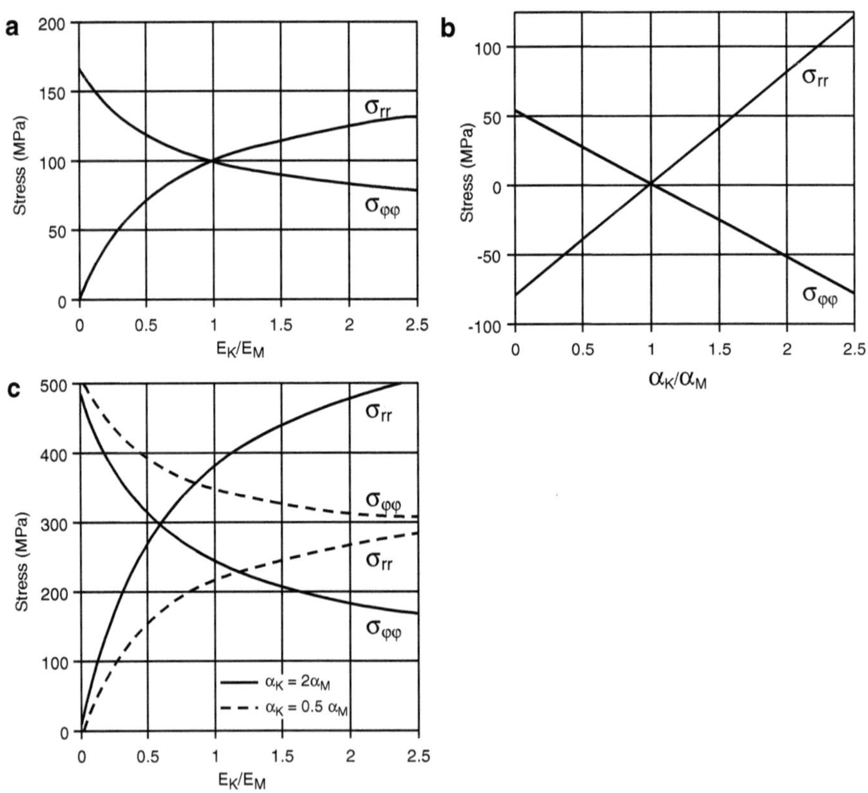

Fig. 4.13 Interface stresses in the self-consistent model of inclusions for three loading scenarios. In **a** unloading with $\Delta p = 100$ MPa and in **b** cooling with $\Delta T = -100°C$ is modelled. Scenario **c** with $\Delta p = 300$ MPa and $\Delta T = -200°C$ simulates interface stresses expected in a bimaterial after stress-relief from a depth of 10 km within the Earth's crust

tial stress $\sigma_{\varphi\varphi}(r = R_K)$ decreases from 165 to 79 MPa (Fig. 4.13a). To account for external loads, both curves must be reduced by 100 MPa. At the intersection of the reduced curves, no stress concentrations occur at $E_K = E_M$. For $E_K << E_M$ (*soft inclusion*), the heterogeneity absorbs the total magnitude of the compressive stress field (−100 MPa) before unloading, whereas the matrix material undergoes circumferential stresses (165 − 100 = 65 MPa). For $E_K > E_M$ or $E_K >> E_M$ (*hard inclusion*), stress components change in sign (radial tension, circumferential compression).

In the cooling scenario (Fig. 4.13b), stress magnitudes vary linearly with the ratio of the thermal expansion coefficients of inclusion and matrix. For this calculation, $0 \leq \alpha_K/\alpha_M \leq 2.5$ varied with $\alpha_M = 10 \times 10^{-6}\ \mathrm{K}^{-1}$, and elastic modulus $E_K = E_M = 100$ GPa, $\nu_K = \nu_M = 0.25$ and inclusion concentration $c_K = 0.1$ are fixed. Again, both stress components are zero for $\alpha_K = \alpha_M$. For $\alpha_K < \alpha_M$, the inclusion tends to shrink less than the matrix producing, most likely, a radial crack pattern in the matrix. For $\alpha_K > \alpha_M$, the sign of stress components are reversed producing, most likely, a coaxial crack annulus at the interface.

In the coupled Δp–ΔT scenario in Fig. 4.13c, the interface stresses are plotted versus the modulus ratio for two sets of mismatches of thermal expansion coefficients ($\alpha_K = 2\alpha_M = 10 \times 10^{-6}\mathrm{K}^{-1}$, $\alpha_K = 0.5\alpha_M = 5 \times 10^{-6}\mathrm{K}^{-1}$). The elastic modulus is chosen according to the unloading scenario. Figure 4.13c demonstrates that small differences in thermo-elastic properties of neighbouring grains result in tensile stresses on the order of 100 MPa, which represents a value about 10 times the tensile strength value of most rock materials. Solving for inclusion models, like in Fig. 4.12, provides an insight into the understanding of stress changes in rock around boreholes, tunnels and mine stopes, as is discussed at the very end of this section. This should be the reader's motivation to tackle the following exercise.

Exercise 4.1 Calculate interface stresses at $r = R_K$ in the self-consistent model for quartz (inclusion) surrounded by feldspar (matrix) embedded in an effective continuum versus the quartz volume fraction $0 \leq c_k \leq 1$ for the three loading scenarios discussed in Fig. 4.13. Use the quasi-isotropic, thermo-elastic properties of the minerals from Table 4.2.

(a) Why do both stress components increase with the quartz volume fraction in the unloading scenario, and decrease in the cooling and coring scenario?
(b) At what critical quartz concentration in the cooling scenario, is grain boundary cracking suppressed, i.e., $\sigma_{rr}(r = R_K) < 10$ MPa?

Table 4.2 Quasi-isotropic, thermo-elastic constants of minerals used for inclusion models. Young's modulus *E* and Poisson ratio ν from Landolt-Börnstein (1982) and linear thermal expansion coefficient α from Skinner (1966)

Mineral	E (GPa)	ν	α ($10^{-6}\mathrm{K}^{-1}$)	K/M-model
Garnet	232	0.27	7	GAR
Diopside	165	0.26	7	DIO
α-Quartz	95	0.08	12	QTZ
Calcite	84	0.32	5	CAL
Albite	70	0.28	7	FSP

(c) Which pressure difference, Δp is needed in order to balance the value of grain boundary stress as resulting from the $\Delta T = -100°C$ cooling scenario in the quartz-feldspar self-consistent system?

If the inclusion has an ellipsoidal contour and anisotropic properties, analytical calculations become more difficult (Eshelby 1957). The state of stress in the ellipsoidal inclusion is homogeneous however if the stress applied to the matrix at infinity is homogeneous. Effective moduli of solids with various shapes of cavities (holes) are compiled, e.g., by Kachanov et al. (1994). Effective media average schemes are given by Mavko et al. (1998). A recent review on techniques in numerical modelling of fractured rock is given in Jing (2003).

Structural stresses can be the result of the combined effect of anisotropy and heterogeneity, as demonstrated in Fig. 4.14. The local stress around a borehole (Fig. 4.14, circle) produces a low stressed region (Fig. 4.14a, dotted area) with respect to the externally applied stress (Fig. 4.14, open arrows indicating far-field stress). In the presence of a cohesion-less discontinuity cross-cutting the hollow inclusion (Fig. 4.14b, fracture), the low stress region becomes a de-stressed region (Fig. 4.14b, dotted area) sustaining even less stress and aligning with the orientation of the fracture (anisotropy).

Structural stresses are essential for other sections in this book. For example, the inclusion stress incompatibility Eq. (4.10) will be discussed in modified form when

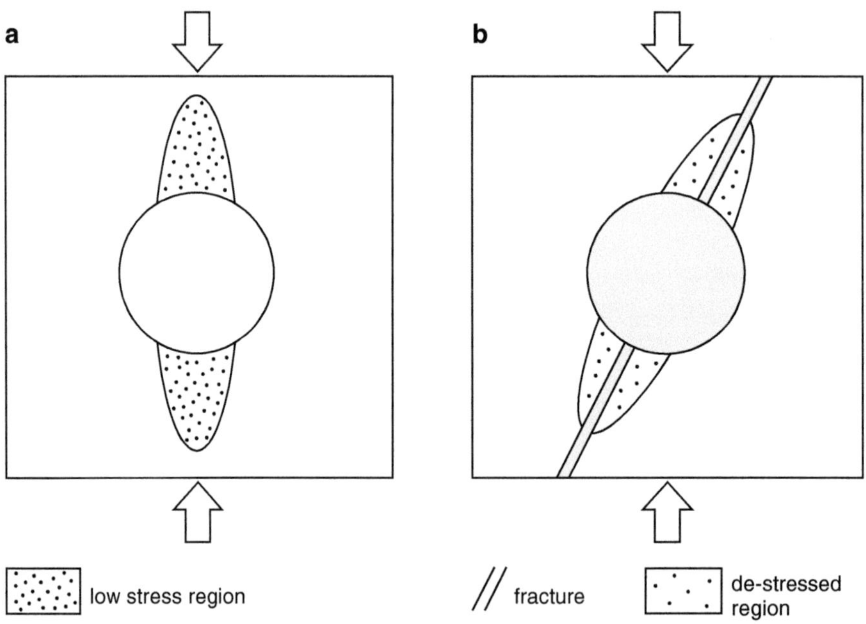

Fig. 4.14 **a** Low stress zone and **b** de-stressed zone around a borehole (*circle in plan view*) under uniaxial remote load (*open arrows*). In homogeneous rock material (**a**), two diametrically opposed low stressed zones develop. In the presence of a fracture intersecting the borehole (**b**), two de-stressed zones align with the fracture

borehole stresses are analyzed (Chap. 7). For this purpose, the borehole is regarded as a cylindrical inclusion with zero-modulus. Other examples of structural stress include the progressive advancement of stopes (mining sciences) and tunnels (geotechnics) in the subsurface. In this context, the stope is regarded as a horizontal, cylindrical inclusion. Calculated cavity-induced stress perturbations then can be related to mine seismicity and mine safety (McGarr et al. 1975). A recent example of the influence of rock heterogeneity on rock bursts in the Lucky Friday Mine in northern Idaho is discussed in Whyatt et al. (2000). A third type of hollow inclusion perturbed stresses is the near-field stress ahead of an advancing tunnel face as computed by 3D finite elements (Eberhardt 2001). As the tunnel passes through the rock mass, spatiotemporal changes in deviatoric stress and stress rotation control fracture propagation and failure near the circular cavity. Today, a 57 km long cylindrical (diameter 10 m) double inclusion (40 m apart) crosses the Alps as the Gotthard base tunnel, and will be used as a fast rail link between Zürich and Milan (Zbinden 2006).

Note-Box The diagram illustrating our rock stress terminology (Fig. 4.1) shows that the in-situ stress field can be subdivided into gravitational, tectonic, residual and terrestrial components. Topographic and terrestrial stress account for in-situ stresses only at shallow depth. The most important contribution to in-situ stress is the tectonic part plus gravitation. Tectonic stresses are subdivided into first order (plate scale), second order (mountain range scale), and third order stresses (fault scale). Residual stresses are locked-in while the rock surface is stress free. As palaeo-tectonic events, residual stresses can contribute to the remnant tectonic stress field. Near-field and far-field stresses are tied to structural stresses, which are the result of rock anisotropy (discontinuity) and heterogeneity (inclusion). Rock excavations (borehole, tunnel) are treated as hollow inclusions and perturb the stress field in the close vicinity (three defect diameters). The true state of in-situ stress in the Earth's crust is always a superposition of several stress components.

Chapter 5
Crustal Stress Models

The stress field of the Earth's crust is often described by three compressive principal stress components, namely the minimum horizontal, S_h, the maximum horizontal tectonic stress, S_H and the vertical stress due to the weight of the overburden, S_V. Depending on the depth variation of principal stresses assumed, different, simplified crustal stress models can be distinguished. When the vertical direction is not a principal stress direction, assuming that one principal component of the stress tensor is equal to the weight of overburden is of course wrong (cf. Sect. 4.1).

5.1 Lithostatic Stress

The inability of rock to support large differential stresses together with the effect of time-dependent deformation of the rock mass can cause horizontal and vertical stress to equalize over periods of geological times (Chap. 1, Heim's rule). This assumed lithostatic state of stress throughout the Earth's crust is governed by

$$p(z) = S_H(z) = S_h(z) = S_V(z) = \rho g z, \qquad (5.1)$$

where g is the gravitational acceleration and z is the depth within the Earth's crust. Strictly speaking, the lithostatic stress is isotropic, having no directional dependence of stress components. The increase of lithostatic stress with depth is linear with a slope of about 27 MPa km^{-1}, assuming an average rock density of $\rho = 2750\,\mathrm{kg\,m^{-3}}$ (Fig. 5.1a). The hydrostatic state of stress exists in liquid (geo) materials (e.g., magma), which are characterized by zero shear strength. It is also used as a proxy for weak rocks like coal, evaporites (rock salt) and shales. In these materials, the lithostatic state of stress gives a good approximation of the in-situ stress field (Hoek and Brown 1980). Since all rocks (in the solid state) support small differential stresses for very long periods (Kirby 1983), they approach, but never reach, the lithostatic state of stress. Below a certain depth in the Earth's crust, the deformation mechanism within the rock changes from brittle fracturing (Chap. 3) to ductile processes such as creep (Sect. 5.5).

A. Zang, O. Stephansson, *Stress Field of the Earth's Crust*,
DOI 10.1007/978-1-4020-8444-7_5, © Springer Science+Business Media B.V. 2010

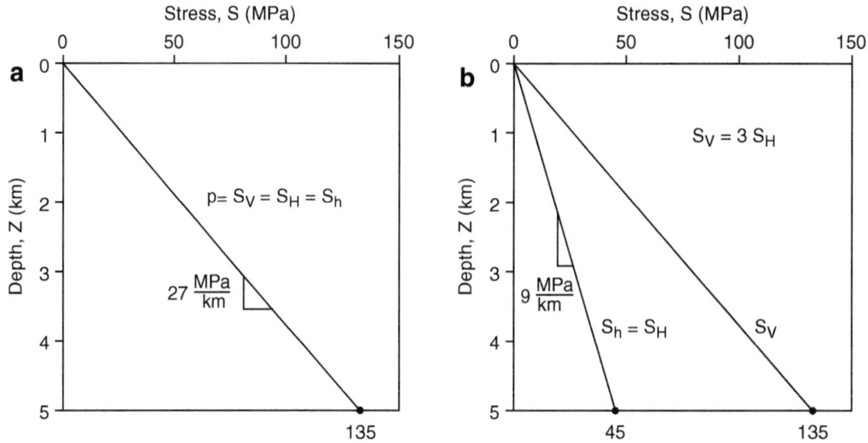

Fig. 5.1 Simple stress models: **a** lithostatic, and **b** biaxial state of stress in the Earth's crust

5.2 Biaxial State of Stress

Terzaghi and Richart (1952) suggested that in the case of sedimentary rocks in geologically undisturbed regions, the strata are built up in horizontal layers (sedimentary basin) assuming that the horizontal dimensions of the basin remain unchanged. Consider a cube of rock mass in the Earth's crust with vertical stress σ_z and horizontal compressive stresses (σ_x, σ_y), it follows from Eq. (4.8), after neglecting thermal stresses, that

$$\begin{aligned} \varepsilon_{\mathrm{x}} &= \frac{1}{E}\left\{\sigma_x - \nu\left(\sigma_y + \sigma_z\right)\right\} \\ \varepsilon_{\mathrm{y}} &= \frac{1}{E}\left\{\sigma_y - \nu\left(\sigma_x + \sigma_z\right)\right\}. \\ \varepsilon_{\mathrm{z}} &= \frac{1}{E}\left\{\sigma_z - \nu\left(\sigma_x + \sigma_y\right)\right\} \end{aligned} \tag{5.2}$$

Suppressing horizontal expansion $\varepsilon_x = \varepsilon_y = 0$, we find from Eq. (5.2) that

$$\begin{aligned} \sigma_x &= \nu\left(\sigma_y + \sigma_z\right) \\ \sigma_y &= \nu\left(\sigma_x + \sigma_z\right). \\ \varepsilon_{\mathrm{z}} &= \frac{1}{E}\left\{\sigma_z - \nu\left(\sigma_x + \sigma_y\right)\right\} \end{aligned} \tag{5.3}$$

Equating the horizontal stresses, $\sigma_x = \sigma_y$, and using our notation for principal crustal stresses ($\sigma_x = S_H$, $\sigma_y = S_h$, $\sigma_z = S_V$), it follows that

$$S_{\mathrm{H}}(z) = S_{\mathrm{h}}(z) = \frac{\nu}{1-\nu} S_{\mathrm{V}}(z). \tag{5.4}$$

For rock having Poisson ratio of $\nu=0.25$, the biaxial state of stress model states that the horizontal stresses are each equal ($S_h=S_H$) to one-third of the vertical stress (Fig. 5.1b). While stresses increase with depths by 27 MPa km^{-1} in the lithostatic case (Fig. 5.1a), the slope of horizontal stress with depth is only 9 MPa km^{-1} in the biaxial case (Fig. 5.1b). We have three remarks concerning Eq. (5.4). First, in the end member case of fluids ($\nu=0.5$), Eq. (5.4) simplifies to the lithostatic expression (5.1). Second, one application of Eq. (5.4) is the calculation of changes in horizontal stress during sedimentation (positive Δp) and erosion (negative Δp). These estimates can be compared to stress changes as inferred from unloading of simple inclusion models (Sect. 4.4.2). Third, one refinement of Eq. (5.4) would be to incorporate thermal stresses as indicated by Eq. (4.8). Thermal stresses in the crust develop in the horizontal direction with the vertical direction being relieved due to the lack of a fixed boundary at the Earth's surface ($z=0$).

Measured in-situ stresses are very seldom as low as predicted by the biaxial state of the stress model. Hast (1958) reported measured horizontal stresses of several times the vertical stress in Scandinavia. It took the rock mechanics community some time to accept Hast's stress data and reject the assumptions made by Terzaghi and Richards. The biaxial stress model where the rock is laterally constrained was severely criticized by McGarr (1988) for being geologically unrealistic. In Sect. 10.1, we discuss more realistic stress depth variations in the Earth's crust.

5.3 Tectonic Stress Field

All phenomena of brittle fracture studied on a laboratory scale (Chap. 3) appear to be reproduced in nature by geological structures. *Faults* form in the upper 10–15 km of the Earth's crust (e.g. Meissner 1996) and are defined as planar discontinuities along which displacement (slip) occurs. Anderson (1951) developed the modern mechanical concepts of the origin of faults and emphasized their important role in tectonics. His key contribution was to recognize that faults result from brittle fracture, and to apply the Coulomb criterion Eq. (3.2) to this problem. Using Eq. (3.3) and applying the condition that near the free surface of the Earth's crust one of the principal stresses is vertical, Anderson showed that the three major classes of faults (normal, strike-slip, reverse) result from the three principal classes of inequality that may exist between the principal stresses.

The three types of *Anderson faulting* are summarized in Fig. 5.2 (reverse or thrust faulting (a), normal faulting (b), and strike-slip faulting (c)). Fracturing and faulting takes place in one or both of a pair of conjugate planes which pass through the direction of the intermediate principal stress and are equally inclined at angles (dip) less than 45° to the direction of the greatest principal stress (about 30° for $\mu=0.6$, see Sect. 3.1). Under these assumptions, *reverse faulting* in the Earth's crust occurs if the least principal stress is vertical, $S_3=S_V$, and the relationship between principal stress magnitudes is $S_V<S_h<S_H$ (Fig. 5.2a). Reverse faults with a low angle of dip are called *thrust faults*. In this case, the upper surface of

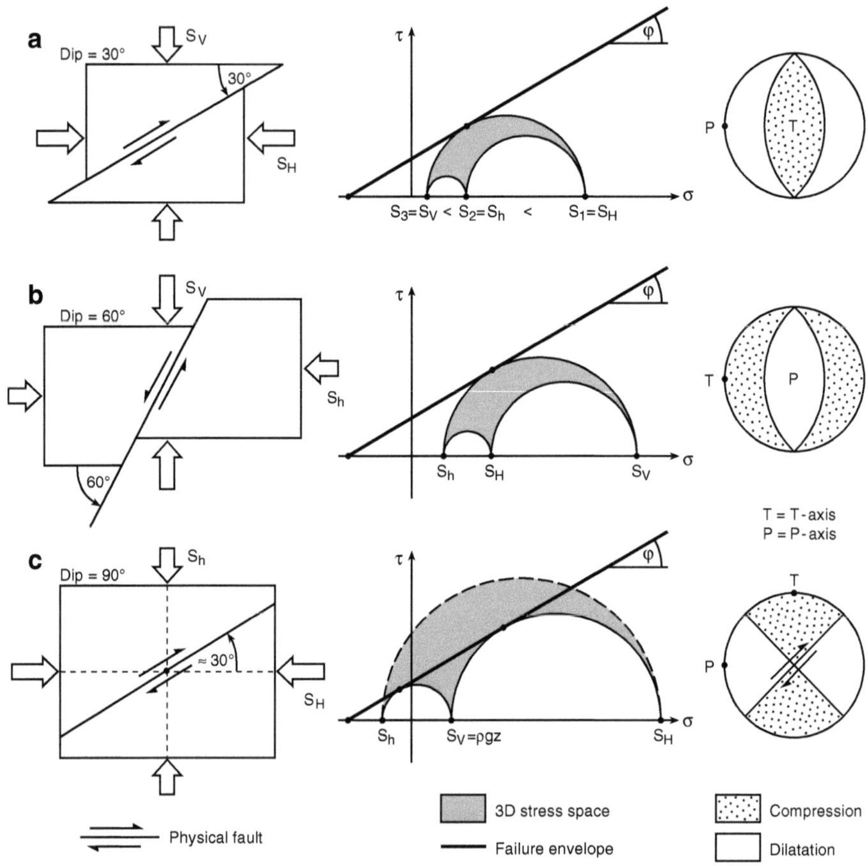

Fig. 5.2 Anderson fault types in physical space (*left*), associated states of stress in Mohr space (*middle*), and corresponding solutions from focal mechanisms of earthquakes (*right*). **a** For thrust faults, the vertical stress is the smallest principal stress. **b** For normal faults, the vertical stress is the largest principal stress. **c** For strike-slip faults the vertical stress is the intermediate principal stress. Earthquake focal mechanisms (*right*) lead to an ambiguity in fault planes (two nodal planes in the lower hemisphere plot) resulting from conjugate fracture planes of the Coulomb fracture criterion

the fault (*hanging wall*) moves upwards relative to the lower surface (*foot wall*). The fault plane develops parallel to S_h and is dipping $<45°$ (30° for $\mu=0.6$) with respect to S_H (Fig. 5.2a). *Normal faulting* occurs if the greatest principal stress is vertical, $S_1=S_V$, and $S_h<S_H<S_V$ is fulfilled (Fig. 5.2b). In physical space, the hanging wall moves downwards relative to the foot wall. The fault plane develops parallel to S_H, and is dipping $>45°$ (about 60°) with respect to S_h. If *strike-slip faulting* occurs, the vertical stress is intermediate, $S_2=S_V$ and the relationship $S_h<S_V<S_H$ applies (Fig. 5.2c). The physical fault plane is vertical (dip 90°) and parallel to S_V. With $\mu=0.6$, the fault plane is tilted 30° from the largest principal stress, $S_1=S_H$. All considerations discussed above depend on the relative mag-

nitudes of principal stresses and are therefore much affected by the depth and tectonics at which faulting is initiated.

In Mohr stress space, the Anderson types of faulting are visualized in Fig. 5.2 (middle column) using the 3D Mohr circle approach (Sect. 2.4). We choose the upper part of the Mohr circle and exclude the ambivalence of conjugate fault planes as discussed in Chap. 3. In terms of the Coulomb criterion of brittle fracture, the three types of faulting (reverse (RF), normal (NF), strike-slip (SS)) in the Earth's crust can be formulated as

$$\begin{aligned} RF &: S_1 = S_\mathrm{H}, S_3 = S_\mathrm{V}, S_\mathrm{H} = qS_\mathrm{V} + C_0 \\ NF &: S_1 = S_\mathrm{V}, S_3 = S_\mathrm{h}, S_\mathrm{V} = qS_\mathrm{h} + C_0 \ , \\ SS &: S_1 = S_\mathrm{H}, S_3 = S_\mathrm{h}, S_\mathrm{H} = qS_\mathrm{h} + C_0 \end{aligned} \tag{5.5}$$

where the constant q is determined with the help of Eq. (3.3) by

$$q = \tan^2\alpha = \tan^2\left(\frac{\pi}{4} \mp \frac{\varphi}{2}\right) = \frac{1+\sin\varphi}{1-\sin\varphi} \tag{5.6}$$

and the uniaxial compressive strength, C_0, is computed from the cohesion term C in the Coulomb criterion Eq. (3.2) by

$$C_0 = 2C\mu^+ \ \text{ with } \ \mu^+ = \sqrt{\mu^2+1} + \mu. \tag{5.7}$$

As we know from Chap. 3, each fracture criterion can be represented in Mohr stress space (σ, τ) and in principal stress space (S_1, S_3). Our Coulomb criterion in Mohr stress space (Eq. (3.2)) reads in principal stress space

$$S_1\mu^- - S_3\mu^+ = 2C \text{ with } \mu^- = \sqrt{\mu^2+1} - \mu. \tag{5.8}$$

In both representations, the terms cohesion C and friction coefficient μ occur. The uniaxial compressive strength C_0, however, is indicated as zero offset in principal stress space (Fig. 5.3a), and has to be computed from Eq. (5.7) using the failure slope $\mu = \tan\varphi$ in Mohr space (Fig. 5.3b). The terms (μ^+, μ^-) we know from discussing

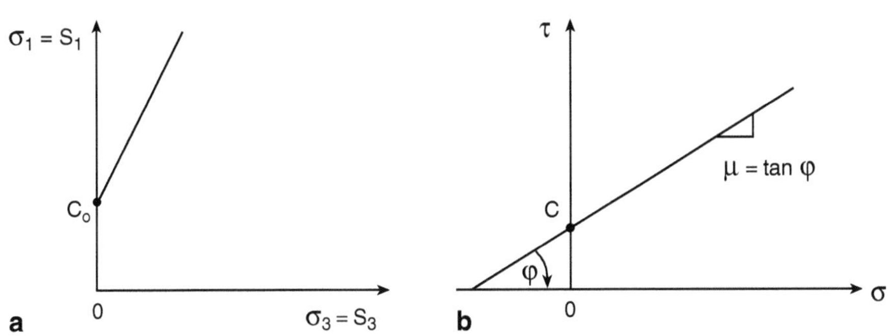

Fig. 5.3 Coulomb fracture criterion in **a** principal stress space, and **b** Mohr stress space

fracture criteria in principal stress space, such as the modified Griffith criterion (Eq. (3.12)), and the wing crack criterion (Eq. (3.14)).

The uniaxial compressive strength of the crust, as appearing in Eq. (5.5), is a cause of controversy. In light of in-situ stress measurements, this value could not be more than 40 MPa (Seiki et al. 1997). However, it can be assumed to be almost zero when inferring the state of stress from different fault types. Using $C_0=0$, Eq. (5.5) simplifies and the ratio of greatest and least principal stress reads

$$\begin{aligned} RF &: S_1/S_3 = S_H/S_V = q \\ NF &: S_1/S_3 = S_V/S_h = q \text{ with } q = \tan^2\alpha = \left(\mu^+\right)^2. \\ SS &: S_1/S_3 = S_H/S_h = q \end{aligned} \tag{5.9}$$

It is obvious from these equations that both the maximum difference and ratio in principal stress magnitudes depend on depth as given by S_V, and knowing whether the study area is an NF, SS or RF regime. Correcting the principal stresses for pore-pressure effects (Sect. 5.4), Zoback et al. (2003) computed depth variations of stress magnitudes for different stress regimes assuming hydrostatic pore-pressure conditions (Fig. 5.4). For their computations, they used a friction coefficient of 0.6 (resulting in $q=3.12$ from Eq. (5.6)), a pore pressure increase of 10 MPa km^{-1} (Fig. 5.4, p_p) and a vertical stress gradient of 23 MPa km^{-1} appropriate for sedimentary rocks. According to Eqs. (5.7) and (5.9), the bold lines in Fig. 5.4 indicate the limiting value

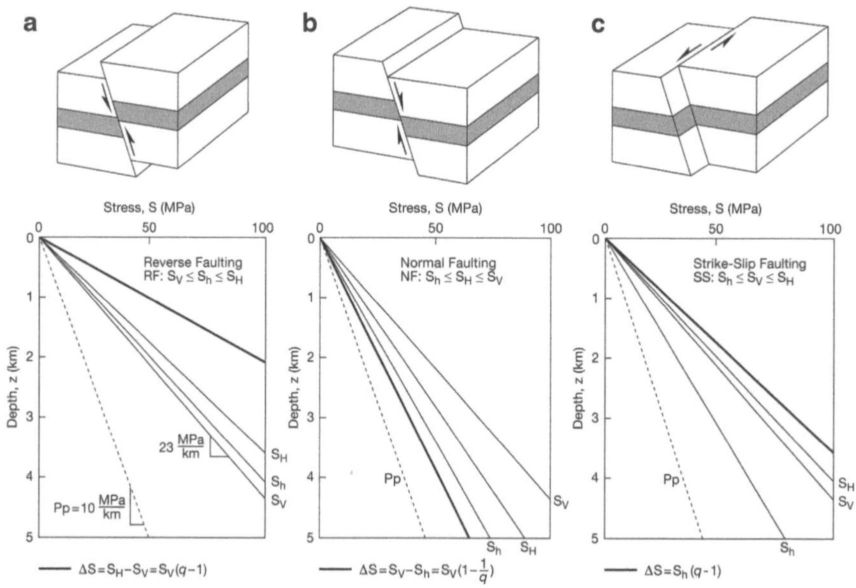

Fig. 5.4 Variations of stress magnitudes with depth in RF (**a**), NF (**b**), and SS tectonic stress regime (**c**) according to Anderson faulting for hydrostatic pore pressure. Assumptions are $\mu=0.6$ for the friction coefficient, 23 MPa km^{-1} for the depth variation of S_V and pore pressure of 10 MPa km^{-1} (after Zoback et al. 2003)

of the principal stress differences at depth. This difference increases with depth due to the increase of crustal strength with depth. Analogous to Fig. 5.1, the stresses vary linearly with depth. Depending on the relationship between the principal stresses, however, a RF regime (Fig. 5.4a, $S_V < S_h < S_H$), a NF regime (Fig. 5.4b, $S_h < S_H < S_V$) and a SS regime (Fig. 5.4c, $S_h < S_V < S_H$) can be distinguished. The cohesive strength of faults is neglected ($C_0 = 0$) because at significant depths (e.g. a seimogenic depth of 15 km), the effect of a cohesive fault (~40 MPa) is relatively unimportant in terms of stress magnitudes and results obtained by using Eqs. (5.5) and (5.9) should be nearly the same (error < 10%). At shallower depths, e.g., in civil and mining engineering applications, a finite value of C_0 must be taken into account.

Anderson (1951) regards the standard state of stress at depth z as lithostatic (Fig. 5.1a, Eq. (5.1)) and assumes that the stress causing faulting is superposed on this state (Figs. 5.2, 5.4, Eq. (5.5)). We close this section by pointing out four limitations of the Anderson faulting theory. (1) Faults can be regarded as discontinuities with zero or non-zero modulus, different from the host rock material (cf. Fig. 4.10). As we know from Sect. 4.4.1, stresses close to discontinuities can change drastically compared to the case in homogeneous materials (cf. Figs. 4.7, 4.11). Therefore, extrapolation of stress orientations into regions for which geological fault data are lacking is not always justified. (2) Anderson faulting cannot accommodate any deformation parallel to the intermediate principal stress axis (Fig. 5.2). If faults are formed in true-triaxial stress systems, they cannot be of Anderson type. Faults that do form under polyaxial stress conditions have lower, orthorhombic symmetry (Reches 1987a, Krantz 1989). (3) The frictional strength of faults is less than the fracture strength of faults. While the fracture strength of faults is indicated by a single point on the Mohr circle, the frictional strength of pre-existing faults is characterized by a range of critical points on the Mohr circle where slip occurs. The stress necessary to form a virgin (young) fault is higher compared to the stress necessary to reactivate a pre-existing (old) fault. This leads to ambiguity in inferring stress-field orientations from the orientation of active faults or earthquake focal mechanisms (Fig. 5.2, right). It seems more likely that an old fault (although not optimally oriented) slips before the condition is met to form a new, young fault. In general, one could use the Coulomb criterion as before, but with lower values of strength and friction coefficient for the planes of weakness. (4) Anderson's theory of faulting assumes isotropic and homogeneous strength of rocks. Instead of worrying about the conditions of failure (conditions necessary to cause slip on a fault plane), in the *Wallace-Bott hypothesis* the shear stress resolved onto the fault plane is assumed to be parallel to the slip vector produced (Wallace 1951; Bott 1959). This empirical law is a fundamental assumption for most inversions of fault-slip data for determining the stress field. Celerier (2008) recently reviewed Anderson faulting in seismology.

Stress-tensor inversion from slip on fault planes (structural geologist perspective) is addressed in Reches (1987b), Pollard et al. (1993) and in the textbook of Ramsay and Lisle (2000). Mapping tectonic stress from earthquakes (geophysicist perspective) is discussed in Arnold et al. (2005). Stress inversion techniques from earthquake focal mechanisms use either a misfit criterion (Angelier et al. 1982; Michael 1984; Albers and Gephart 2001; Angelier 2002; Yamaji and Sato 2006) or a

Bayesian approach (Arnold and Townend 2007). All seismological estimates of the principal stress directions represent at best kilometre-scale volumes of the Earth's crust (Chap. 7). World Stress Map guidelines for stress inversions from earthquakes are summarized in Barth et al. (2008), see DVD Chap. 10. The two most obvious assumptions in making stress inversions, however, are that the stress field is uniform and invariant over time. As we know, this is not true since a discontinuity is able to break stress uniformity (Chap. 4) and if an earthquake occurs, i.e. a slip is activated on a fault plane oriented critically, the Coulomb stress changes in the near-field (Stein 1999). Notice that any stress inversions from earthquake data resolve the stress field at slip occurrence time not after.

Note-Box Simple Earth stress models require a linear increase of stress with depth. In the lithostatic stress model, all principal stresses are equal to the weight of the overburden. In the biaxial elastic stress model, horizontal stresses are each equal to approximately one-third of the vertical stress. In the triaxial stress model, the tectonic stresses are computed based on Anderson faulting and the Coulomb criterion. Depending on the relationship between principal stress magnitudes, reverse faulting (vertical stress=minimum), normal faulting (vertical stress=maximum), and strike slip faulting (vertical stress=intermediate) can be distinguished.

5.4 Effective Stress

The existence of fluids in the Earth's crust has been proven to approximately 10 km depth (Chap. 9). Due to the presence of fluids, the fracture mechanical behaviour of rock from Chap. 3 has to be modified. To do so, elastic stresses must be generalized to *poroelastic stresses* taking into account the effect of a fluid phase circulating or trapped in the pore space of rock. Isothermal poroelasticity was first presented by Biot (1941) and later refined by Rice and Cleary (1976). A good textbook in linear poroelasticity is Wang (2000). Let p be the pressure defined by the mean normal stress of the stress tensor (Sect. 2.2, hydrostatic stress) and p_p denote the fluid pressure inside the connected pores (Fig. 5.5), then the *effective pressure* p^* is defined through the following constitutive relation of linear poroelasticity:

$$p^* = p - \alpha p_p \text{ with } \alpha = 1 - \frac{\mathrm{K}}{\mathrm{K_S}}, \tag{5.10}$$

where the coefficient, α is called *Biot's coefficient* and is derived by Nur and Byerlee (1971). K_S is the bulk modulus of the rock without pores and K is the bulk modulus of the rock with (empty) pores, also called the *modulus of skeleton* or modulus

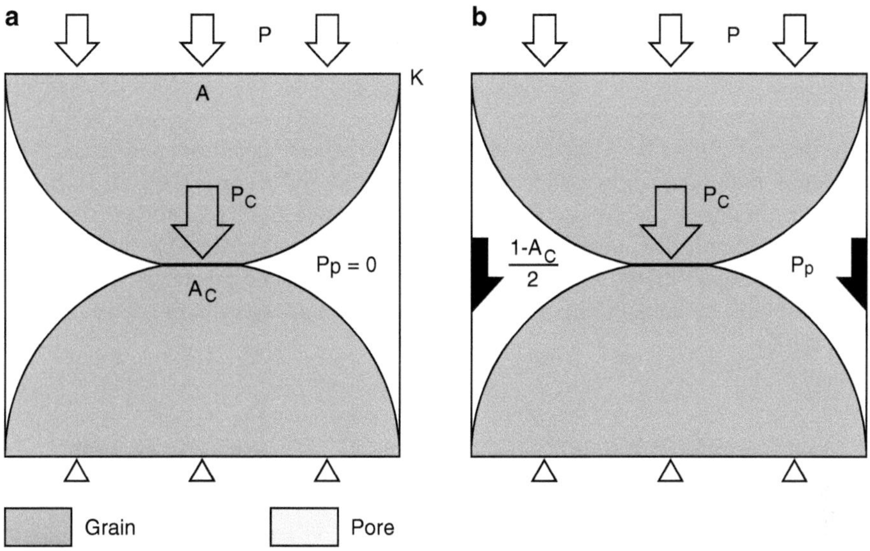

Fig. 5.5 Stress partition in case of a single grain contact for **a** zero, and **b** non-zero pore pressure, p_p

of the frame. Note that the effective pressure equals the differential pressure, p_D (Sect. 2.2) only, if $\alpha = 1$ $(K_s \gg K)$ is valid:

$$p_D = p - p_p. \tag{5.11}$$

The effect of stress partition in the case of a single grain contact is illustrated in Fig. 5.5. In the zero pore-pressure scenario (Fig. 5.5a), the remote stress p is linked to the intergranular stress, p_C, which is transmitted through the grain skeleton by

$$p = ap_C. \tag{5.12}$$

The ratio $a = A_C/A$ is the area of contact between grains A_C normalized to the total grain area, A. In the pore-pressure scenario (Fig. 5.5b), the space outside the area of contact, $1 - A_C$ is being filled with fluid at pressure p_p. In this case, the remote stress is given by

$$p = ap_C + (1 - a)p_p, \tag{5.13}$$

a superposition of intergranular stress and stress resulting from an overpressurized fluid in the pore space. Using different notation, Eq. (5.13) is rigorously proven in Zimmerman et al. (1994). When pore pressure exceeds the hydrostatic stress, the rock fails.

The existence of a fluid phase in rock implies that there are two distinct deformation regimes to be considered (Gueguen and Palciauskas 1994). In the *drained regime*, the fluid pressure remains constant while the amount of fluid within the rock changes. In the *undrained regime*, fluid mass remains constant and the fluid pressure is variable. Stresses in the drained regime are the same as in elasticity, modified by substituting effective stresses σ^*_{ij} for the stresses σ_{ij}:

$$\sigma^*_{ij} = \sigma_{ij} + \alpha p_p \delta_{ij}. \tag{5.14}$$

The drained regime is characterized by three poroelastic constants K, μ and α. The skeleton modulus is only a function of porosity Φ and K_S ($K=K$ (Φ, K_S)). In the undrained regime, the porous rock does not exchange fluid with the outside (i.e., a thermodynamically closed rock-fluid system), can be described by an equivalent effective medium theory and is characterized by four elastic constants (K, μ, α, K_u). K_u is the bulk modulus of the undrained conditions and can be expressed as $K_u=K_u$ (Φ, K_S, K_f).

$$K_u = K + \frac{\alpha^2}{\frac{\Phi}{K_f} + \frac{\alpha - \Phi}{K_S}} = \begin{cases} K + \frac{\alpha^2 K_S}{\alpha - \Phi} \\ K \end{cases} \text{for} \begin{cases} K_f \to \infty \\ K_f \to 0 \end{cases} \tag{5.15}$$

K_u is the effective bulk modulus of a porous rock which contains a second-phase fluid having a bulk modulus K_f. The end member cases of K_u are incompressible fluid ($K_f \to \infty$, e.g. oil saturated sandstone) and rock saturated by gas ($K_f \to 0$). The limiting case $K_f \to \infty$ is useful for plotting results but is not practically meaningful for rocks as pore fluid is always more compliant than the pore space of rock (Zimmerman 1991). The concept of effective pressure, although still valid, is not useful in undrained conditions, since p is no longer an independent parameter. Introducing the dimensionless coefficient B, the change in fluid pressure Δp_u can be expressed in terms of the increase in confining pressure $-\sigma_{KK}/3$ for undrained conditions

$$\begin{aligned} \Delta p_u &= -B \tfrac{\sigma_{KK}}{3} \\ B &= \tfrac{1}{K_u} \left(\tfrac{K_u - K}{\alpha} \right) \end{aligned} . \tag{5.16}$$

B is called *Skempton's coefficient* and is one of the most often-measured parameters in the oil industry. B varies between 0 and 1 (Table 5.1). In porous, unconsolidated rock B approaches 1, since the skeleton modulus approaches zero, $K \to 0$. In porous rock filled with fluid of very high compressibility, the skeleton modulus approaches the modulus of the undrained conditions $K \to K_u$ and, therefore, $B \to 0$.

Skempton (1960) concluded that the use of the differential stress as the "effective stress" in saturated soils is valid if the soil grains are incompressible and the yield stress of the grains is independent of pressure. If this is not the case, Eq. (5.14) has to be used with α depending on the contact area between grains and their mechanical properties.

The effect of pore pressure on rock strength can be demonstrated by combining the Coulomb fracture criterion Eq. (5.5) with the concept of effective stress Eq. (5.14), assuming $\alpha=1$

Table 5.1 Laboratory data on poroelastic constants for rock

Rock	Φ	K, GPa	K_s, GPa	α	B
Granite, Westerly	0.01	25	45	0.47	0.85
Granite, Barre	0.03	15	53	0.72	0.62
Marble, Tennesse	0.02	40	50	0.19	0.51
Limestone, Tonnerre	0.13	19	41	0.53	0.20
Limestone, Lavoux	0.22	14	59	0.77	0.30
Chalk, Lixhe	0.43	4	43	0.91	0.35
Sandstone, Ruhr	0.02	13	36	0.65	0.88
Sandstone, Fontainebleau	0.06	31	35	0.12	0.25
Sandstone, Berea	0.19	8	36	0.79	0.62

Φ porosity, K skeleton modulus, K_s bulk modulus of rock without pores, α Biot's coefficient, B Skempton's coefficient (after Paterson and Wong 2005)

$$S_1^* = qS_3^* + C_0$$
$$\left(S_1 - p_p\right) = q\left(S_3 - p_p\right) + C_0. \tag{5.17}$$

The way in which pore pressure affects failure is seen in Fig. 5.6. The Mohr envelope is that for rock failure at zero pore pressure (Fig. 5.6, straight lines). Curve I is the Mohr circle with zero pore pressure $p_p = 0$ which is seen to lie inside the Mohr envelope representing a stable region of rock (no failure). As the pore pressure increases (curve II, p_p), the Mohr circle is moved by $-p_p$ and now touches the Mohr envelope for $p_p = 0$ and failure takes place. In general, a unique Mohr envelope, not depending on the pore pressure, is obtained by working with the effective stresses, S_{ij}^*.

The criteria for failure in terms of effective stress hold reasonably well if (1) the pore system is interconnected and (2) the pore fluid is inert (Handin et al. 1963). If chemical effects dominate mechanical effects in the fluid-rock interaction, discrepancies occur (Serdengecti et al. 1962). In the latter case, crack propagation is controlled by stress corrosion (Atkinson and Meredith 1987). For the relation to

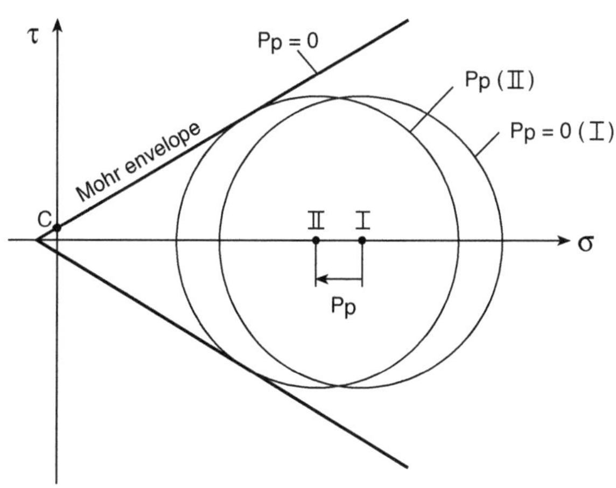

Fig. 5.6 Effect of pore fluid pressure on rock strength. An increase of pore pressure by p_p shifts the Mohr circle towards the Mohr envelope and failure occurs earlier compared to the case of zero pore pressure

swelling in shales, an *adsorptive pore pressure* is introduced (Huang et al. 1986), which is negative, that is, of the same sign as tensile stresses. These "negative" pressures are used in the same way as the pore pressure in a saturated rock to derive conventional effective stresses.

Exercise 5.1 The pore pressure in Aue granite is increased by 20 MPa before the triaxial test from Exercise 3.1 is carried out.

(a) Draw a Mohr circle of the wet test conditions in comparison to the dry conditions (see Fig. 5.6).
(b) In which way will the cohesion and friction coefficient of the rock change when pore fluid pressure is present?
(c) Does pore pressure influence the orientation of final shear rupture when triaxial failure occurs?

Now we are able to estimate the range of possible stress states at any given depth and pore pressures within the Earth's crust. Providing Anderson faulting and the Coulomb failure criterion in effective stress notation (5.17) are valid, we illustrate the principal stresses in the Earth's crust (S_h, S_H, S_V) for NF, SS and RF environments. Figure 5.7 delineates the range of possible stress magnitudes at a particular crustal depth (here $z = 3$ km), pore pressure for a particular coefficient of friction (here $\mu = 0.6$) and S_V computed analogously to Fig. 5.4 from an average rock density of 2350 kg m^{-3} (Zoback et al. 2003). The construction of such figures is discussed in Zoback et al. (1986) and Moos and Zoback (1990). $S_H \geq S_h$ requires all stresses to be above the diagonal line of the unit slope. The lowest value of principal stresses,

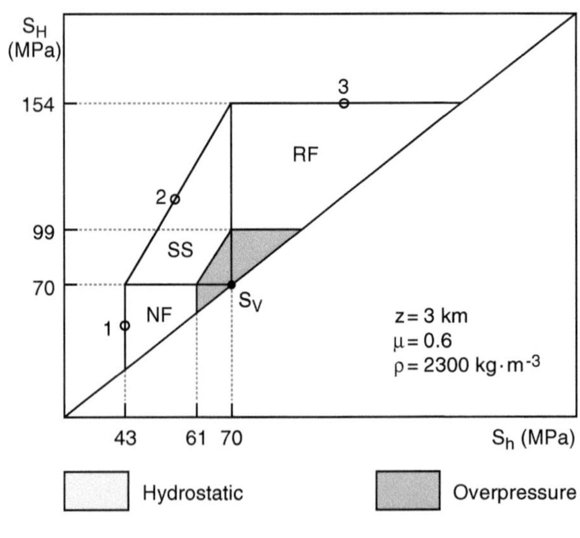

Fig. 5.7 Range of possible principal stress magnitudes (S_h, S_H) with S_V fixed at 3 km depth as defined by Anderson faulting regimes (*NF*, *SS*, *RF*) and effective Coulomb failure. The size of the stress polygon is larger in the hydrostatic pore-pressure scenario compared to the overpressure scenario (modified after Zoback et al. 2003)

S_h=43MPa defines the left border of the NF regime. A Mohr circle (S_h=43, S_H) would touch the frictional failure envelope with a slope of 0.6 (Fig. 5.7, point 1). Similarly, the horizontal line defining the top of the stress polygon (S_h, S_H=154) corresponds to the conditions at which RF would occur (Fig. 5.7, point 3). The third boundary of the stress polygon corresponds to principal stress values between (S_h=43, S_H=70) and (S_h=70, S_H=154) where strike-slip faulting would occur (Fig. 5.7, point 2). The size of the stress polygon for principal stresses (S_h, S_H) with S_V fixed at 3 km depth, is larger in a scenario with hydrostatic pore pressure (Fig. 5.7, hydrostatic p_p=0.3 S_V). In the case of overpressure, where the pore pressure is assumed to be equal to 80% of the overburden p_p=0.8 S_V, the difference among principal stresses at depth becomes smaller (Fig. 5.7, overpressure). If the state of stress is in effective frictional failure equilibrium, it falls on the outer boundary of the polygon.

Note-Box There is no simple universal definition of effective stress which permits uniform treatment of all phenomena affected by pore fluid pressure. For most of the applications related to crustal stress, the conventional effective stress law formulated in linear poroelasticity, Eq. (5.14) with α=1 is appreciable. The effect of pore pressure on rock strength is demonstrated by the shift of the Mohr circle by the amount of pore pressure towards the Mohr envelop until failure occurs. Anderson faulting in terms of effective Coulomb failure can be understood by principal stress polygons with S_V fixed at a certain depth. Each faulting regime (NF, SS, RF) reflects a principal stress triangle. Increasing pore pressure from hydrostatic (p_p=0.3 S_V) to overpressure (p_p=0.8 S_V), the difference in principal stresses at depth is reduced. At very high pore pressures, p_p=S_V small changes in principal stresses are sufficient to cause faulting.

5.5 Laboratory Stress Profiles

The last simplistic technique to estimate the variation of stress magnitudes with depth in the Earth's crust (i.e. stress profile) is to extrapolate laboratory data to crustal conditions. Starting from the surface and penetrating into the deeper crust, three types of experiments can be distinguished. Brittle-fracture tests ranging from low differential (tensile) stress $(\sigma_1-\sigma_3)\leq 0$ (mode I cracking) to high differential (compressive) stress $(\sigma_1-\sigma_3)>0$ (mode II and mode III cracking) simulate the behaviour of the brittle upper crust at low temperatures (T<200°C, z<10 km). Friction tests on rocks with pre-existing planes of weakness at high differential stress (Paterson 1978; Spetzler 1987) simulate the behaviour of rocks in the lower part of the crust. Finally, high-pressure (p<1500 MPa) high-temperature (T<1300°C) experiments simulate the creep behaviour of rocks from the lower crust and upper mantle (Evans and Kohlstedt 1995; Karato 2008). In this section, bounds on the variation of crustal stress (S_H, S_h, S_V) and crustal strength (S_1-S_3) with depth are discussed in the light of pressure, temperature, sample volume, rock material, strain rate and water content.

In the laboratory, the *fracture strength* of rock is commonly measured by means of uniaxial and triaxial compression tests (Chap. 3). The effect of pressure on the strength of rock is quite dramatic and is illustrated in Fig. 5.8 for Westerly granite. In Mohr space (Fig. 5.8a), the maximum shear stress at failure is plotted versus normal stress, while in principal stress space (Fig. 5.8b), the differential stress, $S_1 - S_3$, is plotted versus confining pressure, S_3, in the compression tests. According to Lockner (1995), the following factors reduce the strength of a "typical" granite, saturated with water at room temperature and subjected to confining pressure of 100 MPa, by about 10%: (1) a decrease in confining pressure (or increase in pore pressure) by 25 MPa, (2) an increase in temperature by 200°C, (3) a decrease in strain rate of the experiment by three orders in magnitude or (4) an increase in sample characteristic length by a factor of about two. As we know from Chap. 3, confining pressure affects brittle failure strength by suppressing the growth of dilatant cracks.

Pore fluids affect fracture strength through a direct, mechanical pressure effect and also through chemical interactions with the rock matrix. Fluid-assisted mechanisms will often dominate at intermediate temperatures but, over geologic time scales, they may also play an important role in determining rock strength at room temperature (Rutter and Mainprice 1978). In the brittle field, a strength increase is observed with increasing strain rate. Lockner (1995) compiled brittle fracture-strength values of rocks for strain rates varying from $10^{-9}\,s^{-1}$ to $10^3\,s^{-1}$. Linking the energetics of crack extension to pseudo-viscous response of rock in creep, the rate sensitivity of rock strength in the range from $10^{-9}\,s^{-1}$ to $10\,s^{-1}$ is expressed as

$$\sigma_C = A\left(\frac{d\varepsilon}{dt}\right)^n \text{ with } 0.02 \leq n \leq 0.04, \tag{5.18}$$

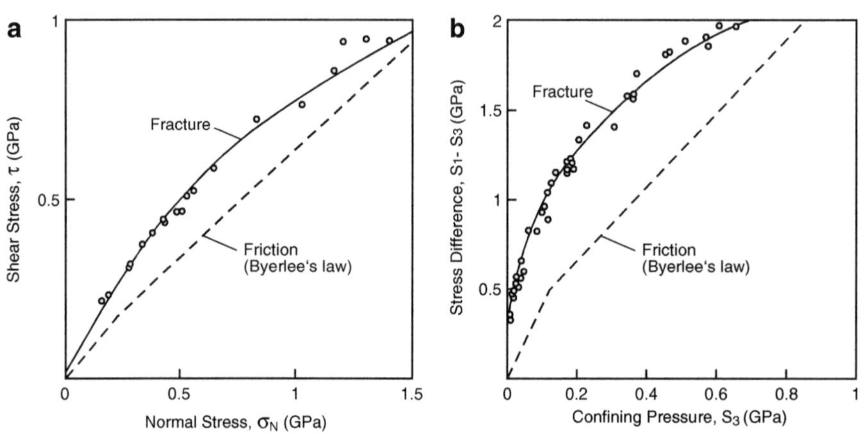

Fig. 5.8 Effect of pressure on rock strength. Fracture and friction of Westerly granite **a** in Mohr stress space (Lockner 1995), and **b** in principal stress space (Scholz 1990)

where A is a constant. Above $10\,s^{-1}$, a significant larger effect of the strain rate on fracture strength, σ_c, is observed. According to Whittaker et al. (1992) the rate sensitivity of the dynamic fracture strength for oil shale in the range from 10 to $10^4\,s^{-1}$ can be described by a cubic root power law (Eq. (5.18) with $n \approx 1/3$). The strength of rock cannot be described by a single size-invariant parameter (such as K_{ic} in Chap. 3) because subcritical crack growth can occur at much lower stress intensity factors. Subcritical tensile crack growth has been identified as the principal mechanism responsible for static fatigue in rock (Kranz 1980; Atkinson 1984; Lockner 1993). In mining engineering, when a load-bearing structure will fail after some time under constant load, this effect is defined as *static fatigue*.

The strain-rate dependence of rock strength leads to the condition that rock fracture strength at geologic strain rates (10^{-10}–$10^{-14}\,s^{-1}$) is less than that measured in the laboratory (10^{-5}–$10^{-9}\,s^{-1}$). From Eq. (5.18) it follows that the strength increase for a 10^9-fold increase in strain rates is $(10^{-5}/10^{-14})^{0.03} \approx 1.86$. A pragmatic way to limit this factor is to look for a natural rock mass deformed close to a fault at a high strain rate ($10^{-11}\,s^{-1}$) and compare it with fault rocks deformed in the laboratory at a low strain rate ($10^{-9}\,s^{-1}$). In this way for the strain rate, only two orders of magnitudes needs to be bridged by extrapolation.

Besides scaling of time, also scaling of sample size is of concern in any application of laboratory studies to geological processes. If L is a characteristic length of the test specimen, it is empirically found that the compressive strength, σ_c, decreases with L according to (Scholz 1990)

$$\sigma_c = AL^{-m} \text{ with } m \leq 1 \text{ for } 0.05 < L < 1\text{m}. \tag{5.19}$$

This effect can be as much as 70% strength loss per decade sample size for a weathered diorite investigated (Pratt et al. 1972). They found that a value of $m = 0.5$ fits the diorite experimental strength over an application range of $0.05\,\text{m} < L < 1\,\text{m}$ best. This behaviour is usually attributed to the dependence of crack size on sample size. As we know from Chap. 3, the critical stress that is needed to fracture a body with a crack of initial size c scales with $c^{-0.5}$, which would lead to $m = 0.5$ in Eq. (5.19). In laboratory specimens nominally free of cracks, strength is often found to be dominated by grain size. This scaling (*Petch's law*) can be understood in terms of the largest cracks limited by the size of the mineral grains in rocks. In natural samples containing flaws, the lower limit on rock strength determined by scale is the frictional strength. Since friction is size-invariant, the lower limit on strength is given by laboratory friction values (Scholz 1990).

The empirical friction law used (Fig. 5.8, dashed lines) is based on *frictional strength* data obtained in the laboratory for a wide variety of rocks (Byerlee 1978) and is described by two straight lines

$$\begin{array}{lcl} \tau = 0.85\sigma_N & \text{for} & 3 \leq \sigma_N \leq 200 \text{ MPa} \\ \tau = 0.6\sigma_N + 50 & & 200 \leq \sigma_N \leq 1700 \text{ MPa} \end{array}. \tag{5.20}$$

Equation (5.20) has become known as *Byerlee's law* (Kohlstedt et al. 1995). It is independent of lithology and holds over a wide range of rock hardness and ductility. The equation is not valid for clay minerals and clay-rich rocks. To first order, it is independent of sliding velocity and surface roughness of fault ligaments (Scholz 1990). Because of its universality we can use it to estimate the strength of natural faults and, therefore, strength profiles within the Earth's crust. For most purposes of estimating rock strength, one can assume a constant friction coefficient. Then the orientation of the fault plane depending on the value of friction coefficient (Chap. 3) is also constant. Note that in the adhesion theory of friction, contacts of surfaces play a key role in understanding the microphysical aspects of friction. In this book, contact mechanics (Johnson 1985) is referred to in three different sections: Hertzian fracture (3.3.1), specific joint roughness (4.4.1) and effective stress law (5.4). The universal importance of friction is worth studying in more detail for geoscientists. Common techniques in laboratory studies of friction include triaxial compression, direct shear, biaxial loading and rotary shear (Paterson and Wong 2005). Analogue to the fracture strength, the frictional strength is affected by several parameters at second order or higher (deviations from Byerlee's law). These factors include the temperature (Stesky et al. 1974), the fraction of contact area during friction (Logan and Teufel 1986), the sliding velocity (Shimamoto 1986) and, if present, a fine-grained gouge material lubricating the friction process in rock between the fault plane ligaments (Scholz 1987).

The third type of laboratory experiment to simulate conditions in the deeper Earth's crust is called creep test. At constant load, a *creep test* requires the measurement of the time-dependent deformation of a rock under high pressure ($p<500$ MPa) and high temperature ($T<1300$°C) conditions, e.g. in a gas-medium *Paterson apparatus*
M5.1, 5.2 (Paterson 1970). Experiments on small rock cylinders (volume 1–2 cm^3) are carried out at temperatures much higher ($T\approx1200$°C) than expected under the geologic conditions ($T>600$°C at a depth of 30 km) in order to obtain measurable strain rates. Creep and flow laws are necessary to be obtained in tests with well-controlled chemical environments (Bai et al. 1991) using apparatus with sufficient stress resolution for accurate extrapolation (Kohlstedt et al. 1995). To estimate crustal rheology, i.e. the creep behaviour of rock, different materials are investigated. Quartz rheologies are assumed to be representative for the upper crust with quartz-rich rocks (Paterson 1989, Paterson and Luan 1990), while feldspar rheologies are used for the lower crust with feldspar-bearing rocks (Rybacki and Dresen 2000, 2004). For mantle rheologies, olivine is used as the standard material (Karato 1986, Mei and Kohlstedt 2000a, b). *Creep-strength* values of a broad range of crustal and mantle minerals and rocks have been published (Karato 2008).

Without recapitulating the fundamentals of crystalline plasticity (Poirier 1985; Ranalli 1995), we introduce the steady-state creep equation, a so-called *creep law*, relating the strain rate of the rock $\dot{\varepsilon}$ to the differential stress $\sigma_D=\sigma_1-\sigma_3$ applied

$$\dot{\varepsilon} = \frac{d\varepsilon}{dt} = A\sigma_D^n \exp\left\{-\frac{E}{RT}\right\} d^{-m}, \tag{5.21}$$

which is commonly used to describe diffusion creep, dislocation creep or superplastic creep in crystalline materials (Weertman 1978; Tsenn and Carter 1987).

The *Arrhenius* relationship in Eq. (5.21) is characterized by an exponential drop of strain rate with an increase in absolute temperatures T while the constant A is the material-dependent pre-exponent and n is called the *stress exponent.* E is the activation energy of the microphysical process involved and R is the molar gas constant in thermodynamics. For diffusion creep, d is the grain size of the rock and m is a constant ranging from 2 to 3 for different diffusion creep mechanisms.

Armed with laboratory results of the three tests types (fracture, friction and creep strength), we are able to construct simple strength profiles of the Earth's crust. Vertical strength profiles in the crust are first computed and displayed by Goetze and Evans (1979), Kirby (1980) and Brace and Kohlstedt (1980). The latter model is summarized in Fig. 5.9. The upper part of the model uses a friction law (with $\mu=0.75$) for the limiting strength and assumes a hydrostatic pore pressure gradient. Anderson faulting reveals stress gradients for reverse faulting (RF) much greater than for normal faulting (NF), depending on principal stress relations (cf. Fig. 5.4). The lower part of the model is based on the extrapolation to low temperatures of the high-temperature creep test of wet quartzite (Jaoul et al. 1984). The 1D strength curves give a lower limit of the strength, $\sigma_D=S_1-S_3$, of the crust. The strength predicted in the upper frictional part of the crust (z<15 km) depends only on the pore pressure assumed, whereas the strength of the lower crust depends strongly on the assumed rock type, temperature, strain rate and water content used in creep tests.

The transition from frictional strength (Fig. 5.9, straight lines) to creep strength (Fig. 5.9, curved line) cannot occur at a single point. The transition is more likely to occur gradually from purely brittle to semi-brittle and ductile material behaviour. In

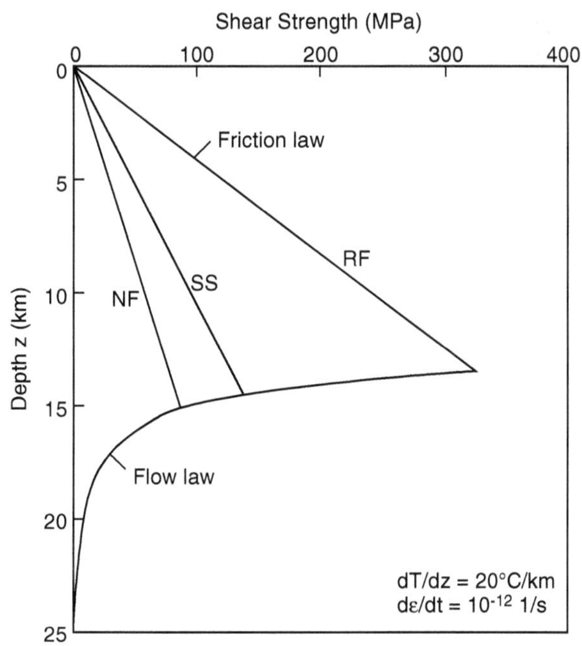

Fig. 5.9 Simple model of crustal strength versus depth. In the upper part, Anderson faulting with Coulomb friction ($\mu=0.75$) and hydrostatic pore pressure is assumed. For the lower part, an experimental flow law for wet quartzite is extrapolated assuming a strain rate $10^{-12}\,s^{-1}$ and a temperature gradient of $20°C\,km^{-1}$ (after Scholz 1990)

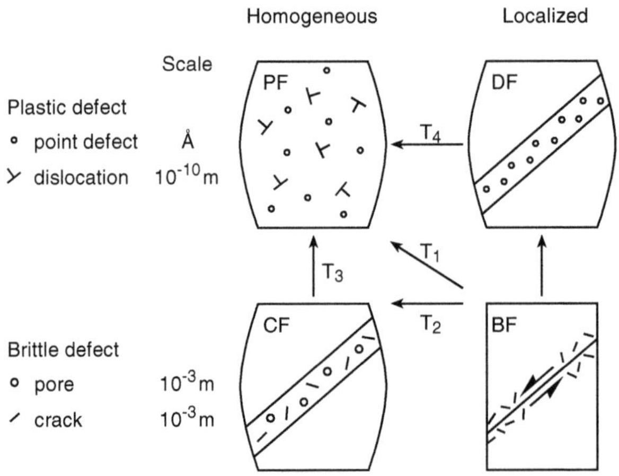

Fig. 5.10 Classification scheme of deformation according to defect scale (row) and homogeneity of deformation (column). Transformations between different regimes are represented by arrows (visualized from Table in Karato 2008). PF= Plastic Flow, DF= Ductile Faulting, CF= Cataclastic Flow, BF= Brittle Fracture, T1=BPT= Brittle Plastic Transition.

this context, the classification of non-elastic rock deformation into four categories by Karato (2008) is helpful (Fig. 5.10). This diagram is based on simple representation of mode of failure transitions in rocks published earlier (Rutter 1986). Depending on the scale of defect and the homogeneity of deformation involved, Karato (2008) distinguishes plastic flow, cataclastic flow, ductile faulting and brittle fracture. Non-elastic deformations change from brittle fracture to plastic flow as one goes from shallow to greater depth within the Earth's crust. The transition from brittle fracture to plastic flow (Fig. 5.10, transition T_1) corresponds to the condition at which the stress needed for plastic flow is the same as the stress needed for frictional sliding. The brittle plastic transition T_1 is bounded in differential stress space by the *Goetze criterion* (Evans and Kohlstedt 1995, $\sigma_1-\sigma_3=p_c$). At confining pressures greater than the rock strength, $\sigma_1-\sigma_3<p_c$, deformation proceeds without cracking. All transitions from Fig. 5.10 (arrows T_1, T_2, T_3, T_4) are discussed for Earth materials in detail by Karato (2008) in the light of the micromechanism behind. What is still missing is a quantitative model for the interaction between fracture and flow.

In the friction regime with hydrostatic pore pressure, the strength predicted by Byerlee's law provides upper limits to the in-situ stress measured in boreholes (Hickman 1991). If chemical effects dominate, the friction coefficient is reduced and Byerlee's law fails (Chester 1994; Blanpied et al. 1995). Rate and state variables are used (Dietrich 1979; Ruina 1985) to quantify the friction coefficient, μ versus temperature, sliding rate, pore and total pressure, and also with respect to the seismicity at faults (Hillers et al. 2006).

In the fracture regime, the strength of rock mass provides an upper bound of crustal strength and stress. The actual failure mode of rock, however, changes when confining pressure is increased. Dilatant mode I fracture is replaced by mode II shear failure in Man-nari granite at *p-T*-conditions (0.8–1 GPa, 100–300°C) as described by Shimada (1992). Based on its low and high pressure fracture types of granite, Shimada (1993) constructed a strength-depth diagram for a granitic crust (Fig. 5.11). We follow Shimada's (1993) argumentation in demonstrating how strength profiles of the Earth's crust are generated. In the brittle regime, the vertical stress is taken to be equal to the lithostatic overburden ($\rho = 2750\,\mathrm{kg\,m^{-3}}$) for a dry, granitic crust. If pore pressure is present, an effective lithostatic stress is used and computed from reduced density values ($\rho = (\rho - \rho_w) = 1750\,\mathrm{kg\,m^{-3}}$) for a wet granitic crust under hydrostatic pore pressure.

$$\begin{aligned} S_V &= \rho g z = 27z \quad [\mathrm{MPa\,km^{-1}}],\ \mathit{dry} \\ S_V &= (\rho - \rho_w) g z = 17z \quad [\mathrm{MPa\,km^{-1}}],\ \mathit{wet} \end{aligned} \tag{5.22}$$

Lithostatic stresses are represented as principal compressive stresses according to Anderson faulting (Eqs. (5.5)–(5.9)). When friction strength is lower than fracture strength, Eq. (5.20) with $\mu = 0.85 = \mathrm{cotan}(2\varphi)$ is used for input into the Anderson equations and the ratio of principal stresses, S_1/S_3, as well as the difference in principal stresses $S_1 - S_3$ (strength) is calculated. For a thrust (reverse) faulting regime (RF), Shimada obtains

$$\begin{aligned} S_1 - S_3 &= S_H - S_V = 99z \quad [\mathrm{MPa\,km^{-1}}],\ \mathit{dry} \\ S_1 - S_3 &= S_H - S_V = 63z \quad [\mathrm{MPa\,km^{-1}}],\ \mathit{wet} \end{aligned}. \tag{5.23}$$

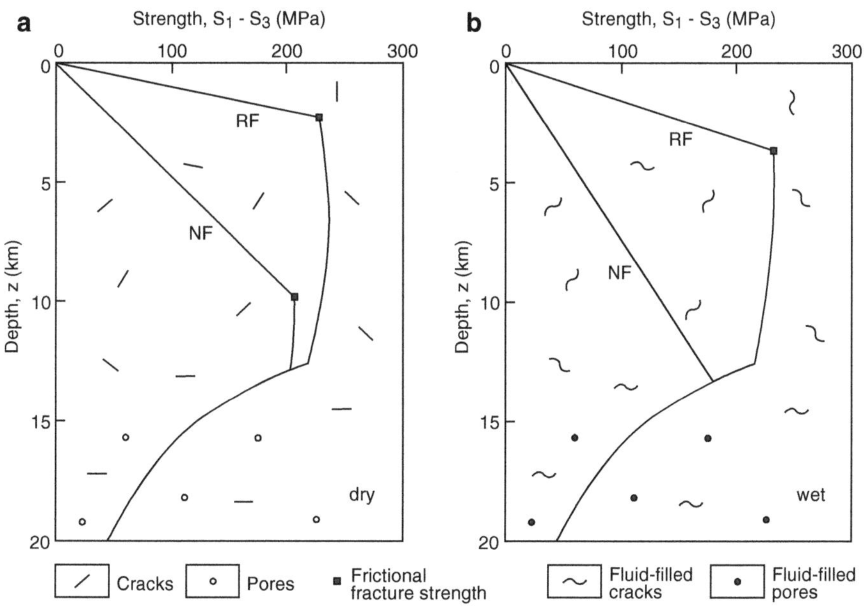

Fig. 5.11 Estimated strength-depth diagram in **a** dry and **b** wet granitic crust (modified after Shimada 1993)

The straight lines of the RF and NF regimes defining the upper part of the crustal strength profile, are plotted for a dry (Fig. 5.11a) and a wet granitic crust (Fig. 5.11b). While in the upper part of the crust, open cracks (faults) exist. Cracks partially close at a depth greater than 15 km and degenerate into more stable, stress-resistant pore space (Fig. 5.11a). In the wet granitic crust, cracks and pores are fluid-filled (Fig. 5.11b). When the frictional strength becomes equal to the fracture strength, the fracture strength of the high-pressure type failure of Man-nari granite is used to limit the crustal strength value (Fig. 5.11, black squares). For this purpose, the fracture strength of Man-nari granite is corrected for pressure, temperature and large grain size, while the geotherm of Goetze and Evans (1979) is used. The cut-off value of strength, e.g. for the dry RF regime, is 2.3 km, as shown in Fig. 5.11a (black square). In the ductile regime, Shimada (1993) used laboratory creep parameters (*A*, *n*, *E* in Eq. (5.21)) obtained on dry and wet granite (Kirby and Kronenberg 1987)

$$\begin{aligned} A &= 10^{-8.6}\ \text{MPa}^{-n}\text{s}^{-1} \quad & n &= 3.4 \quad & E &= 139\ \text{kJ mol}^{-1},\ \textit{dry} \\ A &= 10^{-3.7}\text{MPa}^{-n}\text{s}^{-1} \quad & n &= 1.9 \quad & E &= 137\ \text{kJ mol}^{-1},\ \textit{wet} \end{aligned} \tag{5.24}$$

to compute the strength, $\sigma_D=\sigma_1-\sigma_3$, from (5.21) in the lower part of the crust. For this purpose, again the Goetze and Evans (1979) geotherm is used and a crustal strain rate of $10^{-14}\,s^{-1}$ is assumed. It is evident from Fig. 5.11 that the peak strength of the crust occurs in the brittle regime. Commonly, the 350°C isotherm is thought to be the onset of plasticity in rocks of granitic composition. For gradients of 27–30°C km^{-1}, this temperature corresponds to 11–13 km depth.

One shortcoming of this approach and all other extrapolations of laboratory creep data to natural conditions (Paterson 1987) is that the microphysical process observed in small laboratory test specimens must be assumed to hold on a larger scale in nature. The approximate maximum size of the rock samples used in the laboratory to infer rock-strength values is shown versus crustal depth, for assumed pressure and temperature-conditions simulated in the experiment (Fig. 5.12a). While a rock core 15 cm in diameter and 30 cm in length is suitable to investigate rock fracture in the brittle crust (Fig. 5.12b, triaxial test with stiff loading frame, confining pressure 200 MPa, rock volume $V=5\times10^{-3}$ m, room temperature), a rock core 1 cm in diameter and 2 cm in length is suitable to analyze rock creep in the lower ductile crust (Fig. 5.12c, creep test with Paterson apparatus, confinement 500 MPa, $V=1.6\times10^{-6}$ m, high (1300°C) temperature). In the schematic diagram of Fig. 5.12b, grain-scale cracking (mm $=10^{-3}$ m) can be resolved using acoustic emission techniques (Fig. 5.12b, AE). In the diagram of Fig. 5.12c, diffusion processes at an atomistic scale (Å $=10^{-10}$ m) can be resolved by investigating the specimen post mortem "in the light" of a transmission electron microscope. It is the experimentalist's responsibility to make sure that the sample size inspected is appropriate to resolve the physical process involved.

In Fig. 5.13, we discuss three important factors influencing crustal strength, namely pore fluid pressure (Fig. 5.13a), rock material (Fig. 5.13b) and strain rate

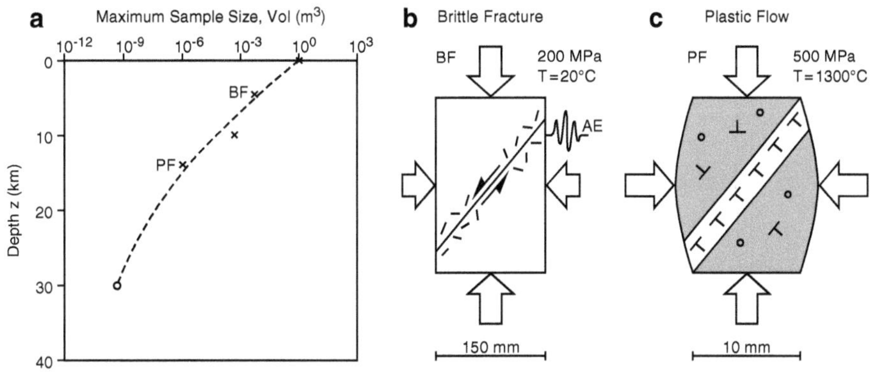

Fig. 5.12 **a** Maximum size of rock sample tested in the laboratory used to infer crustal strength data versus actual depth in the Earth's crust (*z*, kilometre) simulated in the laboratory experiment. **b** End-member brittle fracture, *BF* test specimen representative for upper crustal conditions and **c** end-member plastic flow test specimen representative for lower crust and upper mantle conditions are shown for reference

(Fig. 5.13c). Pore fluid pressure p_p can substantially reduce the maximum strength of the crust (Fig. 5.13a), where

$$\kappa = \frac{\lambda}{\lambda_H}, \; \lambda_H = \frac{\rho_w}{\rho}, \; \lambda = \frac{P_p}{S_V}, \tag{5.25}$$

where λ_H is the hydrostatic pore pressure ($\kappa = 1$). As fluid pressure builds up, rock strength decreases and the depth at which the maximum strength is reached increases. When faulting occurs, pore pressure drops due to the increase in rock permeability. If the fault is sealing over geologic times, pore pressure builds up again.

In Fig. 5.13b, creep laws from recent laboratory investigations are extrapolated to nature assuming $\dot{\varepsilon} = 10^{-14}\,\mathrm{s}^{-1}$ and $dT/dz = 27°\mathrm{C\,km}^{-1}$ (Rybacki 2009, personal communication). Two creep curves of wet quartz are shown (Fig. 5.13b, 1 = Hirth et al. (2001) and 2 = Rutter and Brodie (2004)), representative for the rock rheology of the upper crust. Two creep curves of feldspar are shown, one for wet anorthite (Fig. 5.13b, 3 = Rybacki et al. (2006)) and one for dry anorthosite (Fig. 5.13b, 5 = Rybacki et al. (2006)), both representative for the rheology of lower crustal rocks. In addition, one creep curve of wet olivine is added (Fig. 5.13b, 4 = Hirth and Kohlstedt (2003)), which is assumed to be representative for the upper mantle rheology of the Earth. The onset of creep in nature relying on geologic observations is bar-coded at the right margin of Fig. 5.13b (Passchier and Trouw 1996). Note that dry rocks are considerably stronger than wet rocks as shown for feldspar (Fig. 5.13b, curve 3 (wet anorthite) and curve 5 (dry anorthite)).

The third factor influencing crustal strength is the strain rate (Fig. 5.13c), shown at $\kappa = 1$ and $dT/dz = 30°\mathrm{C\,km}^{-1}$ for wet quartz rheology at two-end member strain

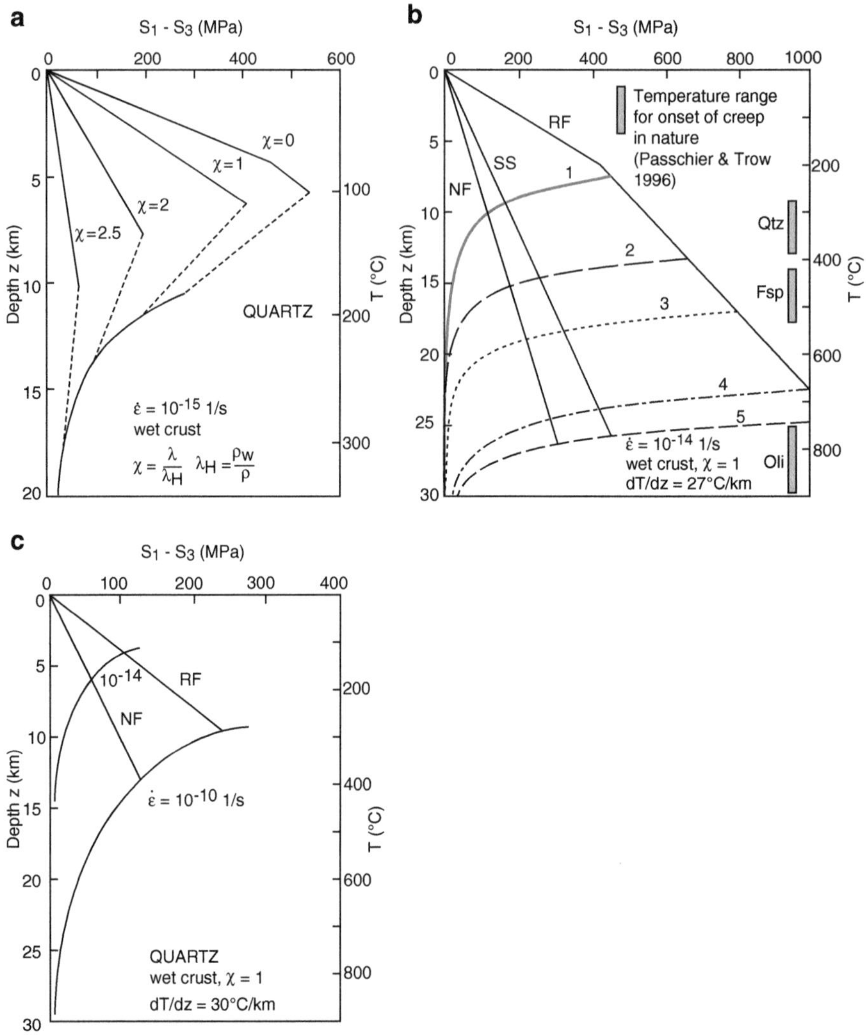

Fig. 5.13 Factors influencing the shape of crustal strength profiles: **a** pore fluid pressure, **b** materials and **c** strain rate assumed. In **a**, the strain rate and the material (wet quartzite) is fixed (after Kohlstedt et al. 1995). In **b**, the strain rate and $\kappa = 1$ is fixed. Materials: *1*+*2*=wet quartz, *3*=wet anorthite, *5*=dry anorthite, and *4*=wet olivine. References to curve 1=Hirth et al. (2001), 2=Rutter and Brodie (2004), 3+5=Rybacki et al. (2006), 4=Hirth and Kohlstedt (2003). (after Rybacki 2009, personal communication). In **c**, the material (wet quartz) and $\kappa = 1$ is fixed (after Passchier and Trouw 1996)

rates, $10^{-14}\,s^{-1}$ and $10^{-10}\,s^{-1}$. The former strain rate is representative of "slow", homogeneously deforming lithospheric plates, while the latter is found when approaching narrow banded fault zones with "fast", highly inhomogeneous deformation.

Large uncertainties in simple differential stress profiles (Fig. 5.13) result from the water content and temperature in the actual crust as well as uncertainties in the experimental data on the influence of water fugacity. The trade-off between the influence of temperature and water on strength-depth profiles is discussed in Karato (2008, his Sect. 19.3.2) for a crust made of coarse-grained basalt (diabase). Also in here, the reader interested in computing strength-depth diagrams will find an excellent compilation of flow-law parameters of relevant crustal and upper mantle minerals and rocks (Karato 2008, his Table 19.1).

In constructing crustal strength profiles, Kohlstedt et al. (1995) assumed that brittle failure and plastic flow are independent processes. The smaller strengths calculated for each mechanism is chosen to be the strength of the crust. As pointed out by Karato (2008), in reality a broad regime of *semi-brittle* behaviour is observed, where the interaction of brittle defects (cracks) with plastic defects (dislocations) becomes significant (Fig. 5.10). There is no accepted constitutive law for the semi-brittle regime. Consequently, the strength profiles near the peak strength (Fig. 5.13) have a large uncertainty and are likely to overestimate the actual crustal strength. Also, strength computations based on steady-state rheology must be applied with caution to real crustal conditions.

Note-Box Differential stress-depth diagrams of the Earth's crust are based on laboratory fracture, friction and creep tests. Fracture strength has to be corrected for (pore) pressure, temperature, deformation rate and sample size. Friction strength is well described by the universal Byerlee law. Only for second-order effects it is necessary to consider rate and state variables in friction laws. In the deeper crust, creep tests bound the strength of rock by placing a flow law, which intersects the Byerlee straight lines for different faulting regimes. Besides errors from the extrapolation of laboratory data to nature, large uncertainties in simple strength profiles result from differences in water content and temperature in the actual crust.

Part II
Measuring Stress

Chapter 6
Physics of Stress Measurements

Stresses cannot be measured directly. Stress determination is made indirectly, e.g. by the measurement of strain. Deformation values obtained from an unbalanced body approaching equilibrium in combination with theoretical knowledge about constitutive behaviour (stress-strain relationship) allows us to evaluate the state of stress existing in any deformable body. The physics of stress measurements can be subdivided into six methodologies (see Table 6.1). For rocks and minerals, some modifications of these techniques used in material sciences are required.

6.1 Mechanical Methods

A mechanical procedure gives stress data with knowledge of Young's modulus E and measurements of the change in length ΔL of a linearly elastic body normalized to its initial length L_0

$$\sigma = E\varepsilon = E\frac{\Delta L}{L_0} = E\frac{L - L_0}{L_0}. \tag{6.1}$$

The accuracy of length measurements (L_0 and L) required in the 1D Hooke's law from Eq. (6.1) is about 1 µm to gain stresses with an accuracy of 0.1 MPa, assuming a Young's modulus of rock in the order of 100 GPa. The vertical stress component (Fig. 6.1, $\sigma=\sigma_{zz}$) inside a test specimen compressed in the laboratory can be computed from the relative vertical shortening measured (Fig. 6.1, $\varepsilon=\varepsilon_{zz}$) and the Young's modulus determined in a separate experiment (e.g. dynamically from wave velocity data). For rock material, the effect of cracks on Young's modulus has been found to be important. Walsh carried out uniaxial (1965) and hydrostatic compression tests (1980) on rock in the laboratory to quantify the effect of cracks on elastic rock properties. O'Connell and Budiansky (1974) also considered fluid-filled cracks. The highly non-linear relationship of elastic rock properties with crack closure has been discussed in Chap. 4 (e.g., Fig. 4.9).

A. Zang, O. Stephansson, *Stress Field of the Earth's Crust*,
DOI 10.1007/978-1-4020-8444-7_6,

Table 6.1 Physical methods for determining stress in deformable materials with application to rocks and minerals

	Method	Physical Parameter	Mathematical Relation	Reference
1	Mechanical	Change in relative length, ΔL; effect of cracks on moduli	$\sigma = E\frac{\Delta L}{L_0}$	Walsh (1965, 1980), O'Connel and Budiansky (1974)
1A	Fracture Mechanical	Brittle tensile fracture strength, T_0	$T_0 = \frac{K_{IC}}{F(c/R)\sqrt{c\pi}}$	Abou-Sayed et al. (1978)
2	Strain Gauges	Relative electrical resistance, ΔR	$\frac{\Delta R}{R_0} = k\frac{\Delta L}{L_0}$	Nichols (1975)
2A	Differential Strain Analysis	Crack closure strains from ΔR	Analogue to 2 with 3 strain gauge rosettes	Simmons et al. (1974), Dey and Brown (1986)
3	X-Ray Diffraction	Change in angle of reflection, $\Delta\theta$	$\varepsilon(d) = -\cot(\theta)\Delta\theta$	Krawitz (2001)
3A	Neutron Diffraction	(a) Angle dispersive method, $\Delta\theta$ (b) Time-of-flight *TOF*, Δt	Analogue to 3 $\varepsilon_{hkl} = \frac{d-d_0}{d_0} = \frac{t-t_0}{t_0}$	Hutchings et al. (2005), Daymond (2006)
4	Photoelasticity	Number n of isochromatic fringes	$nS = D(S_1 - S_2)$	Frocht (1941), Hawkes (1971)
4A	Holographic Interferometry	Laser light intensity, I with number of fringes	$I = UU^*$	Hecht and Zajac (1979), Bass et al. (1986), Schmitt et al. (2006)
5	Ultrasonic Method	Ultrasonic wave velocities, v	$\frac{\Delta v}{v_0} = \frac{4G+n}{8G^2}(S_2 - S_3)$	Murnaghan 1951
5A	Wave Velocity Analysis	Effect of cracks on wave velocity in pressure vessel	$v = v(\varphi, p_c)$	Birch (1960), Ren and Hudson (1985)
6	Micromagnetic Method	Magnetic field *B(H)* hysteresis	*B(H)*-Hysteresis	Peiter (1992), Theiner (1997)

Another approach to incorporate the effect of cracks into stress determination is the use of fracture mechanic principles (Chap. 3). Hardy and Fairhurst (1974) applied this approach to the results from hydraulic fracturing tests (Sect. 7.2). Abou-Sayed et al. (1978) quantified the effect of pre-existing cracks in the borehole wall and calculated for zero confinement the critical pressure for the well bore to fail (Table 6.1, method 1A)

$$p_{crit} = T_0 = \frac{K_{IC}}{F(c/R)\sqrt{c\pi}}. \tag{6.2}$$

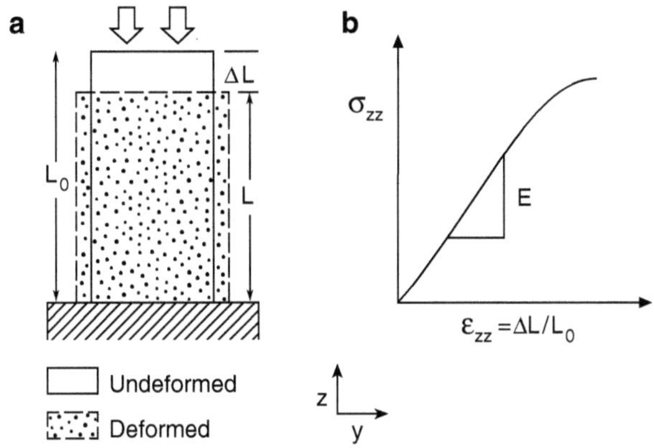

Fig. 6.1 **a** Uniaxial compression test of a rock in the laboratory and **b** determination of vertical stress component from measured vertical shortening of the core

The rock breakdown pressure turns out to be equal to the rock tensile strength T_0 and is computed from the fracture toughness of rock K_{IC}, the initial length of borehole wall cracks c and the radius of the drill hole R. The term $F(c/R)$ is a complex function whose values were determined by Paris and Sih (1965).

6.2 Strain Gauges and Overcoring

Strain gauges are able to detect the change in length ΔL of a rock specimen by measuring the change in electrical resistance ΔR using the linear relationship

$$\frac{\Delta R}{R_0} = k\varepsilon = k\frac{\Delta L}{L_0}, \tag{6.3}$$

where R_0 is resistance at rest and k is a constant calibration factor for different types of strain gauges (Table 6.1, method 2). Strain gauges are used in the majority of overcoring stress measurement methods and equipments (Sect. 7.1). A schematic sketch of a strain gauge with a single, active measuring grid is shown in Fig. 6.2a. A copper wire is embedded into a support foil which is directly glued to the rock surface of interest. When the rock is deformed, the support foil as well as the wire follows and the change in the length of the copper wire causes the electrical resistance of the wire to vary. To increase the sensitivity of ΔR-values, the copper wire meanders and forms an active measuring grid (Fig. 6.2a). Soldering contacts allow the connection of the active grid to a Wheatstone bridge circuit. Dimensionless strain values ε [$\mu m\, m^{-1} = 10^{-6}$] are computed from measured ΔR-values using Eq. (6.3). Like in the mechanical method (Sect. 6.1), measured strain values are converted to stress by assuming a linear rock constitutive relationship.

To obtain principal strains in space (ellipsoid) or principal strains in a plane (ellipse), strain-gauge rosettes are used (Fig. 6.2b). Typical strain-gauge arrangements glued on rock cores for stress analysis are shown in Fig. 6.3. Depending on

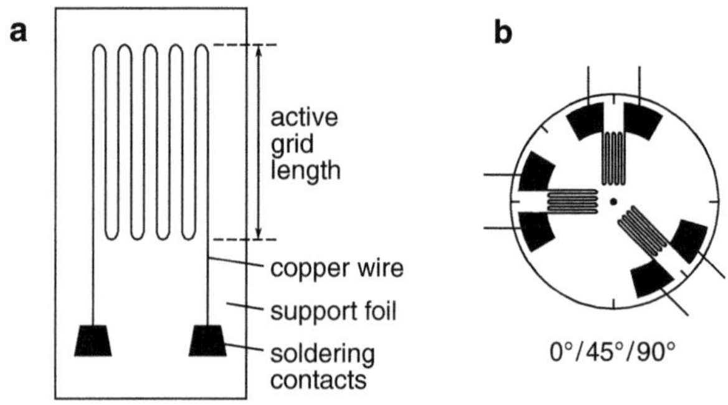

Fig. 6.2 Components of a strain gauge with **a** single active grid and **b** three active grids in a strain-gauge rosette

the relative location of active grids and the drill hole, we can distinguish between *overcoring* techniques (Fig. 6.3a, strain gauges located inside the trace of the drill bit, e.g. Leeman 1964), *undercoring* techniques (Fig. 6.3b, strain gauges located outside the trace of the drill bit, e.g. Duvall 1974), and mixed coring techniques (Fig. 6.3c, e.g. Nichols 1975). All coring techniques belong to relief methods (Sect. 7.1) and are used for estimating the complete in-situ stress tensor. The physical principle behind relief techniques is to isolate a rock sample (fully or in part) from the in-situ stress field by coring and monitor its re-equilibrium deformation response.

Early residual strain measurements in granitic rock are reported, e.g., by Swolfs et al. (1974) and Nichols (1975) using the combined coring technique (Fig. 6.3c). Residual strains can be converted into residual stress (Sect. 4.3) using the elastic parameters of the rock. The overcoring technique is a *destructive method* to analyze residual stresses in rock. In subsequent sections, we will also learn about *non-destructive methods* used in structural and residual stress analysis (e.g., Noyan and Cohen 1987; Hauk 1997). A non-destructive laboratory method in rock physics is the differential strain analysis (DSA) used by Simmons et al. (1974) and Feves et al. (1977). Strains are measured with strain gauge rosettes on orthogonal sides of a rock cube subjected to confining pressure (Table 6.1, method 2A). Stresses are calculated based on the premise that strain-relief cracks are caused by the removal from the in-situ stress field. If the rock is retrieved from great depth, the differential

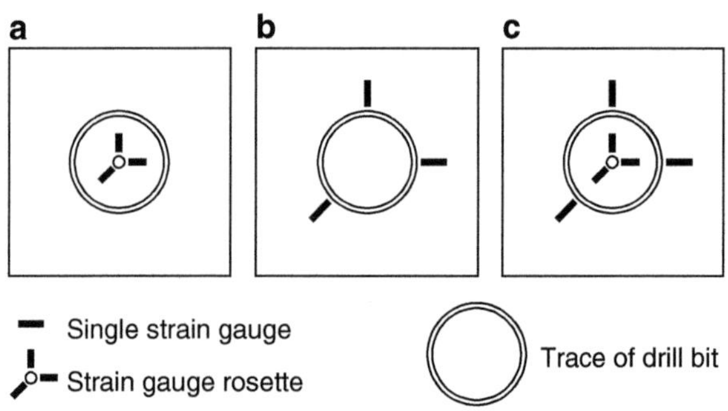

Fig. 6.3 Strain gauge assembly for stress analysis with **a** the overcoring technique, **b** the undercoring technique and **c** the mixed coring technique

strain curve analysis (DSCA) allows the determination of in-situ stresses (Dey and Brown, 1986).

Successful strain-relief results depend on the established constitutive law assumed and instrumentation sensitive enough to capture small strains ($\mu m\, m^{-1}$). Errors in stress analysis from overcoring were discussed by Amadei (1985). The main drawback with relief methods is the small rock volume involved (Sect. 7.1). If strains measured are sensitive to variations in mineral content or grain size of rock, we need to increase the volume of rock for stress determination (Chandler 1993). Details on different overcoring techniques can be found in Sjöberg et al. (2003). General steps in overcoring illustrated by the *Borre Probe* (Hakala et al. 2003) are shown in Reinecker et al. (2008a) and in two accompanying movies on the DVD (M6.1 Overcoring with Borre Probe, M6.2 Overcoring stress analysis). Relief methods are further discussed in Chap. 7.

M6.1, M6.2

Exercise 6.1 The rectangular strain-gauge rosette (Fig. 6.2b) is used to measure plane strains at the flat end of a rock core after overcoring (Fig. 6.3a). The three single gauges of the rosette 45° apart indicate strain values of $\varepsilon_A(0°)=10$, $\varepsilon_B(45°)=15$, and $\varepsilon_C(90°)=20\ \mu m\, m^{-1}$.

(a) Calculate the principal normal strains of the rock plane in question according to $\varepsilon_1=A+B$ and $\varepsilon_2=A-B$ with

$$A=\frac{1}{2}\left(\varepsilon_{\mathrm{A}}+\varepsilon_C\right),\ \text{and}\ B=\frac{\varepsilon_{\mathrm{A}}-\varepsilon_C}{2\cos 2\theta}.$$

(b) Calculate the angle $\theta(\varepsilon_1, \varepsilon_A)$ between the principal strain coordinate system of core $(\varepsilon_1, \varepsilon_2)$ and the strain rosette fixed coordinate system $(\varepsilon_A, \varepsilon_C)$ according to

$$\tan 2\theta=\frac{\left(\varepsilon_{\mathrm{A}}-\varepsilon_C\right)-2\varepsilon_B}{\varepsilon_{\mathrm{A}}-\varepsilon_C}.$$

(c) Calculate the principal strains of a delta rosette where single gauges are 60° apart, $\varepsilon_D(0°)=10$, $\varepsilon_E(60°)=15$, and $\varepsilon_F(120°)=20\ \mu m\, m^{-1}$. Assume the ε_D-direction of the delta rosette to be parallel to the ε_A-direction of the rectangular rosette.

6.3 Diffraction Methods

Diffraction methods measure lattice strains that are the result of internal stresses. These stresses may be applied externally as a load or can be a residual of former deformation or phase transformation. While relief techniques rely on strains stored in several mineral grains of a rock (e.g., residual stress of the first kind, Sect. 4.3) diffraction methods are designed to measure strains stored in single mineral grains (e.g., residual stress of the second kind). Measurements of residual stresses by diffraction methods are discussed in standard books (Hauk 1997; Krawitz 2001; Hutchings et al. 2005). In addition, there is an excellent review on neutron scattering in earth sciences by Daymond (2006).

There are two methods to determine internal stress by diffraction: local stress (Table 6.1, method 3: microscale, synchrotron *X-ray diffraction*), and average stress by peak broadening, peak shift and asymmetry (Table 6.1, method 3A: neutron scattering). Particularly, the former has become very interesting, documenting residual strains that are rather large (synchrotron Laue microfocus, e.g. Tamura et al. (2002), Lynch et al. (2007)). The *neutron diffraction* method (Table 6.1, method 3A) has become an important tool for measuring stress in material sciences (Hutchings and Krawitz 1992; Daymond 2006) and also has been applied to rock (Frischbutter et al. 2000; Pintschovius et al. 2000; Darling et al. 2004). The importance of this method is recognized by the construction of new *stress spectrometers* (SMARTS at Los Alamos, USA (Brown et al. 2003), ENGIN-X at Rutherford Appelton Laboratory, UK (Santisteban et al. 2006) and EPSILON at Dubna (Russia) pulsed reactor (Frischbutter et al. 2006)). These spectrometers measure lattice strains, i.e. the changes in lattice spacing relative to an undeformed perfect crystal. The resolution of these new generation instruments is better than 30 microstrains.

Methodically, X-ray and neutron diffraction are similar. Frozen-in deformations in atomic lattice planes of a crystalline body (mineral grain) are quantified by ray diffraction. For example, in X-ray diffraction a monochromatic X-ray primary beam with wavelength λ of less than 0.1 nm and penetration depth of 20 μm shot onto a crystal is reflected according to *Bragg's law*

$$n\lambda = 2d_0 \sin\theta_0 \quad \text{with} \quad n = 0, 1, 2, \ldots, \tag{6.4}$$

where d_0 is the distance between neighbouring lattice planes, θ_0 is the angle of the incipient and reflected wave (Fig. 6.4a) and n is the order of the interfering fringe pattern. Once external forces are applied to the minerals of a rock, the lattice planes deform (Fig. 6.4b). Residual strains, the relative changes in lattice plane (hkl) distances due to external loads ($\Delta d = d - d_0$), can be determined by detecting the corresponding change in reflection angle ($\Delta\theta = \theta - \theta_0$) of the incident beam

$$\varepsilon(d) = -\cot(\theta)\Delta\theta. \tag{6.5}$$

Since the difference Δd in lattice plane deformations is small (Fig. 6.4, $d < d_0$), also the difference in changes of the reflection angles are small (Fig. 6.4, $\theta > \theta_0$) in the order of about 1°. The accuracy of energy-dispersive reflection angle measurements has to be 0.01°. In general, the resolution of $\Delta d/d$ is about 5×10^{-3} in X-ray stress analysis (Kocks et al. 2000).

A recent example to quantify residual stresses in a quartz vein (Sambagawa metamorphic belt, Japan) by X-ray diffraction is given by Sekine and Hayashi (2009). Compressive residual stresses of −15 to −20 MPa were measured in the direction parallel to the vein. Using a platy vein-like inclusion model, residual stress

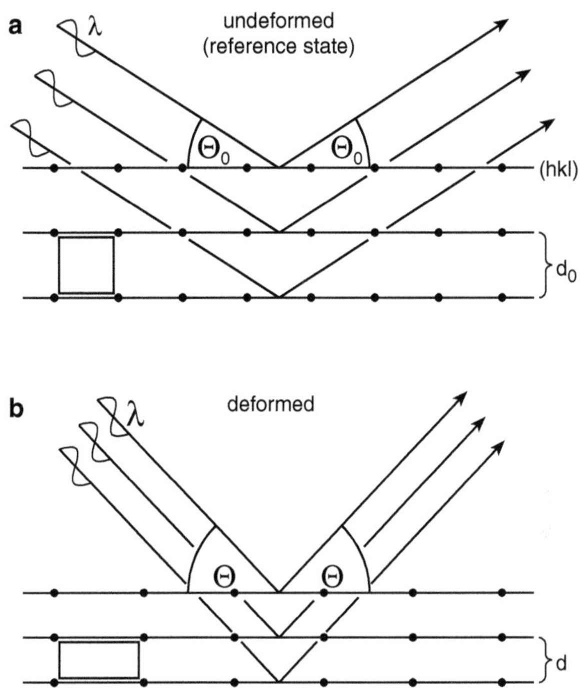

Fig. 6.4 Monochromatic X-ray beam with wavelength λ and angle of incidence θ_0 reflected from atomic lattice planes (*hkl*) of a mineral **a** in the reference state with lattice distance d_0 and **b** in the deformed state with lattice distance d

magnitude and fluid-inclusion micro-thermometric data, the authors estimated the paleostress magnitude at the time of vein formation.

Compared to X-rays with penetration depth of μm, neutron diffraction allows stress analysis within several cubic centimetres of a rock specimen (Hutchings 1992). Two sub-techniques exist for measuring lattice strain: the angle-dispersive and the wavelength-dispersive method (Table 6.1, method 3A). The *angle-dispersive* (monochromatic) neutron diffraction method is based on Bragg's law where the lattice strain ε_{hkl} is related to the observed shift of the Bragg angle $\Delta\theta_{hkl}$ of the maximum diffraction peak (Krawitz 2001). Equation (6.5) holds for this method with $\varepsilon=\varepsilon_{hkl}$, $\theta=\theta_{hkl}$ and $\Delta\theta=\Delta\theta_{hkl}$.

The *wavelength-dispersive* (time-of-flight) neutron diffraction method has the advantage that all permitted *Bragg* reflections can be determined at once. A polychromatic ("white") neutron beam allows the simultaneous observation of multiple lattice spacing. Since no angle corrections are needed, the accuracy of the measurements is increased. Lattice deformations ε_{hkl} are related to the times of flight (TOF) of the neutrons, t:

$$\varepsilon_{hkl} = \frac{d - d_0}{d_0} = \frac{t - t_0}{t_0}, \tag{6.6}$$

where t_0 is the flight time in the stress-free material and t is the time in the stressed material. TOF is measured from the surface of the moderator of the neutron reactor through the rock specimen to the detector. In order to compute the relationship between TOF values and the wavelength or velocity of neutrons unequivocally, the emitting source must be pulsed (Hutchings et al. 2005). The time-of-flight method has become a favourite technique to investigate in-situ structural changes and internal stresses during deformation (Daymond 2006).

Both methods, conventional angle-dispersive and time-of-flight neutron diffraction, were applied to investigate residual stresses in sandstones recovered from groundwater wells in the Elbezone close to Königstein in Germany (Scheffzück et al. 1998; Pintschovius et al. 2000; Frischbutter et al. 2000).

Combining *X-ray synchrotron* diffraction with high-pressure *diamond anvil cells*, the influence of stress on the texture development of minerals can be studied in-situ (i.e. during the deformation experiment). Materials like oxides and silicates, brittle at ambient conditions, become ductile at pressures above 5 GPa, even at room temperature, and develop lattice preferred orientation (texture) when a directional stress field is applied. Using radial, rather than conventional axial X-ray diffraction geometry, the influence of non-hydrostatic stress on lattice strains has been quantified. Merkel et al. (2002) applied this method to MgO polycrystals up to 47 GPa to calculate shear strength, elastic anisotropy and texture development under stress. Currently, the most elegant and powerful method to infer quantitative texture information from radial X-ray diffraction images is the Rietveld technique (Wenk et al. 2006). In this method, the wavyness of diffraction lines is used to quantify the effect of stress on lattice strains. Such experiments provide information about deformation mechanisms and mechanical properties of minerals in the lower Earth mantle that are a prerequisite for realistic geodynamic models where the Earth's crust is one part of the models.

As we know from Chap. 4, paleostresses reflect ancient crustal stresses at the time of tectonic deformation, averaged over the duration time and rock volume involved in the tectonic process. On the other hand, in-situ stress data provide a snapshot of ambient crustal stresses. Paleopiezometry provides information about peak differential stress attained during a particular tectonic event in rock mass history. *Paleopiezometers* rely upon a relationship between differential stress and the development of microstructural elements in a rock, e.g., dislocation density, (Pfiffner 1982), recrystallization of calcite and quartz (Twiss 1977), mechanical twinning of calcite and dolomite (Jamison and Spang 1976; Rowe and Rutter 1990) which has to be calibrated experimentally. Grain-size reduction, dynamic or fluid-enhanced recrystallization can influence paleopiezometers. Both paleostress magnitudes from paleopiezometry and in-situ stress magnitudes from deep drill holes, even at mid-crustal depth, are found to be in agreement with laboratory-derived friction laws (Lacombe 2007). It has been suggested that recystallized grain size in fault-zone rocks rather than being a paleopiezometer, is likely to be a paleowattmeter (Austin and Evans 2007). While empirical paleopiezometer laws where average reduced grain size is inversely related to stress, are independent of temperature and strain rate, kinetic balance models such as the paleowattmeter contain the temperature dependence.

6.4 Optical Methods

Photoelasticity uses transparent (non-opaque), stress-active model materials in order to visualize mechanical stresses. In plexiglass or Araldite, the speed of light changes depending upon loading. If a vertically polarized ray of light falls onto a biaxially stressed disk (Fig. 6.5), it splits along the two principal stress directions and propagates inside the disk (thickness D) with two different velocities. A pattern of interfering lines occurs behind the analyzer. By simultaneous rotation of the polarizer and the analyzer, we are able to determine the principal stress directions at any location within the plexiglass test specimen (Sect. 2.4, isoclines). To analyze stresses, the isoclines are filtered out (by $\lambda/4$-plates) and an isochromatic fringe pattern (lines of the same color) results.

The difference in principal stresses ($S_1 - S_2$) between neighbouring isochromatic lines changes by S the so-called constant of photoelasticity

$$nS = D(S_1 - S_2), \quad n = 0, 1, 2, \ldots \tag{6.7}$$

D is the thickness of the test specimen and n is the order of isochromatics which interfere. A result from photoelastic analysis is shown in Fig. 6.6. A cylindrical sample of phenol resin (D=10 mm) is compressed between two jaws. The vertical load—not important for the final fringe pattern shown—was 2.45 kN (Fig. 6.6a, arrows). The pattern of isochromatics in the Brazilian test body (cylindrical disk) and the force transmitters (upper and lower jaw) are displayed (Fig. 6.6b). The number of isochromatics is n=15 in the disk and 11 in the lower jaw leading to local higher stresses in the test disk compared to the loading platens.

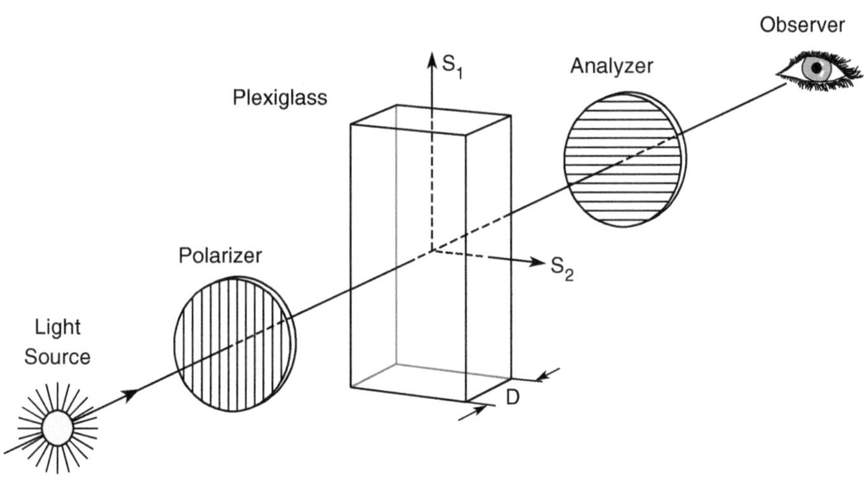

Fig. 6.5 Photoelastic setup for determining isoclines in a biaxially stressed plexiglass specimen

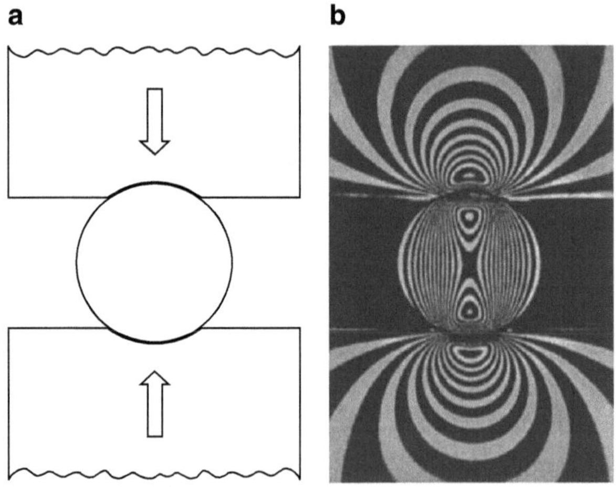

Fig. 6.6 Brazilian test configuration showing **a** the experimental setup with vertical load (*arrows*) and **b** the corresponding fringe pattern in the phenol resin disk and loading platens (modified after Bergmann-Schaefer 1978)

Photoelasticity in rock mechanics is used in an indirect way to record stresses, since rock material is opaque. The photoelastic tracer material is bonded onto the rock surface of interest (Hawkes and Moxon 1965). A review on interpreting measured isochromatic fringe pattern of overcored rock for stress measurement purposes is given by Hawkes (1971). The advantage of this quick and fairly reliable method is reduced by cementing problems in the usually wet conditions in boreholes when stresses at depth in the Earth's crust need to be determined.

Holographic interferometry is analogue to photoelasticity an optical procedure, but uses laser light instead of visible light (Hecht and Zajac 1979). Like in relief methods, holographic images are taken from an excavated rock surface before and after coring to determine the in-situ stress level in boreholes (Bass et al. 1986). Stress is relieved locally by drilling a small borehole (slave pilot hole) into the existing wall of the master borehole. The resultant entire displacement field due to stress relief is obtained.

Highly coherent, monochromatic light from a laser is divided into two beams. One of the beams is used to illuminate the object (stressed rock), while the other is directed at the holographic film without hitting the object. The complex amplitude of these two monochromatic wave fronts of light can be represented by

$$\begin{aligned} U_o(r) &= A_o(r)\exp\{i[\Phi_o(r)+\omega t]\} \\ U_r(r) &= A_r(r)\exp\left\{i[k(r\sin\alpha + r^2/2Z)+\omega t]\right\} \end{aligned} \tag{6.8}$$

where the subscripts o and r refer to the object and reference beams, respectively. Each light wave is described by an amplitude A, phase Φ and angular frequency ω. The term $k(r\sin\alpha+r^2/2Z)$ accounts for changes in phase due to the angle α between the wave vector k (which equals 2π/wavelength) and the film plane normal, and also for the curvature of the wave front (Fig. 6.7).

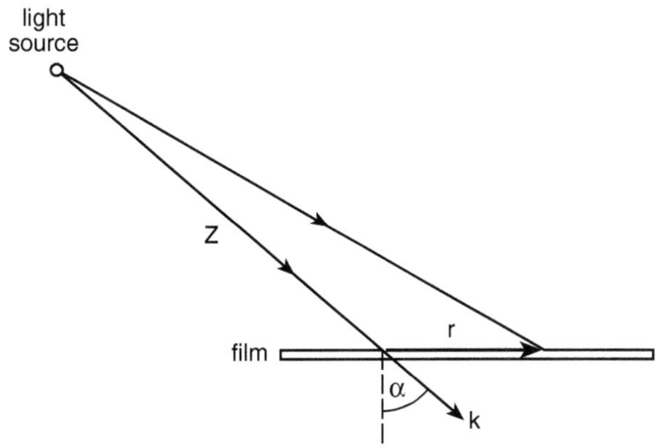

Fig. 6.7 Light source with wave vector k striking a film at angle α. The light at position r will be phase-shifted due to α between k and the film plane and the curvature of the wave front (after Bass et al. 1986)

At the film surface, the resultant wave amplitude U is the sum $U_o + U_r$. The film will record the intensity of light I, which is given by

$$I = UU^* = A_o^2 + A_r^2 + 2A_oA_r \cos\left\{\Phi_o(r) - k[r \sin\alpha + r^2/2Z]\right\}, \quad (6.9)$$

where U^* is the complex conjugate of U. Interference of the two beams at the film surface (Fig. 6.7) leads to variation in the recorded light intensity as described by the cosine term in Eq. (6.9). Development of the photographic plate results in fringes on the film. If the developed film is placed in its original position and illuminated only by the reference beam, the fringes on the film will act as diffraction gratings which serve to reconstruct the original object beam wavefront. The exposed and developed holographic film does not look anything like the recorded object, as in the case of a conventional photographic negative; rather it is the fringes, closely spaced and normally unobservable, which carry all of the information about the recorded object and a reconstruction beam is needed to recover the original scene from the hologram (Bass et al. 1986).

One example of a doubly exposed hologram that yields topographic maps showing surface deformation of rocks between two exposures is shown in Fig. 6.8a. The hologram was taken as a pyrophyllite sample with an artificial fault (saw cut) was stressed to a near stick-slip event (Spetzler et al. 1991). The deformation pattern on both sides of the fault is illustrated in Fig. 6.8b. The vertical scale is in the fringes. The space between two neighbouring dark lines represents a deformation magnitude which equals half the wavelength of the illuminating laser light (here $\lambda = 314$ nm). The peak of deformation shown in Fig. 6.8b is about 8 μm and is located close to the fault discontinuity.

Schmitt et al. (2006) used dual beam digital electronic speckle interferometry to record stress-relief displacements induced by drilling blind holes into acrylic blocks subjected to uniaxial compressive stress. Stress values inferred from interferograms agreed with known stresses better than 70%.

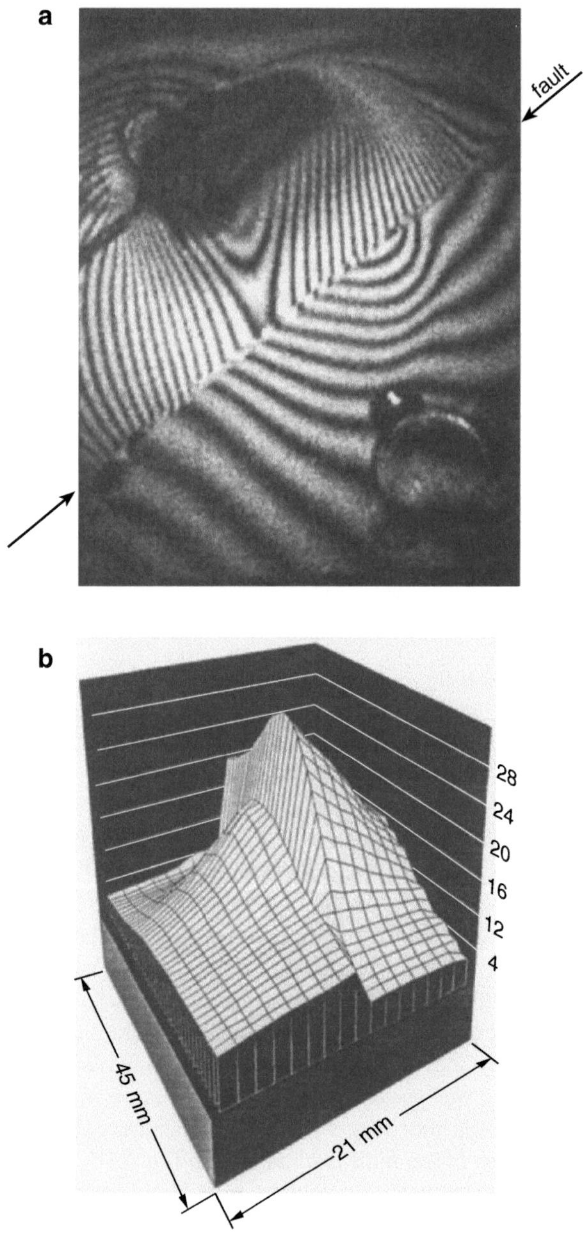

Fig. 6.8 Fringe pattern of hologram **a** from a pyrophyllite sample with an artificial fault (*arrows*) stressed biaxially in the laboratory to a near stick-slip event. The surface deformation on both sides of the fault is displayed in **b** for the central part (21 × 45 mm) of the rock specimen (modified from Spetzler et al. 1991)

6.5 Ultrasonic Wave Speed

Direction-dependent measurements of ultrasonic waves supply information about anisotropy in the material, our method No. 5. Quadratic terms in the analysis of strain are considered in the stress-strain-relation of the material (‰-effects). High-resolution velocity measurements permit estimates about stored strains and stress within the material.

A quasi-isotropic polycrystal has three independent constants of third order (*Murnaghan* constants l, m, n) which describe the stress sensitivity of wave velocities v_{ij} using the elastic Lame constants (λ and μ) (Murnaghan 1951).

$$\begin{aligned} \rho v_{11}^2 &= \lambda + 2\mu + (2l + \lambda)\Theta + (4m + 4\lambda + 10\mu)\varepsilon_1 \\ \rho v_{12}^2 &= \mu + (\lambda + m)\Theta + 4\mu\varepsilon_1 + 2\mu\varepsilon_2 - \frac{1}{2}n\varepsilon_3 \\ \rho v_{13}^2 &= \mu + (\lambda + m)\Theta + 4\mu\varepsilon_1 + 2\mu\varepsilon_3 - \frac{1}{2}n\varepsilon_2 \end{aligned} . \quad (6.10)$$

The indices 1, 2, 3 describe a right-angled coordinate system in which compression waves (v_{11}, v_{22}, v_{33}) and shear waves (v_{12}, v_{13}, v_{21}, v_{23}, v_{31}, v_{32}) are propagating. The first index i is the direction of wave propagation, the second index j the direction of polarization (oscillation) of the wave. Principal strains are denoted by ε_1, ε_2, ε_3 and the volumetric strain by $\Theta=\varepsilon_1+\varepsilon_2+\varepsilon_3$. The velocities in all other azimuths result from permutation indices of the above equations. Each ultrasonic procedure for stress estimates must proceed from these three equations (Lubarda and Richmond 1999; Längler and Schneider 2005).

With the birefringence procedure, two transverse waves spread along the same direction (Fig. 6.9, 1-direction) but have a polarization different by 90°. From the *Murnaghan* Eq. (6.10), it follows that

$$\frac{\Delta v}{v} = \frac{v_{12} - v_{13}}{v_{13}} = \frac{4G + n}{8G^2}(S_2 - S_3), \quad (6.11)$$

where G is the shear modulus of the material and $(S_2 - S_3)$ is the stress difference. In the uniaxial state of stress, one measurement of wave propagation perpendicular to the stress direction leads to the stress value. In the biaxial state of stress, two waves propagating perpendicular to each other are needed and in the triaxial stress case, three wave propagation paths are required.

In rock physics, the effect of cracks on the ultrasonic velocities is utilized (Table 6.1, method 5A). Theoretical background is found in Crampin et al. (1980) and Hudson (1981) where crack populations are analyzed according to their velocity anisotropy generated in the rock. Application of differential wave-velocity analysis (DWVA) for stress estimates in deep boreholes of the petroleum industry was given by Ren and Hudson (1985). Zang et al. (1996a) used wave-velocity analysis (WVA) to determine stress in the scientific ultradeep KTB well. Hypotheses for crustal stress estimates with both methods is (1) that the closure pressure of cracks preserved in rock cores from deep wells equal the in-situ stress magnitude, and (2) that the normal of stress-relief cracks is assumed to point towards the direction of minimum horizontal compression stress, S_h (Sect. 4.4). In some approaches, also high precision phase velocity measurements are used to infer the state of stress

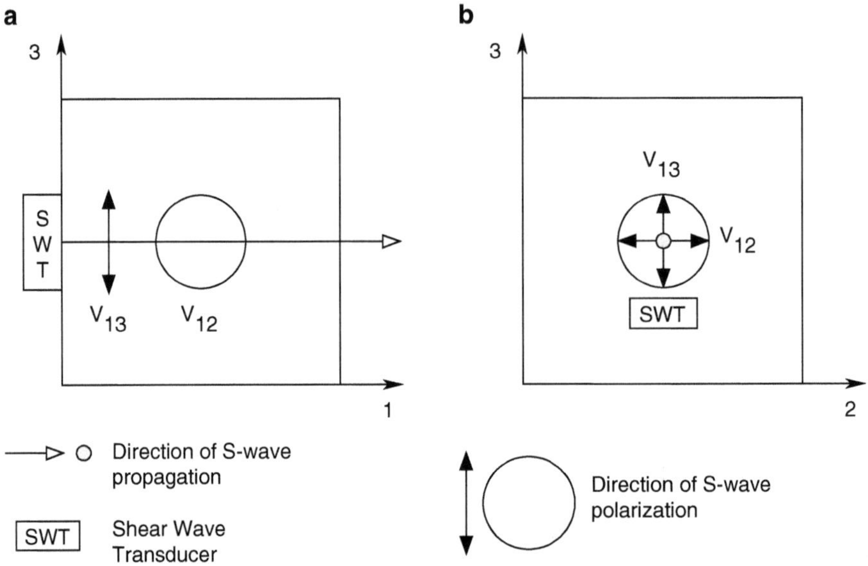

Fig. 6.9 Laboratory setup of birefringence velocity stress measurements in **a** the lateral view and **b** the view from above (modified after Peiter 1992)

(Kobayashi et al. 1997). Core-based methods for stress estimates in the crust are listed in Sect. 7.1 and are presented in Chap. 8.

6.6 Micromagnetic Method

Micromagnetic stress measurements are limited to ferromagnetic material like magnetite. *Weiss* domains are regions of the same saturation magnetization and are sensitive to external loads applied to ferromagnetic material. *Weiss* domains take preferred orientation when large stresses are applied without the magnetic field (magnetostrictive effect). In the case of large tensile stresses without the magnetic field, the domains re-orient parallel to the stress direction; in the case of large compressive stress, the domains orient perpendicular to the stress direction (Fig. 6.10).

Changes in the size and rotation of ferromagnetic domains can be detected with the help of *Barkhausen noise*. The sudden transitions in the magnetization of the material produce current pulses in a detector coil which can be amplified to produce a series of clicks in a loudspeaker. This so-called Barkhausen noise can be measured as a function of applied external magnetization of the test specimen (e.g., Theiner 1997). The characteristic frequency required to obtain magnetic hysteresis curves is typically about 100 Hz, resulting in a penetration depth of the method of about 1 mm. Therefore, the micromagnetic method ranges somewhere between X-ray diffraction and neutron diffraction for stress analysis (Peiter 1992). Although this is a

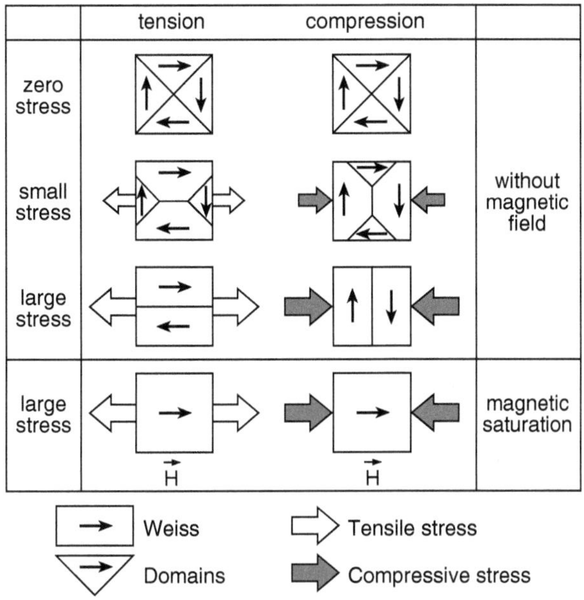

Fig. 6.10 Preferred orientation of *Weiss* domains due to large stress (modified after Peiter 1992)

well-known stress measuring method in material science, it—to our knowledge—has never been applied to rocks. A possible test site would be the large underground iron ore mine in Kiruna, Sweden.

Note-Box Six different physical methodologies for determining stress are presented. The indirect parameters to measure stress are changes of sample length and brittle failure (mechanical methods), electric resistance (strain gauge), atomic lattice plane deformations (X-ray and neutron diffraction), fringes of interference (photoelasticity and holographic interferometry), ultrasonic velocities and magneticially preferred orientated *Weiss* domains. From the indirect parameter, measured stress values are computed by making assumptions about the elasticity of rocks, the wave velocity and lattice distance of minerals. While most methods discussed refer to near-surface stresses (holography, X-ray (micrometer), micromagnetics, neutron (centimetre)), mechanical overcoring and crack-related ultrasonic and strain techniques allow estimating stresses in rock volumes of decimeter-scale.

Chapter 7
Measuring Crustal Stress: Borehole Methods

Stresses within the Earth's crust are "measured" indirectly by coring, slotting and loading a piece of rock with subsequent analysis of re-equilibrium deformations. This action requires assumptions about constitutive behaviour of the rock, i.e. the relationship between measured strain and inferred stress (Eq. (4.2) for anisotropic, Eq. (4.8) for isotropic rock). In addition, Eq. (4.8) includes the effect of temperature on mechanical stresses. If the location of stress measurement is chosen to be close to natural discontinuities (fracture, fault) or excavation (borehole, tunnel) boundaries, near-field stresses are determined (Sect. 4.4). The virgin or in-situ stress field can only be observed at distances of two to three times the size of the excavation discontinuity or any other stress concentrator (inclusion).

The accuracy of rock stress measurements is a superposition of errors due to (1) the influence of structural stress in different rock types (*intrinsic errors*), (2) the bias in stress-measurement techniques (*measurement errors*) and (3) errors in the statistical analysis of various stress data (*statistical errors*). The effect of rock structure on the accuracy of stress data is demonstrated in Fig. 7.1. While in homogeneous material (Fig. 7.1a), stress magnitudes can be determined as precise as ±10%, and stress orientations with angles as precise as ±10°, the errors in non-homogeneous rock material can be much larger (Cai et al. 1995). Leijon (1986) estimated errors in rock stress in homogeneous granite (magnitude ±14%, angle ±15°), and in anisotropic leptite (magnitude ±35%, angle ±40°) at the same depth of 600 m in the Malmberget mine in Sweden.

In anisotropic rock material (Fig. 7.1b), local stress data (deviating from far-field stress) can be determined with small errors. In the case of heterogeneous material (Fig. 7.1c), local stress data (indicating the far-field stress) are inferred with large standard deviations. As we know from Sect. 4.4, anisotropic rock can force the measured stress field to rotate towards, e.g. bedding planes, foliation or preferred oriented fracture networks. In Fig. 7.1b, the in-situ stress direction, S_H (far-field stress) is overprinted by the orientation of rock anisotropy resulting in measured stresses, $<S_H>$ which deviate by $\alpha=50°$ from the in-situ stress direction.

In heterogeneous rock, local stresses inside rock heterogeneities (near-field stress) may vary drastically both in magnitude and orientation (Fig. 7.1c, circular inclusions). Nevertheless, the overall determination of in-situ stress is possible with

A. Zang, O. Stephansson, *Stress Field of the Earth's Crust*,
DOI 10.1007/978-1-4020-8444-7_7, © Springer Science+Business Media B.V. 2010

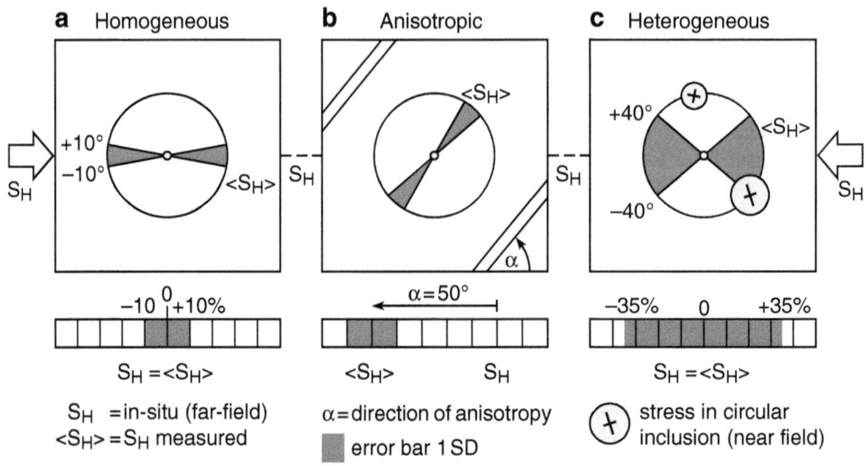

Fig. 7.1 Effect of structural stress on measured stress data. **a** homogenous, **b** anisotropic and **c** heterogeneous rock material

large standard deviations. In order to decide which case or combination of cases in Fig. 7.1 is realized at a certain location for stress estimation, the stress analyst has to be experienced with the effect of structural stresses. The effect of rock elastic anisotropy on underground excavations is made systematically available in Tonon and Amadei (2003).

Type (2) errors in stress measurements are discussed in the appropriate section of the specific technique presented (Sects. 7.2, 7.3, 8.1, 8.2). Statistical type (3) errors are relevant to stress data presented in Chaps. 10 and 11.

7.1 Classification of Measurement Techniques

Classical crustal in-situ stress measurement techniques require either a well bore (*borehole methods*, e.g. breakouts, hydrofrac) or core material (*core-based methods*, e.g. overcoring, strain relief). Therefore, the ultimate depth of crustal stress data is restricted to the final depth of the deepest hole in the Earth's crust (Sect. 1.2). Below the ultimate borehole depth, only focal mechanisms of earthquakes or models contribute to the knowledge of the stress field. No earthquake stress data are available, if the seismogenic zone is missing in a particular tectonic setting.

Rock stress measurements in the Earth's crust can be classified according to their underlying physical principle (Dyke 1988), or according to the rock volume involved in the measurement technique (Amadei and Stephansson 1997). In Table 7.1, we group crustal stress techniques according to the first classification scheme. Category M7.0 (1) mechanism is related to rock fracture (Chap. 3) as applied to boreholes. The most important method of this category in Earth sciences is *hydraulic fracturing* (HF) launched in the 1940s to stimulate production in reservoirs (Hubbert and Willis

Table 7.1 Physical mechanism, experimental technique and ultimate borehole depth (UBD) for crustal stress estimates published in a classical and current article

Category	Mechanism	Technique	References: ■ classical, current	Depth (m)
1	Rock fracture in borehole	Hydraulic Fracturing	■ Hubbert and Willis (1957)	
			te Kamp et al. (1995a, b)	9066
		HTPF	■ Cornet and Valette (1984)	
			Cornet (1986)	973
		Sleeve Fracturing	■ Stephansson (1983)	
		Borehole Breakouts	■ Bell and Gough (1979)	
			Zoback et al. (1986)	11600
2	Elastic strain relief by coring	Surface Relief, SR	■ Leeman (1964)	0
		Borehole Relief, BR	■ Leeman and Hayes (1966)	
			Herget (1986)	2100
		Relief of Large Rock Volumes, RLRV	Sakurai and Shimizu (1986)	UBD
			Sakurai and Akutagawa (1994)	
		Flat Jack	■ Tincelin (1951)	UBD
		Borehole Jack	■ Bock (1993)	
3	Crack-induced strain relief in drill cores	ASR	■ Teufel (1983)	1685
			Lin et al. (2006a)	4544
		DRA	■ Yamamoto (1990)	
		DSA	■ Simmons et al. (1974)	
		DSCA	Ren and Roegiers (1983)	2438
			Dey and Brown (1986)	3791
			Lin et al. (2006b)	4545
		DWVA	■ Ren and Hudson (1985)	1810
		WVA	Zang et al. (1996a)	8080
		DIF	■ Pendexter and Rohn (1954)	
		Core Disking, CD	Haimson (1997)	3582
			Li and Schmitt (1998)	
		Kaiser Effect, KE	■ Kaiser (1953)	
			Holcomb (1993a, b)	1600
			Villaescusa et al. (2006)	
4	Rock properties and stress	Shear wave polarisation	Crampin (1978)	
			Crampin et al. (1986)	
			Bonness and Zoback (2006)	
		Stonely wave -	Zoback et al. (1986)	1670
		Electrical resistivity	Goodman (1980)	UBD
			Stopinski and Dmowska (1984)	
5	Fault properties and earthquakes	Fault plane solutions	■ Byerly (1928)	UBD-Moho
		Natural seismicity, NS		
		Induced-, IS	Trifu (2002)	
		- Mining-, MIS	Knoll (1990)	
		- Fluid-, FIS	■ Healy et al. (1968)	
		- Reservoir-, RIS	■ Carder (1945)	0
		Fault slip striations	Spudich et al. (1998)	

ASR Anelastic Strain Recovery, *DIF* Drilling Induced Fractures, *DRA* Deformation Rate Analysis, *DSA* or *DSCA* Differential Strain (Curve) Analysis, *DWVA* or *WVA* (Differential) Wave Velocity Analysis, *HTPF* Hydraulic Tests on Pre-existing Fractures

1957) but at that time not intended for stress measurements. A sealed-off section of a borehole in an oil-producing well is pressurized until fracture develops in the borehole wall. The minimum horizontal stress S_h is inferred from the pressure at which the hydraulic fracture closes in the pressure-time record. For brief estimation of maximum horizontal stress S_H the peak fluid pressure and the rock tensile strength T_0 is needed. In cased holes with perforations, which are more common in the oil industry, only S_h can be determined accurately. In very porous rock, hydraulic fracturing is difficult to perform. The world deepest hydraulic fracturing test was intended at a depth of 9066 m (te Kamp et al. 1995a, b) at the drilling site of the *Kontinentales TiefBohrprogramm*, KTB (Sect. 9.1). The method is presented in Sect. 7.2.

M7.3 HTPF

One modification of HF is HTPF, so-called *hydraulic tests on pre-existing fractures* (Cornet and Valette 1984). The fluid pressure in HTPF balances exactly the normal stress across the pre-existing fracture (Sect. 4.4, joint stiffness). Having made tests on non-parallel pre-existing fractures, the stress field can be determined without making any assumption with regard to the principal stress orientation and constitutive behaviour of rock. The 2D stress state can be determined theoretically with three pre-existing fractures, the 3D state with six pre-existing fractures. In practice, the number of isolated, pre-existing fractures required is five up to 10 in 2D, and nine up to 20 in 3D to obtain reliable stress data (Ljunggren et al. 2003). HTPF does not work well if structural stresses dominate (anisotropic, heterogeneous rocks). As compared to HF, HTPF has the advantage of less limitation as regards geologic structures. The method does not require the determination of T_0. However, HTPF is more time-consuming as the down-hole equipment must be positioned at the exact location of each discrete fracture to be tested. This requires a good accuracy in the depth calibration of the tools.

M7.2 Sleeve

During *sleeve fracturing* (Stephansson 1983), no fluid penetrates the rock (advantage), but the breakdown pressure for stress estimates is not often well defined. The technique was developed to minimize the problems of hydraulic fracturing associated with fluid-penetration into pre-existing fractures and the build-up of pore pressure in the vicinity of the borehole. No acceptable theory exists with which sleeve fracturing alone can be used to determine the horizontal stress state from a vertical borehole (Ljunggren 1987). Nevertheless, a refined method using urethane sleeves was further developed using single and double fracturing and applied to boreholes in Japan to determine in-situ stresses (Serata et al. 1992). Sleeve fracturing generates two fractures parallel with the axis of the borehole. Later, these fractures were used in hydraulic fracturing and HTPF stress measurements.

The last technique belonging to category (1) mechanism in Table 7.1 is the analysis of borehole breakouts (Bell and Gough 1979). *Borehole breakouts* (BBO) are defined as zones of broken and fallout rock material at the borehole wall occurring 180° apart. The major axis of the diametrically faced breakout zones is an indicator of the direction of minimum horizontal compressive stress. While in hydraulic fracturing, the instantaneous tensile strength of rock is relevant for the pressurized borehole wall to fail, breakouts are the result of mixed-mode failure processes in the borehole wall related to the strength and deformability of the rock involved and time. Breakouts can have a length between centimetres up to several hundred metres in the borehole wall. Attempts have been made to estimate stress magnitudes from breakout geometry (Haimson and Herrick 1986; Zoback et al. 1986; Vernik et al. 1992; Haimson and Lee 1995; Haimson et al. 2009). Breakout analysis is quick to use and requires only the measurement of the diametrical changes of the borehole wall to obtain the S_h-direction. The method reveals stress information valid at very great depth. The deepest breakout located at a depth of 11.6 km in the Kola Peninsula hole, has been evaluated by Zoback et al. (1986). The major limitation of breakout analysis is that they do not appear in every borehole. For example, breakouts are rare in Swedish crystalline rocks above 1 km (Ljunggren et al. 2003) so that the method is inapplicable in shallow holes. The method is discussed in Sect. 7.3. Drilling-induced tensile fractures (DITF) is mostly used as an additional technique (constrain) to breakout analyses.

Category (2) mechanisms (Table 7.1) are related to *elastic strain relief* due to coring. *Relief methods* rely on converting measured strains to in-situ stress through Hooke's law (Sect. 6.2) and were originally developed for hard rocks (Leeman 1964). The physical principal of relief methods was expanded in the overcoring movies from Chap. 6. Relief methods are the most widely used techniques in the engineering application of stress measurement. The technique can be further subdivided into *surface relief* (SR) (e.g., undercoring, Duvall 1974), *borehole relief* (BR) methods (Lee et al. 2006) and techniques that involve *relief of large rock volumes* (RLRV) with subsequent analysis of re-equilibrium deformation (Sakurai and Shimizu 1986).

Variations of borehole relief methods can be sub-classified according to the type of strain analysis at the borehole wall: (1) diametral strains (Leeman 1959; Obert et al. 1962; Crouch and Fairhurst 1967; Suzuki 1969), (2) strains at the flat end of the borehole (Mohr 1956; Leeman 1971; Oka et al. 1979), (3) strains at the hemispherical end of the hole (Sugawara et al. 1986), and (4) strains at the conical end of the hole (Kobayashi et al. 1987; Sugawara and Obara 1995). Disadvantages with the conical or hemispherical cell are that they require preparation of the bottom hole. Another limitation is their poor success in water-filled boreholes. In addition, we can distinguish between strain analysis in shallow holes (e.g., CSIR South Africa cell (Leeman and Hayes 1966)) and strain analysis in deep boreholes (e.g., Borre probe (Sjöberg et al. 2003), US Bureau of Mines gage (Merrill 1967), CSIRO hollow inclusion cell (Worotnicki and Walton 1976)). According to Sect. 6.2, all strain cells except US Bureau of Mines gage allow the determination of the 3D state of stress from one single measurement. The physical principle of stress determination with the Borre Probe has been demonstrated in the accompanying DVD, Chap. 6. A compilation of methods for rock-stress
1, 6.2 estimation by overcoring suggested by the International Society of Rock Mechanics (ISRM) can be found in Sjöberg et al. (2003). Guidelines for stress estimates from overcoring techniques are summarized in Reinecker et al. (2008a), DVD Chap. 11.

The rock volume involved in overcoring stress measurements is small (Table 7.2, BR) and depends on the drilling technique used (overcoring, undercoring at bottom hole, pilot hole or sidewall holes of the main borehole). The *doorstopper* is a method of gluing a strain rosette (Sect. 6.2) directly on to the flattened bottom of a borehole. The technique is so named because the circular rubber or epoxy assembly

Table 7.2 Rock volumes involved in crustal stress measurements and sketch of technique

Rock Volume	Abb.	Stress Inversion Technique, Dimension of Stress Field Analyzed, Estimation of Size	Sketch of Technique	Category from Table 7.1
$10^{10}m^3$	FPS	Focal mechanism of natural earthquakes Fault Plane Solutions, 2D, Plane 10 x 100 km x 10 m thick		(5)
10^9m^3	FSS	Fault Slip Striations, 1D, Lines on planar surface	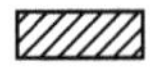	(5)
10^7m^3	MEX	Mine Excavation, 3D		
10^6m^3	FIS	Fluid Induced Seismicity, 2D, Pressure front between two boreholes 3 km deep, 100 m apart, thickness 10 m		(5)
10^5m^3	IS	Induced Seismicity (4<*M*<6), 2D Fault Plane 1 x 10 km x 10 m thick	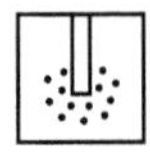	(1,5)
10^4m^3	MIS	Mining Induced Seismicity, 2D Rock Bursts (2<*M*<4) Fault Plane 100 x 100 m x 1 m thick		(5)
10^3m^3	RLRV	Relief of Large Rock Volumes, 3D		(2)
10^2m^3	HF	Hydraulic Fracturing, 2D, 0.5-50 m^3		(1)
10^1m^3	HTPF	Hydraulic Testing on Pre-Existing Fractures, 3D, 1 to 10 m^3	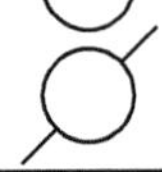	(1)
10^0m^3	SR	Surface Relief, 3D, 1 to 2 m^3		(2)
$10^{-1}m^3$	FJ	Flat Jack, 3D, 0.5 to 2 m^3	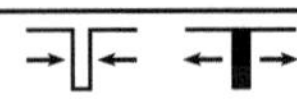	(2)
$10^{-2}m^3$	BBO	Borehole Breakouts, 2D, 0.01 to 100 m^3	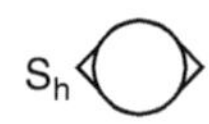	(1)
$10^{-3}m^3$	BR OC	Borehole Relief, 3D, 10^{-3} to $10^{-2}m^3$ Overcoring, Under-, Multiple-, 2-3D	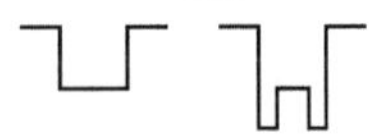	(2)
$10^{-4}m^3$		Core-Based Methods, 2-3D		
	ASR	Anelastic Strain Recovery	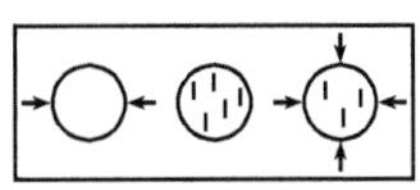	(3)
$10^{-5}m^3$	DSA	Differential Strain Analysis		
	WVA	Wave Velocity Analysis		
	KE	Kaiser Effect, 3D	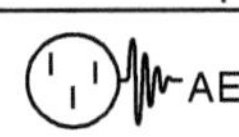	(3)
$10^{-6}m^3$	RS	Single grain residual stress, 3D Diffraction measurements		

Abb.= Abbreviation of method AE= Acoustic Emission *M*= Earthquake Magnitude

with a rosette has the appearance not unlike that of a standard doorstopper in the bathroom. Leeman (1959) indicates that the doorstopper was used as early as 1932 to determine stress in a tunnel below the Hoover Dam (USA). The technique is still used and has recently been adapted for applications in deep boreholes (Martino et al. 1997; Sjöberg et al. 2003). *Deep doorstopper gauge systems* (DDGS, Thompson et al. 1997; Thompson and Martino 2000) allow overcoring measurements at depths as great as 1.2 km. Today, pilot-hole drilling followed by overcoring is used on a regular basis in underground projects. The doorstopper is the next most used technique, mainly performed in highly stressed rock or when the fracturing is too intense to allow pilot-hole drilling (Ljunggren et al. 2003). Doorstoppers include the possibility for continuous monitoring and the application in water-filled boreholes. Disadvantage with doorstoppers is that they require polishing the hole bottom and that they determine planar stresses only.

Overcoring methods are limited by the magnitude of in-situ stress which can only be determined at depths for which the strength of rock near the borehole wall is not exceeded. Herget (1986) reported extraordinary overcoring stress values of 130 MPa in high-strength rocks of the Canadian Shield at a depth of 2.1 km. The success rate of overcoring, however, is less than 50% (Herget 1993). Relief methods are affected by rock damage and weathering (surface relief), and the gauges by humidity and dust (borehole relief). The main drawback is that relief methods involve small rock volumes (Table 7.2). Chandler (1993) suggested multiple, concentric coring with increasing diameter to overcome stress sensitivity from rock microstructures (mineral content, grain size). Other approaches like RLRV, measure stress while excavating large underground openings (Sakurai and Shimizu 1986; Sakurai and Akutagawa 1994). These methods require displacement measurements of an opening following excavation and back analysis by boundary or finite-element modelling to relate strain to stress. The advantage of back-analysis is that it is quick to use, requiring a simple measurement procedure and includes large rock volumes (Table 7.2, RLRV). The limitations include that the technique is tied to underground openings. The method requires numerical analysis and, like any other inversion technique, provides no unique solution.

Lastly, *jacking methods* (Tincelin 1951), also called stress compensation methods, involve cutting a slot into the rock mass and measuring the response. To reconstruct the shape of rock, a jack is pressurized until the original shape of the rock without a slot is obtained. This shape-compensation method does not require rock elastic properties to determine stresses. A review of borehole-slotting methods is given by Bock (1993). The method does not require any overcoring and is quick to use (Ljunggren et al. 2003; ten measurements per day). The borehole, however, must be dry. Mizuta et al. (2004) modified the borehole-jack fracturing method by increasing the contact angle of platens in the borehole from 90° (Azzam and Bock 1987) to 160°. Both Azzam and Bock (1987) and Mizuta et al. (2004) employ tangential strain sensors or crack-opening displacement sensors for detecting fracture formation.

Category (3) mechanism (Table 7.1) is related to *crack-induced strain relief* in drill cores. Microcracking is caused in stress relief when the rock is cut from the in-situ stress field at the bottom or the wall of a deep well. These, also called microcrack-related phenomena (Engelder 1993), strain recovery methods (Amadei and Stephansson 1997) or *core-based methods* (Sano et al. 2005), record core damage following drilling due to elastic (Table 7.1, category (2)) and inelastic material behaviour (Table 7.1, category (3): viscoelastic, cracking). The crucial assumption in all core-based stress measurements is that the dominating portion of the microcracks observed is caused by relief of the present-day stress field. If this assumption is valid, azimuths of crack populations can be related to in-situ stress orientations and crack closing pressures (Sect. 4.3), inferred from repressurized core specimens, can be used to compute in-situ stress magnitudes (Dey and Brown 1986). Core-based methods are discussed in Chap. 8.

Core-based methods further can be subdivided into the analysis of (a) strain data (ASR, DRA, DSA), (b) wave velocity data (DWVA, WVA) and (c) fracture inspection data (CD, DIF) as shown in Table 7.1 (category (3) mechanism). In an *anelastic strain recovery* (ASR) experiment, the time-dependent strain of freshly recovered cores is measured to estimate in-situ stresses (Teufel 1983). *Differential strain analysis* (DSA) on rock cubes (Simmons et al. 1974) and *deformation rate analysis* (DRA) on rock cylinders (Yamamoto et al. 1990) require reloading of the relaxed, cracked material in order to infer in-situ stresses. *Differential wave-velocity analysis* (DWVA) (Ren and Hudson 1985) and *wave velocity analysis* (WVA) (Zang et al. 1996a) also require the investigation of relaxed drill cores in a pressure vessel but, in contrast to the strain analysis techniques, the sensitivity of ultrasonic wave velocities on stress-relief cracks is utilized to estimate in-situ stresses. Even though laboratory tests like ASR and DSA are termed exotic, require costly cores and are hardly ever in use due to practicability in petroleum engineering (Reynolds and Shaw 2006), it is worth understanding the physics behind (Chap. 8), since in severe environments of very deep boreholes, core-based methods are among the important techniques to be applied when classical stress measurement techniques fail. Ljunggren et al. (2003) rate core-based methods of great value while most of them miss their full commercial platform.

Fracture inspection methods usually rely on boreholes drilled in highly stressed rock. The resulting cores often appear as an assemblage of disks (Pendexter and Rohn 1954) which are often shaped like a horse saddle. Much experimental work (Haimson 1997) and theoretical analysis (Li and Schmitt 1998) has been done to understand the micromechanical process involved in *core disking* (CD). Results indicate that the S_H magnitude can be estimated from the average thickness of the disks, and the S_H-orientation from the through axis of the "horse saddle"-shaped disk. Haimson (1997) pointed out that, together with breakouts, core disks can provide upper limits on the in-situ stress levels and help to assess the maximum horizontal stress. Hakala (1999) listed the minimum information needed for the interpretation of disks with respect to in-situ stress. When core disking occurs, rock stresses exceed rock strength. Such information obtained during drilling is, of course, valuable.

Another exotic technique within category (3) mechanisms (Table 7.1), cracking phenomena in drill cores, is the Kaiser Effect (KE). In the Middle Ages, tin casters estimated the quality of plates by listening to "tin-clamour". It was Joseph Kaiser (1953) who first brought to our attention the increase of sub-audible acoustic noise emitted from metals in tension when the previously applied stress level has been exceeded. The phenomenon of recalled maximum stress verified in various materials, is termed *Kaiser Effect*. It has been hypothesized that the stress experienced by a rock in-situ can be inferred from monitoring acoustic emissions (AE) on drill cores from deep wells. Despite encouraging results, applying KE to drill cores from great depth (Dyke 1988; Wenzel 1990), the research by Holcomb (1993a, b) revealed that KE in uniaxial compression tests cannot be used to infer in-situ stresses. The Kaiser effect was applied for in-situ stress estimates by combining Holcomb's damage surface approach with polyaxial test data of rock cores in the laboratory (Pestman and Munster 1996; Pestman et al. 2001, 2002). Reliable stress data applying the Kaiser effect to deep drill cores are obtained to an ultimate depth of 1600 m (Table 7.1, Villaescusa et al. 2006). The Kaiser effect is discussed in Sect. 8.2.

Category (4) mechanisms (Table 7.1), also called borehole (seismic) logging (Engelder 1993) or indirect methods (Amadei and Stephansson 1997), combine the variation of physical rock properties with stress. Research into the estimation of in-situ stress orientations by the analysis of *shear-wave polarisation* was pioneered by Crampin (1978). A shear wave entering a region of anisotropy splits into two phases, each travelling at a different velocity (Sect. 6.5, v_{SH}, v_{SV}). Wave polarisation and the delay between the two phases vary with azimuth, allowing the symmetry of the anisotropy to be identified. Crampin et al. (1986) suggested that stress aligned fluid-filled cracks (with dimensions of less than one wavelength of the *S*-wave) are the cause of *shear wave splitting*. Both Crampin et al. (1986) and Peacock et al. (1988) report observations of temporal variation in shear-wave splitting, over periods of years, interpreted as due to changes in the microcrack population induced by stress change. Within rocks exhibiting wave anisotropy from causes other than the present day in-situ stress field, the technique may yield limited results. Nevertheless, shear-wave anisotropy has been used to map stresses until recent times (Boness and Zoback 2006). Prioul et al. (2007) stated that the fast shear-wave azimuth measured in sonic borehole logs is caused by a combination of stress and fracture effects.

Stonely waves, frequently called tube waves, propagate as surface waves along borehole walls (White 1962) at phase velocities less than the shear-wave velocity of rock (Cheng and Töksoz 1984). In a borehole drilled into isotropic, elastic, homogeneous material, the Stonely wave-particle motion is prograde elliptical, with its major axis along the borehole axis. However, if principal stresses in the plane orthogonal to the borehole are not equal, even in isotropic material wave velocities can be very anisotropic. Based on this, Zoback et al. (1986) proposed a new technique for estimating principal stress directions involving analysis of stress induced polarisation of tube wave particle motion.

Goodman (1980) cites the use of *electrical resistivity* as a means of measuring in-situ stress. As cracks close, the resistivity decreases (Hoerning 1979). Resistivity sensitivity to crack opening and closing allowed Stopinski and Dmowska (1984) to observe stress build-up and release in the vicinity of underground workings. This method, as well as other exotic acoustic, sonic, electromagnetic or holographic methods (Chap. 6), are listed here for completeness (and for students to invent new techniques) but have not yet gained much popularity in stress-measurement practice (Amadei and Stephansson 1997).

The mechanism (5) for stress estimates in Table 7.1 is concerned with physical properties of pre-existing fault zones in the Earth's crust and the related earthquakes. According to Table 7.2, stress determination techniques within this category involve very large rock volumes. The end-members are *fault plane solutions* (FPS), depending on the size of the fault plane activated during an earthquake, and may reach rock volumes of up to $10^{10}\,m^3$ (e.g., planar surface 10×100 km with a thickness of 10 m).

Focal mechanisms of earthquakes can provide the orientation of principal stresses using the Coulomb failure criterion (Sect. 3.1), and the relative magnitudes of the three principal stresses (Sect. 5.3). It is the only method that, first provides stress data within the Earth's crust below the ultimate borehole depth (UBD) down to the Moho (Table 7.1, Category (5)) and, second, dominates the overall entries of stress data in the World Stress Map (DVD, Chap. 11). The portion of focal mechanism to WSM data increases from 54% (Zoback 1992) to 77% (Reinecker et al. 2006), while the total number of stress entries in the same period more than doubled from 7,328 entries in 1991 to 15,969 entries in 2005. There are two big uncertainties in stress determination from earthquake focal mechanisms: (1) the precise value of the friction coefficient of faults in the Coulomb fracture criterion (Chap. 3), and (2) the precise orientation of maximum horizontal stress (30° error) which can lie anywhere in the dilatational quadrant (McKenzie 1969).

From our viewpoint we have to separate between stress inversions from *natural seismicity* (NS) and *induced seismicity* (IS). In contrast to NS, the term IS refers to typically minor earthquakes and tremors that are caused by human activity that alters the crustal stress field (Gupta and Chadha 1995; Trifu 2002, 2010; Cornet 2007). We further refine IS into *mining-induced seismicity* (MIS) and *fluid-induced seismicity* (FIS). MIS includes seismic events and rock bursts arising from stress changes associated with mining excavations (Mendecki 1997) and can be regarded as a large-scale relief test, where the mining excavation process determines the overcored rock volume involved. FIS are caused by the impoundment of large reservoirs (RIS), injection or extraction of fluid in liquid waste disposal or the fracturing of hydrocarbon and geothermal reservoirs. On the other hand, FIS can be regarded as a large-scale hydraulic test where the volume of fluid and pressure front determines the extension of hydraulic fractures. In addition, microseismic monitoring of real hydraulic-fracture tests at the borehole scale can be performed within nearby boreholes (Doe 1981), downhole within injection wells (Albright and Pearson 1982) or on the ground surface (Pine and Batchelor 1984). Despite there being many similarities between IS and NS, there are also notable differences. The most important being the aspect of human intervention in natural processes. With respect to in-situ stress, the most important fact is that IS counts for near-field stress around excavations and reservoirs, while NS is able to determine far-field stress depending on the size of the fault zone under consideration.

In future from our point of view, the WSM will benefit from both stress inversion data of natural and induced seismicity events for two reasons. First, stress inversions from focal mechanisms of earthquakes become more and more reliable (see Sect. 5.3) and, second, the number of human triggered seismicity sites continually increases according to energy supply from (geothermal, hydrocarbon) or waste disposal (radioactive, CO_2 sink) into the Earth's crust. *Large magnitude events* (LME) will play a key role in stress inversion. The largest induced seismicity event ever recorded was a magnitude $M=7.0$ gas withdrawal event in Gazli, Uzbekistan in 1984 (Simpson and Leith 1985). A large MIS event was the $M=5.6$ rock burst that occurred in a potash mine near Völkershausen, Germany in 1989 (Knoll 1990). Significant fluid-induced seismicity events associated with long-term well injection include, the 1966 $M=5.2$ event at the Rocky Mountain Arsenal near Denver, Colorado (Evans 1966; Healy et al. 1968). Significant FIS arising from short-term stimulation operations at geothermal sites include the 2003 $M=3.7$ event at the Cooper basin site, Australia (Baisch et al. 2006) and the 2006 $M=3.4$ microquake near Basel, Switzerland. Enhanced oil and gas recovery events are usually below $M=3$, but possible FIS associated with well operations have local magnitudes as high as 6.5 (Nicholson and Wesson 1992).

Stress inversions from induced seismicity events, together with stress inversions from background natural seismicity, are useful tools to identify stress perturbations triggered by human activity. Due to the fact that induced seismicity is tied to the points of human interaction within the Earth's crust (borehole, dam, mining shaft), local seismic arrays can be operated resulting in high precision locations of IS events, and best estimated fault plane solutions as well as full moment tensor inversions. Compared to NS, in the case of IS we eliminate one degree of freedom (location) in earthquake prediction besides time and magnitude of occurrence. Certainly, induced events can occur a few 100 m away from the source, e.g. mining activity or borehole (Sect. 9.1). Stress inversion techniques from earthquake focal mechanisms are not expanded in this edition of the book, but stress data from focal mechanisms are included (DVD, Chap. 11 WSM).

In the following, we select four stress determination techniques arranged in terms of decreasing rock volume involved (Table 7.2). From category (1) mechanisms (Table 7.1), we choose the classical hydraulic fracturing method where the borehole wall needs to be pressurized instantaneously (Sect. 7.2), and the borehole breakout analysis technique which relies on the time-dependent failure behaviour of the borehole wall (Sect. 7.3). The category (2) mechanism borehole relief was described earlier (Chap. 6). From category (3) mechanisms (Table 7.1, core-based methods), we select the anelastic strain recovery method, which applies to freshly recovered drill cores (Sect. 8.1), and the Kaiser effect methodology which recalls the maximum stress from acoustic emission data obtained on relaxed cores by polyaxial laboratory reloading (Sect. 8.2). It has to be emphasized that for practical rock stress measurements in the field we recommend to use one or several of the three most important methods like overcoring, hydraulic techniques and borehole breakout analysis (see Sect. 10.4).

7.2 Hydraulic Fracturing

An instantaneous rock mass fracturing technique like hydraulic testing, requires a hermetically sealed part of a borehole in the crust which is blasted open by fluid overpressure. The pressure-time-curve of the rock mass response detected allows determining the S_h magnitude. The orientation of the hydraulic fracture plane induced provides the S_H direction of the in-situ stress field. The method was first applied in 1947 (at that time called "hydrafrac" instead of "hydrofrac") when borehole Keppler No. 1 (West Kansas) was stimulated for gas production (Clark 1949).

Today, hydraulic fracturing is a commercial standard procedure for stress estimates and oil and gas stimulation of reservoirs in petroleum industry. The "monograph" in reservoir stimulation by hydrofrac testing is Valko and Economides (1995) or Economides and Nolte (2000).

7.2.1 Theoretical Basis of Hydrofracs

The classical hydraulic fracturing theory by Hubbert and Willis (1957) requires isotropic, elastic, homogeneous and impermeable rock material which is subjected to three principal stresses at depth. One principal stress is assumed to be parallel to the axis of a vertical borehole, S_V, and equals the weight of the overburden. The near-field stress around the pressurized hole with inner pressure p is calculated according to Fig. 7.2. The p is defined as the difference between the fluid pressure injected into the well and the formation pore pressure.

For tangential stress, $\sigma_{\varphi\varphi}$ also called circumferential or hoop stress at the borehole wall ($r=R$), the Kirsch (1898) solution (Sect. 7.3, Eq. (7.9)) leads to

$$\sigma_{\varphi\varphi} = S_H + S_h - 2(S_H - S_h)\cos 2\varphi - p. \tag{7.1}$$

The minimum value of $\sigma_{\varphi\varphi}$ is located at azimuths $\varphi=0$, π (i.e., $\cos 2\varphi=1$), and reads

$$\sigma_{\varphi\varphi}^{\min} = 3S_h - S_H - p. \tag{7.2}$$

It is a tangential tensile stress (negative) when $p>3S_h-S_H$ is fulfilled. If $\sigma_{\varphi\varphi}^{\min}$ equals the tensile strength T_0 (Chap. 3) of the rock mass, a radial tensile fracture develops at the borehole wall propagating bilaterally along the azimuths 0° and 180°, which are parallel to the S_H direction of the pre-existing virgin far-field stress (Fig. 7.2,

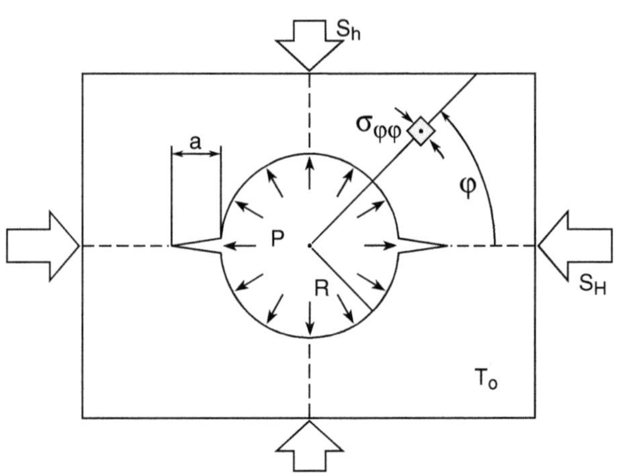

Fig. 7.2 Geometry of hydraulic fracture visualized in a 2D cut of a circular borehole

crack tips). We can formulate our Hubbert-Willis *elastic hydrofrac criterion* based on elastic and impermeable rock as

$$p_{crit}^{HWS} = 3S_h - S_H + T_0, \tag{7.3}$$

where p_{crit} is the critical fluid pressure for fracture initiation at the borehole wall, also called fracture breakdown pressure (*FBP*). Note that it was Scheidegger (1962) who first introduced the rock tensile strength into Eq. (7.3) in order to determine three principal stresses from hydrofrac data in boreholes. Therefore, we extend the superscript in the elastic model to HWS indicating the *Hubbert-Willis-Scheidegger criterion* for the breakdown pressure. Hubbert and Willis (1957) suggested that the minimum horizontal stress S_h is equal to the well pressure required to extend the hydraulic fracture while holding it open. Kehle (1964) was more precise and suggested that $S_h = SIP$, where shut-in pressure *SIP* is the minimum pressure needed to keep the fracture open against fracture-normal stress (Chap. 4) after pumping has been stopped (see Sect. 7.2.2). Also, the fracture breakdown pressure (*FBP*) is discussed in detail in the following section.

One assumption made in the elastic model is that no borehole inspected fluid penetrates into the host rock. This assumption cannot hold in clean, open holes of types commonly used for in-situ stress measurements, since pressure diffusion creates an additional stress perturbation around the borehole. Haimson and Fairhurst (1967) allow the rock formation to be porous and permeable. In their approach, the elastic model is modified for the penetrating fluid (permeable rock), and the *poroelastic hydrofrac criterion* results

$$p_{crit}^{HF} - p_f = \frac{3S_h - S_H + T_0}{2 - \alpha\left(\frac{1-2\nu}{1-\nu}\right)}, \tag{7.4}$$

where p_f is the formation fluid pressure, α is the Biot constant and ν is the rock Poisson ratio. The denominator in *Haimson-Fairhurst breakdown pressure*, $p_{crit}^{HF} = \text{FBP}$ in Eq. (7.4) for isotropic rocks can vary between 1 and 2, since $0 \le \alpha \le 1$ (Sect. 5.4) and $0 \le \nu \le 0.5$. For hard rocks, the denominator approaches 2, for porous sedimentary rocks it approaches 1. The upper limit for the fracture initiation pressure (Eq. (7.3), p_{crit}^{HWS}) is referred to as the "fast" pressurization limit, and the lower limit (Eq. (7.4), p_{crit}^{HF}) as the "slow" pressurization limit (Detournay and Carbonnel 1997). In real situations, the fracture initiation pressure will usually be somewhere between these two limits. The lower limit of fracture initiation applies above a rock permeability of 1 μD (Detournay and Cheng 1992). The real breakdown pressure of the formation will depend on stresses, borehole inclination, rock strength, permeability, borehole fluid properties and operational procedures. Although poroelasticity is used in the derivation of Eq. (7.4), the coupled diffusion-deformation relationship in fluid-saturated porous rock was treated in depth by Detournay et al. (1989) based on the poroelastic theory expanded in Sect. 5.4 (Rice and Cleary 1976). For low porosity rocks, Schmitt and Zoback (1989) modified the rock tensile strength by an effective stress law (see Eq. (5.10)) and calculated the breakdown pressure for both, permeable and impermeable rock. The introduction of the modified stress law lends support to the elimination of the pore pressure term in the interpretation of hydrofrac stress data (Eqs. (7.3) and (7.4)) in low-permeability, hard rocks.

Conventional elastic (Eq. (7.3)) and poroelastic solutions (Eq. (7.4)) assume that the rock subject to hydraulic fracturing behaves as a continuum. The presence of cracks in the test interval, however, can violate these assumptions. Abou-Sayed et al. (1978) introduced a fracture mechanics approach to the hydrofrac criterion, assuming the existence of arbitrarily oriented cracks in rock (Chap. 6). From a fracture mechanics view-point, the breakdown pressure characterizes the end of the elastic regime and the sudden entry into the inelastic regime indicated by the initiation of tensile failure at the borehole wall. Rummel and Winter (1983) and Rummel (1987) derived a fracture mechanics solution that can be used to calculate the maximum horizontal compressive stress. Using the approximation that crack lengths (Fig. 7.2, a) are large compared to the borehole radius ($a>3R$), Rummel (1987, 1989) computed the critical pressure for unstable crack extension at the borehole wall as

$$\begin{aligned} p_{crit}^{RW} &= K_0 \frac{K_{IC}}{\sqrt{R}} + K_1 S_h + K_2 S_H \\ K_0 &= \frac{1}{h+i}, K_1 = \frac{g}{h+i}, K_2 = \frac{f}{h+i} \end{aligned} \quad (7.5)$$

The terms $f(b)$, $g(b)$, $h(b)$ are dimensionless stress intensity functions, which depend on the normalized crack length, $b=1+a/R$ (Fig. 7.2), of the propagating hydraulic fracture. $2R$ is the borehole diameter. The stress intensity function $i(b)$ depends on the pressure distribution along the hydraulic fracture ligaments. The apparent hydraulic fracturing tensile strength (Eq. (7.5), $S_h=S_H=0$) depends on K_{IC}, the fracture toughness of rock in mode I (Chap. 3), the borehole radius R and the crack length a as well as the pressure distribution within the fracture. The latter is related to the pumping rate or fluid pressurization. For rocks without pre-existing cracks, it can be shown that $K_1=3$ and $K_2=-1$. In this case, the fracture mechanics approach (superscript R for Rummel and W for Winter) equals the elastic solution ($p_{crit}^{RW} = p_{crit}^{HWS}$) with $T_0=K_0K_{IC}R^{-0.5}$.

So far we have discussed the case of a vertical fracture (Fig. 7.2, 2D) which occurs when the fluid pressure in the well plus the rock tensile strength is smaller than the vertical stress due to the weight of the overburden. This case corresponds to a breakdown of the borehole wall which is caused by azimuthal tension resulting in a *vertical hydrofracture* plane (Fig. 7.3a, b, $FBP<S_V=\rho gz$). If the pressure in the well plus the tensile strength of rock becomes greater than the lithostatic stress, the breakdown is caused by vertical tension in the borehole wall resulting in a *horizontal hydrofracture* plane (Fig. 7.3c, d, $FBP>S_V$). If the radial stress in the borehole wall becomes smaller than the rock strength plus the well pressure, *concentric failure* of the hole is the result (Fig. 7.3e, f). Concentric enlargement of the borehole diameter may be the result of a low mud weight. As we know from Chap. 3, as long

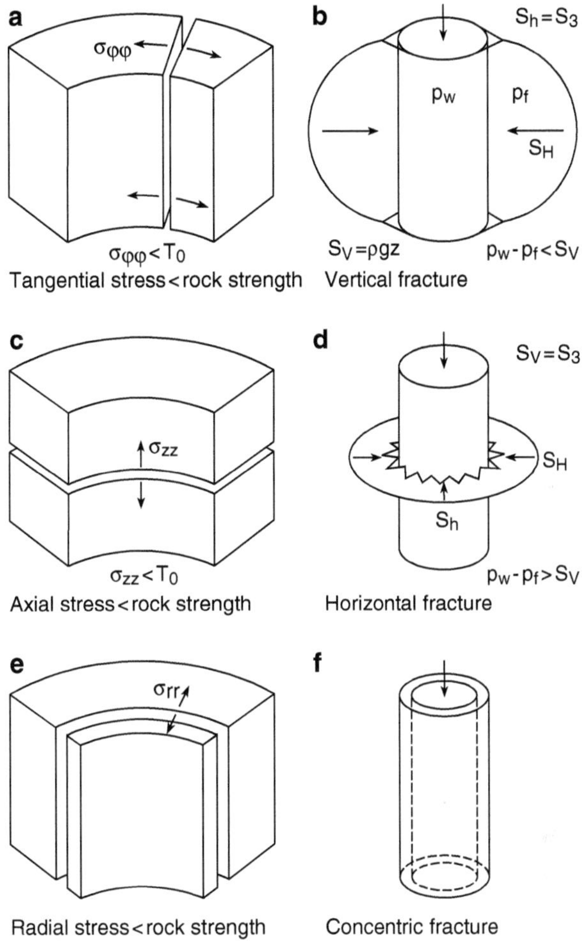

Fig. 7.3 Tensile failure process in the borehole wall from the inner hole perspective (*left column*) and outer hole perspective (*right column*). Depending on the relationship between the radial, tangential and axial stress, we can distinguish three practical cases: **a**, **b** vertical hydraulic fracture, **c**, **d** horizontal hydraulic fracture and **e**, **f** concentric borehole enlargement

as the rock is reasonably homogeneous and isotropic, any fracture in the borehole wall will propagate normal to the least principal stress which is equal to S_h (tangential stress) in Fig. 7.3a, b, equal to S_V (axial stress) in Fig. 7.3c, d and equal to the radial stress in Fig. 7.3e, f.

7.2.2 Practice of Hydrofracturing

Schematically, the hydraulic fracturing operation is visualized in Fig. 7.4. For determining in-situ stresses at depth *z*, a section of the borehole is sealed by inflating (Fig. 7.4a, pressure pipe 1) two packers (Fig. 7.4a, 3). The sealed-off section (Fig. 7.4a, injection interval ~1 m) is pressurized (Fig. 7.4a, pressure pipe 2) until the borehole wall is fractured (Fig. 7.4a, HF). From the fracture breakdown pressure

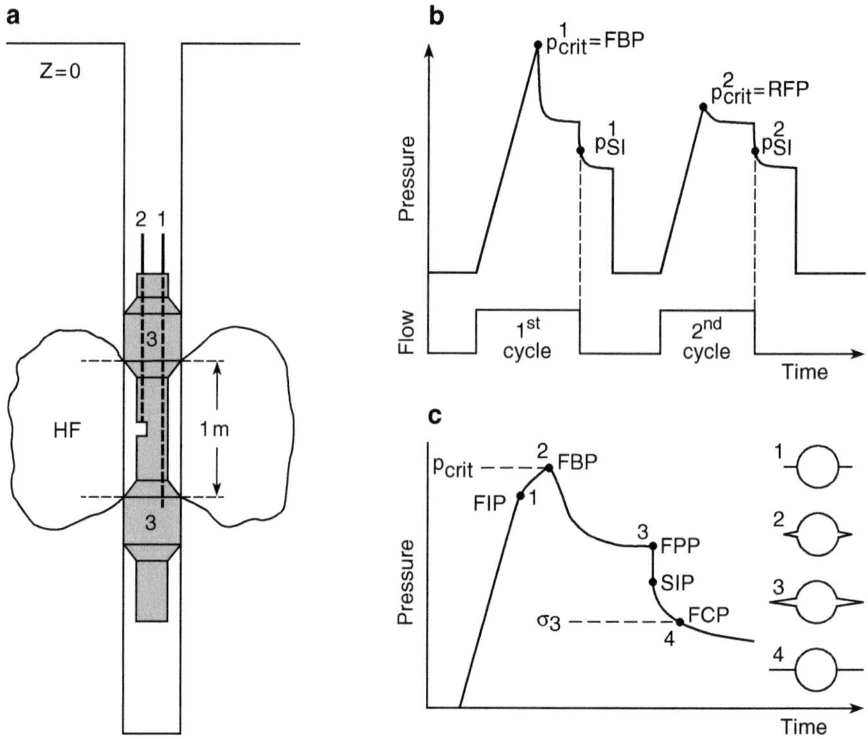

Fig. 7.4 Schematic view of **a** downhole equipment, **b** bottom hole pressure-time-curve of hydraulic treatment, and **c** crack mechanics in hydraulic fracturing demonstrated for the first pressurization cycle. In **a**, *2* is the pressure pipe for initiation of the hydraulic fracture, *1* is the pressure pipe for inflating the packers *3*, and the injection interval is indicated by the *horizontal dashed lines*. In **b**, the peak pressure measured is the fracture breakdown pressure (*FBP*), and *SIP* is the shut-in pressure (virgin loading of the hole, first cycle). Reloading the hole fractured (second cycle) leads to superscript *2* pressures. For reference, the constant flow rate in the injection interval is seen. In **c**, fracture initiation pressure (*FIP*), fracture breakdown pressure (*FBP*), fracture propagation pressure (*FPP*), fracture closure pressure (*FCP*) and the shut-in pressure (*SIP*) related to the different stages of crack growth (initiation, propagation, arrest, re-propagation) are shown

(Fig. 7.4b, $p^1_{crit} =$ FBP) and the pressure necessary to hold the surfaces of a fracture open (Fig. 7.4b, p^1_{SI} =shut-in pressure=SIP), we can calculate in-situ stress magnitudes (elastic solution Eq. (7.3), impermeable borehole wall)

$$\begin{aligned} S_h &= p^1_{SI} = SIP1 \\ S_H &= 3S_h + T_0 - p^1_{crit} = 3S_h + T_0 - FBP. \\ S_V &= \rho g z = p^2_{SI} = SIP2 \end{aligned} \tag{7.6}$$

In a normal faulting stress regime ($S_V<S_h<S_H$), the initiation and propagation of a vertical hydraulic fracture (Fig. 7.3a, b) requires the shut-in pressure (Kehle 1964)

and therefore $S_h = p^1_{SI}$ (Fig. 7.4b, c). The minimum horizontal stress value S_h is the most reliable value in Eq. (7.6) and in hydraulic testing in general (Ljunggren et al. 2003; accuracy about ±5%). The magnitude of the maximum horizontal stress S_H is calculated from the measured breakdown pressure FBP in a bottom-hole pressure chart of the hydraulic fracturing treatment (Fig. 7.4b, c). Computing S_H (Ljunggren et al. 2003; accuracy about ±20%) requires all assumptions from the elastic model in Eq. (7.3) plus an independent determination of the tensile strength of rock T_0, e.g. by laboratory testing. The vertical stress S_V is equal to the measured value of the shut-in pressure in the second fracturing cycle, the so-called refrac cycle (Fig. 7.4b). This is only the case if the minimum principal stress is vertical and aligned with the borehole axis (Fig. 7.3c, d, horizontal fracture). In a scenario where an initially vertical oriented hydraulic fracture "rolls over" at some distance from the hole into a horizontal oriented fracture, all three principal in-situ stress magnitudes (S_H, S_h, S_V) can be determined from one pressure-time record of a single hydraulic test. Note, however, when there is a risk for a "roll-over hydraulic fracture" the shut-in pressures determined from different stages of fracture propagation have clearly to be separated (Baumgärtner and Zoback 1989).

The different pressures involved in a hydraulic fracturing treatment are shown in a high-resolution pressure-time-record of the virgin test cycle together with its
7.1 HF
physical interpretation of crack growth (Fig. 7.4c). Sometimes a leak-off point, the so-called *fracture initiation pressure* (Fig. 7.4c, FIP), is clearly visible in the pressure chart before breakdown. In this stage, a small, closed (zero aperture) starter crack is present in the borehole wall (Fig. 7.4c, stage #1). The physical reason of *FIP* is not yet clear, but it seems to be related to the plastification of the well, or the stress and temperature sensitivity of the host rock (e.g., Fjaer et al. 2008). In this case, the Kirsch solution has to be replaced by the Brady solution (Goodman 1980) in order to characterize the plastic zone around a well (Abass and Neda 1988). Leak-off tests are usually run in non-reservoir rock to determine the lower bound of S_h (Hawkes et al. 2005). As such, the pore pressures in these rocks are unaltered from their natural, in-situ values.

The *fracture breakdown pressure* (Fig. 7.4c, FBP at stage #2) is defined as peak pressure in the pressure-time chart of the virgin hydraulic fracturing cycle. In physical space, FBP is represented by a small starter crack with finite aperture (Fig. 7.4c, stage #2). Fracture-breakdown pressures are usually measured in reservoir formations, produced for extended periods and pore pressures have often decreased significantly. Pressure depletion tends to lower S_h (Addis 1997).

Once the starter crack is initiated (FBP) it continues to propagate (Fig. 7.4c, stage #3). The *fracture propagation pressure* is well recognized at the "knee" of the pressure-time record (Fig. 7.4c, FPP). During field testing one might not observe the FPP knee as pronounced. In physical space, FPP is represented by the hydraulic fracture of final length and finite aperture (Fig. 7.4c, stage #3). The *shut-in pressure* (Fig. 7.4c, SIP) is defined as the pressure at which the hydraulic fracture stops propagating (crack arrest). However, in many situations it is gradual with no obvious kink. Note that also in physical space there is no simple representation of SIP. The *fracture closure pressure* (Fig. 7.4c, FCP) is the last of our four pressure values of each test cycle and is given by the fluid pressure at which the two crack ligaments collapse and close again. In physical space, FCP is shown as a closed (zero-aperture) fracture of final length (Fig. 7.4c, stage #4).

Conceptually, the pressure required to extend a hydraulic fracture can be divided into the pressure required to keep the fracture open towards the least-principal stress, the pressure required to flow the fluid through the fracture and the pressure required to overcome the resistance at the fracture tip creating new surface area. Fischer et al. (2008) analyzed thousands of microquakes induced during hydraulic fracture treatment of a tight gas reservoir in the Canyonsand Formation in west Texas (USA). According to the microseismic signatures found, the whole fluid volume injected was used for creating the new fracture. In other words, the diffusion infiltration of the fluid injected into the reservoir rock was negligible.

Determining S_h from shut-in pressures of hydraulic fracturing tests is not straight-forward, since there is a gradual decrease from FPP to FCP in the pressure-time curve, and SIP has to be computed using different techniques. Tunbridge (1989) gave a review on methods determining SIP from pressure-time records by the inflection point, tangent intersection, and maximum curvature techniques in the second and third refracturing cycles. The physical reason for imprecise determination of SIP is the fact that fracture closure does not occur instantaneously (cf. Figs. 4.6 and 4.9). The pressure measured immediately after the shut-in operation started is referred to as *instantaneous shut-in pressure* (ISIP), and is an upper bound for the least horizontal stress magnitude. To assist in interpretation of fracture response, a number of different graphical methods are used (square root of time, log time, log pressure versus time). For an overview of methods, we recommend Guo et al. (1993) or Amadei and Stephansson (1997). Estimating the smallest principal stress from fracture closure pressures is discussed by Fjaer et al. (2008).

As can be inferred from Fig. 7.4b, pressures from the second fracturing cycle can be interpreted analogously to those of the first cycle using superscripts 2 (e.g. p^2_{crit} and p^2_{SI}). In hydraulic fracturing terminology, the breakdown pressure of the second cycle p^2_{crit}, however, is referred to as reopening or *refrac pressure* (*RFP*). Bredehoeft et al. (1976) suggested that the tensile strength of the rock formation in the field T_0^{HF} can be determined from the difference in the first $p^1_{crit} = FBP$ and second breakdown pressure $p^2_{crit} = RFP$

$$T_0^{HF} = p^1_{crit} - p^2_{crit}, \tag{7.7}$$

assuming that the fracture closes completely between each cycle of pressurization and that RFP is the pressure level at which the pre-existing hydraulic fracture just begins to open. Ratigan (1992) calculated the reopening pressure by a fracture mechanics approach for the case of permeable rock. He concluded that Eq. (7.7) is valid only for two specific cases, namely (1) slow pumping and low-viscosity fluid $a>4R$ and $S_H=2S_h$, and (2) fast pumping and high viscosity fluid $a<R$. Otherwise rock strength has to be obtained from laboratory tests.

Laboratory minifrac tests predict that the hydraulic fracture tensile strength depends on the inverse of the square root of the borehole radius. This is evident from two fracture mechanics approaches, like the one of Abou-Sayed et al. (1978) in Eq. (6.2) and the one of Rummel (1987) in Eq. (7.5), by setting $S_h = S_H = 0$. While the former model incorporates the tensile strength decrease with an increase in the central hole diameter (Eq. (6.2)), $F(c,R)$ at zero confinement), the latter approach predicts the variation of minifrac pressure with confinement. Laboratory results and in-situ data have been compared by Rummel and Winter (1983) for sandstones and granites. Such data show that the critical crack length for HF in the laboratory are in the order of the grain size of rocks, while the critical "in-situ" crack length for a hydraulic fracture to occur is one order of magnitude larger.

Exercise 7.1 The stress in a granitic rock mass (tensile strength 10 MPa, density 3000 kg m^{-3}) has been measured with the hydraulic fracturing method. Two tests were conducted in a vertical borehole at depths of 1 km (SIP = 15 MPa, FBP = 25 MPa) and 2 km (SIP = 30 MPa, FBP = 35 MPa)

(a) Calculate the principal stresses S_h, S_H, S_V at the two depth levels using hydrostatic pore pressure conditions.
(b) Draw a stress magnitude-depth diagram and determine the style of Anderson faulting according to the relation of principal stresses found.
(c) A geothermal reservoir is expected to operate at 3 km depth in the granitic rock mass. What effective stress conditions do you compute by extrapolating hydraulic stress data to the reservoir depth?

The orientation of the induced hydraulic fracture (Fig. 7.2, S_H direction) can be determined by (1) conventional impression-packer fingerprinting the fracture trace, (2) borehole televiewer (BHTV) mapping the acoustic properties of borehole wall rocks, (3) formation microimager (FMI) mapping the electrical resistivity of rocks in the borehole wall (Zemanek et al. 1970) and (4) Mosnier's azimuthal laterolog (Cornet 1993). The impression packer starts the downhole measurement oriented absolutely with respect to geographic North. The trace of the hydraulic fracture is mapped by inflating the packers to the fractured borehole wall. The fracture trace is imaged onto the packer recording (Fig. 7.5). Electrical images from FMI are well suited for investigation of fine structures like fractures and bedding planes. Acoustical images from BHTV are best suited for detection of borehole breakouts (Sect. 7.3). Mosnier's wireline method represents a hybrid between the laterolog technique and micro-electric imaging. The laterolog principle is based on measurements of azimuthal distribution of formation resistivities around the borehole. The different geologic well-logging tools are described in detail by Luthi (2001).

Images of vertical hydraulic fractures obtained by impression packers, BHTV or FMS are rarely found in the form of two straight lines 180° apart (Fig. 7.5a). More common are discontinuous sets of lines (Fig. 7.5a, en-echelon-type cracks) tilted in small angles with respect to vertical fracture traces. Lee and Haimson (1989) introduced circular statistics to delineate coaxial hydraulic fracture traces. For an inclined fracture intersecting the borehole (Fig. 7.5b), they applied a sinusoidal regression method to the photographs yielded from any fracture detection system. The dip direction of the fracture (Fig. 7.5, α = azimuth) can be determined from the

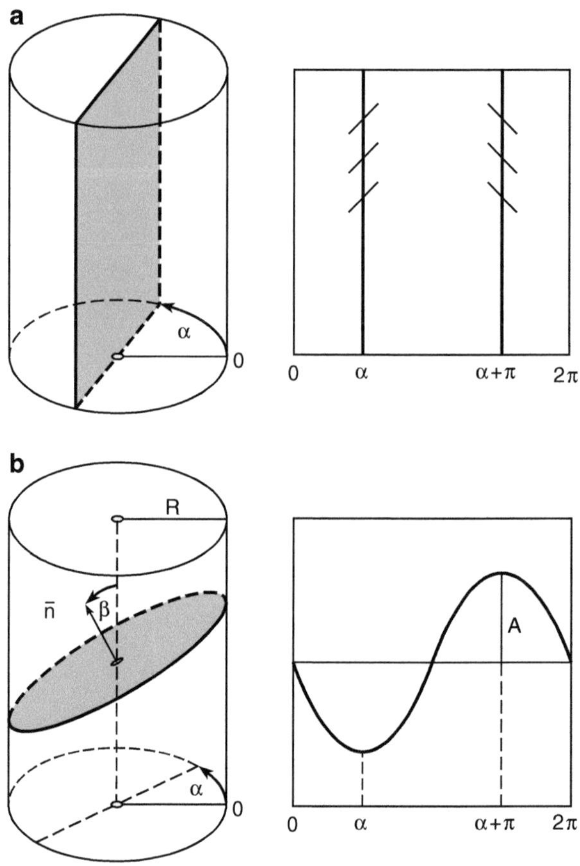

Fig. 7.5 Borehole geometry for imaging **a** vertical and **b** inclined hydraulic fracture planes

minimum of the sinusoidal curve, while the dip angle (Fig. 7.5, β=dip) can be computed from the amplitude of the sinusoidal curve and the borehole radius.

Historically, three early hydraulic fracturing field tests were reported by Von Schonfeldt and Fairhurst (1970) in Minnesota, Haimson (1973) in the Rangely oil field, Colorado, and by Rummel and Jung (1975) in Germany. Starting from the 1980s, hydraulic fracturing stress data were collected world wide (e.g., Pine et al. (1983) UK, Cornet (1983) France, Tsukahara (1983) Japan, Li et al. (1983) China, and Enever and Wooltorton (1983) Australia). An early field-test system realized was the Bochum Wireline Hydrofrac System (Rummel and Baumgärtner 1985, Rummel 2005) allowing both in-situ stress and in-situ permeability measurements (Fig. 7.6a schematic, 7.6b nature). A second early field system was the Swedish multihose hydrofrac field system (Bjarnason et al. 1989). One advantage of hydraulic fracturing is that the deformation behaviour of the rock (constitutive equations) is not necessary to calculate stresses. A second advantage of this technique is that we are able to obtain more or less "continuous" stress versus depth diagrams within the Earth's crust provided deep boreholes exist (e.g. Rummel 1979).

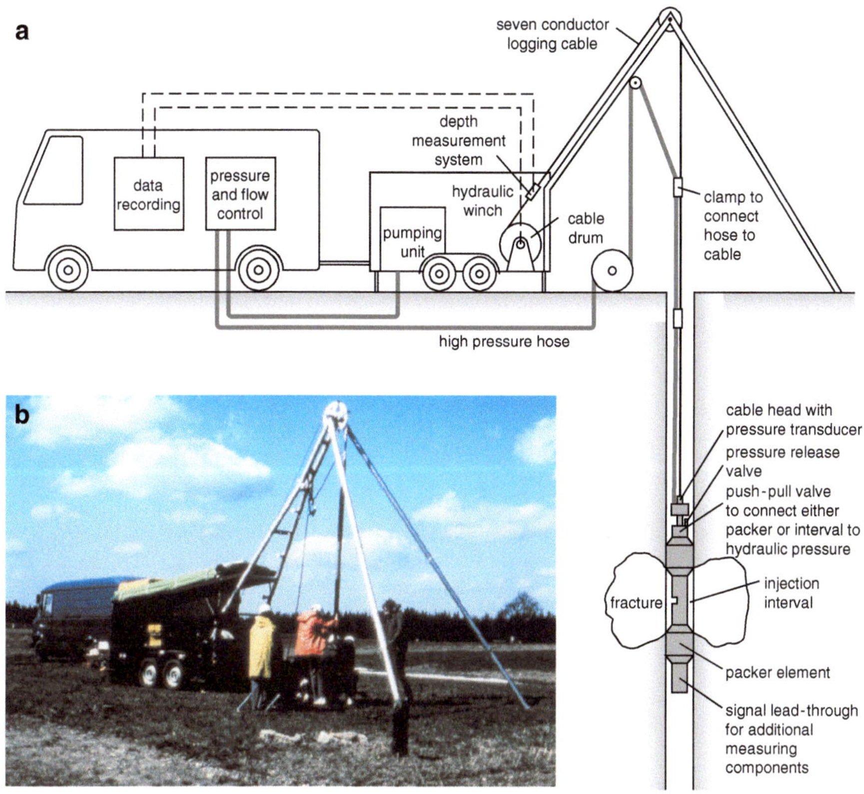

Fig. 7.6 The Bochum Wireline Hydraulic Fracturing system for in-situ stress and permeability measurements **a** schematically (after Rummel and Baumgärtner 1985), and **b** in the field (courtesy of Fritz Rummel)

An historical pressure-time-chart (dated December 1994) of one of the world deepest (9066 m) hydraulic fracturing test at the KTB drilling site, is shown in Fig. 7.7. The upper figure displays the flow chart of the virgin fracturing cycle and the first two refrac cycles; the lower viewgraph indicates the surface pressure in the borehole versus time. For reference, the pressure terminology introduced in Fig. 7.4 is used to mark breakdown pressure (FBP), shut-in pressure (SIP), and refrac pressure (RFP). According to Fig. 7.7b, the field tensile strength of the KTB rock is FBP–RFP = 8 MPa applying the Bredehoeft approximation from Eq. (7.7). Brazilian laboratory test data on crystalline drill cores from the KTB pilot hole lead to average tensile strength values of 19 MPa for amphibolites and 23 MPa for metagabbros, while gneisses showed a strength anisotropy between 12 and 16 MPa (Zang and Berckhemer 1993). Hydraulic fracturing stress data are discussed in the context in Sect. 9.1, where the integrated KTB stress strategy is presented.

Haimson and Cornet (2003) discuss assumptions, limitations and technical aspects for both hydraulic fracturing and hydraulic testing on pre-existing fractures. The S_H value in Eq. (7.6) is full of uncertainty. An important aspect is the

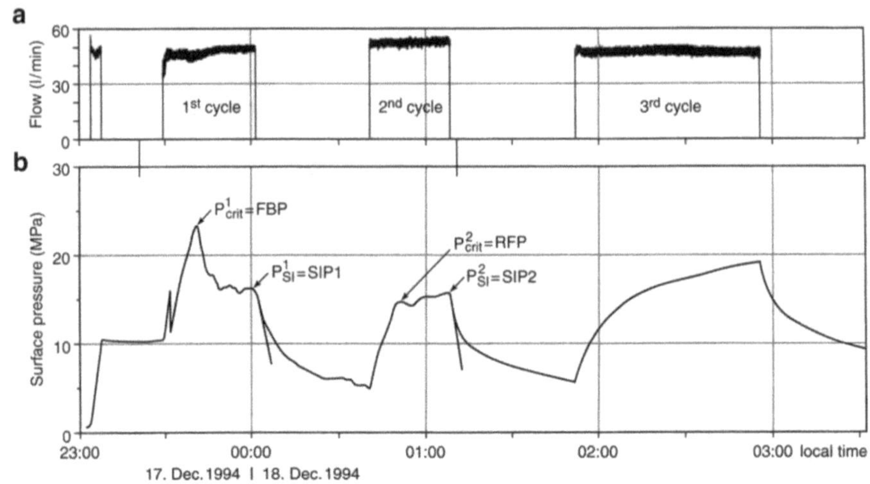

Fig. 7.7 Virgin fracturing (first cycle) and refracturing (second, third cycle) of the KTB main hole at a depth of 9066 m recorded at midnight December 17–18, 1994. The flow rate is shown in **a** and the subsurface pressure in the well versus time is displayed in **b** (after te Kamp et al. 1995a, b)

combination of HTPF with real hydraulic fracturing, for solving the maximum horizontal stress magnitude, without requiring the use of the fracture breakdown pressure. A recent review of stress measuring techniques by Sano et al. (2005) classifies pressure-time-curves according to the three types published by Hickman and Zoback (1983) and raises questions about hydraulic techniques and theory have which not yet been conclusively answered. In their opinion, the permeation process of fluid into the fracture plays a key role. This process is influenced by the residual aperture and the length of rupture (Ito et al. 1999; Rutqvist et al. 2000). Both sets of authors suggest that the conventional refrac pressure RFP can only be detected when the water permeates a fracture several times larger than the borehole radius. Second, they propose using a high-stiffness hydraulic fracturing measurement system. The crucial effect of the system compliance on S_H estimation from HF data has recently been evaluated by Ito et al. (2006) in theory and field tests. For the case of inclined wells (Zoback et al. 2003) and boreholes drilled through anisotropic rocks (Sect. 4.4), Eq. (7.1) needs to be modified. A special HF borehole tool was developed by Manthei et al. (1998) using two rings of acoustic emission sensors to map the fracture during the hydraulic test in-situ. The hydraulic pressurization unit in the middle of the borehole tool (length 2 m) is bordered by two rings of four acoustic emission transducers at both ends. The AE-HF tool is applicable in boreholes with diameters between 98 and 104 mm and has been applied in German salt mines (Manthei et al. 2003). The latest ongoing developments in HF and HTPF techniques including the new *quadruple packer* system of *Vattenfall* designed to fracture high stress formations by ~70 MPa fluid overpressure (Ask et al. 2009) can be viewed on the lecturing DVD, movies Chap. 7.

M7.1 QPT

Note-Box Hydraulic fracturing (HF) is an important borehole technique for stress estimates in the Earth's crust. A sealed-off section of the borehole is pressurized until a fracture develops in the borehole wall. From the pressure at which the induced hydraulic fracture closes in the pressure-time record the magnitude of the minimum horizontal in-situ stress is inferred. Fracture-breakdown pressure and the rock tensile strength are needed to obtain the value of maximum horizontal stress. The orientation of the induced hydraulic fracture is inferred from oriented impression packers or borehole logging tools, and provides the azimuth of the maximum in-situ compressive stress. In very porous sedimentary rock and highly anisotropic or fractured rock, hydraulic fracturing stress data are of limited value.

7.3 Borehole Breakouts

In 1970, a Schlumberger field engineer, working in the Alberta basin (Canada) noticed that four-arm dipmeter caliper logs measured borehole cross-sections that were systematically elongated parallel to the trend of the foothills (Cox 1970). Initially it was thought that the breakout indicated the direction of natural fractures (Babcock 1978). Later Bell and Gough (1979) argued that the elongation was due to wellbore stress concentration.

The effect that the stress in the vicinity of an inclusion or a cavity rises above the far-field stress is defined as *stress concentration* (Chap. 4). The German engineer *Ernst Gustav Kirsch* (1841–1901) was the first to quantify stresses around a circular hole in an infinite elastic plate under uniaxial tension (Kirsch 1898). Applied to the case of biaxial horizontal stresses in the Earth's crust, the Kirsch solution predicts rock failure perpendicular to the direction of the maximum horizontal compressive stress (Bell and Gough 1979). The zones of broken and fall-out rock material, diametrically faced in the borehole wall, are called *borehole breakouts* (BBO). Borehole imaging tools are able to map breakouts and identify the azimuth of S_h in a well or field as a function of depth.

Leeman (1964) used borehole breakouts for stress determination. Bell and Gough (1979) explained borehole elongations through shear fracturing in zones of amplified stress differences in the wall. Plumb and Hickman (1985) emphasized the need to differentiate between actual breakouts and other forms of well bore enlargements such as washouts and key seats. Plumb and Hickman define a *washout* as complete failure of the borehole wall indicated by a 360° enlargement of the drill bit diameter (Fig. 7.3e, f). They define a *key seat* as asymmetrical notching of the well caused by mechanical wear of the borehole at top and/or bottom. While it is accepted that excessive stresses near the borehole wall are the reason for breakouts, Zoback et al. (1985) favour shear failure (mode II cracks) in the breakout zone, while Zheng et al. (1989) suggest extension mode failure (mode I cracks) in the breakout zone

(Guenot 1989). Rutqvist et al. (1990) used a damage model to simulate breakout initiation and propagation into the rock mass parallel to the S_h direction. By drilling under stress into cubes of Lac du Bonnet granite, a laboratory loading path closer to the actual field conditions was simulated (Lee and Haimson 1993; Haimson and Lee 1995). They found tensile cracking to be the basic mechanism of breakout initiation, followed by progressive detachment of rock flakes bounded by cracks leading to the final 'V'-shaped cross-section of the breakout zones. Since the depth and angular width of breakouts are found to increase linearly with S_H magnitude for fixed S_h and S_v values, the geometry of breakouts also stores information on in-situ stress magnitudes. The extrapolation from laboratory to field conditions, however, is difficult due to an apparent increase in rock strength around smaller diameter boreholes. The strength-scale dependence seems to disappear for boreholes with diameters larger than about 75 mm (Martin et al. 1994). Physical aspects of breakout micro-failure mechanisms were recently reviewed by Haimson (2007).

Borehole breakout analysis has become routine work in deep continental drilling programs such as Cajon Pass, USA or KTB, Germany (Baumgärtner et al. 1993). Today, breakout analysis is often interpreted together with hydraulic fracturing stress results from the same drill site (Galera 2006), or breakout analysis is used as part of integrated stress estimation strategies as in the case of KTB (Sect. 9.1). Haimson (1997) evaluated breakouts and core disks as tools for estimating in-situ stress in deep holes from the laboratory perspective. Shen et al. (2002) used a 2D boundary element method code with a displacement discontinuity method to simulate breakouts by simultaneous propagation of mode I, mode II, and mixed mode cracks. Cook et al. (2004) applied the discrete-element method to laboratory simulation of near-borehole-wall mechanisms. A review on borehole breakouts and in-situ stress from the viewpoint of reservoir applications was given by Prensky (1992). Acock et al. (2004) discussed borehole breakouts in the context of hydrocarbon exploitation problems.

7.3.1 *Theory of Breakouts*

Consider a vertical hole in homogeneous, isotropic and linear elastic rock mass subjected to effective stresses S_h and S_H acting at infinity (Fig. 7.8, open arrows indicating far-field stress). According to *Kirsch's solution for the biaxial stress case*, e.g. Scheidegger (1962) found for the stress components in polar coordinates in the plane perpendicular to the hole

$$\begin{aligned}\sigma_{rr} &= \left(1-\rho^2\right)\frac{S_H+S_h}{2}+\left(1-4\rho^2+3\rho^4\right)\frac{S_H-S_h}{2}\cos 2\varphi+\Delta p\rho^2\\ \sigma_{\varphi\varphi} &= \left(1+\rho^2\right)\frac{S_H+S_h}{2}-\left(1+3\rho^4\right)\frac{S_H-S_h}{2}\cos 2\varphi-\Delta p\rho^2\\ \sigma_{r\varphi} &= -\left(1+2\rho^2-3\rho^4\right)\frac{S_H-S_h}{2}\sin 2\varphi\end{aligned}\quad, \qquad (7.8)$$

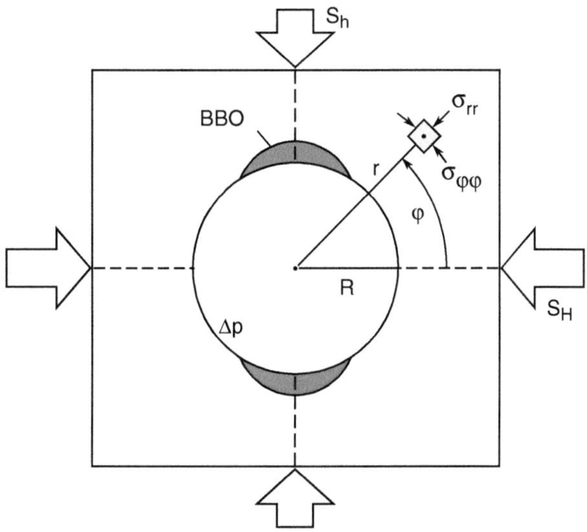

Fig. 7.8 Borehole breakout (*BBO*) of a circular hole in an infinite elastic rock mass under biaxial far-field stresses, S_h and S_H

where R is the borehole radius, r is the distance from the centre of the hole ($\rho = R/r$), φ is the angle measured from the direction of S_H and Δp is the difference between the fluid pressure in the borehole and that in the rock formation (Fig. 7.8). We corrected the stress component $\sigma_{r\varphi}$ in the Scheidegger (1962) solution for the minus sign. The reason for the sign error in the biaxial Kirsch solution in the literature is addressed in Fjaer et al. (2008, p. 145).

Note that perturbations in the stress field caused by the presence of the hole drop off at least as fast as ρ^2 and are negligible at distances greater than $10R$ from the hole (Jaeger et al. 2007). From Sect. 4.4.2 findings, we estimate this distance as three inclusion (borehole) diameters which is $6R$.

At the wall of the borehole, $\rho = R/r = 1$ Eq. (7.8) reduces to

$$\begin{aligned} \sigma_{rr} &= \Delta p \\ \sigma_{\varphi\varphi} &= S_H + S_h - 2\left(S_H - S_h\right)\cos 2\varphi - \Delta p. \\ \sigma_{r\varphi} &= 0 \end{aligned} \tag{7.9}$$

The tangential stress, $\sigma_{\varphi\varphi}$ relevant for the breakout hypothesis follows (Fig. 7.8)

$$\begin{aligned} \sigma_{\varphi\varphi}(r = R, \varphi = 0) &= 3S_h - S_H - \Delta p \equiv \sigma_{\varphi\varphi}^{\min} \\ \sigma_{\varphi\varphi}(r = R, \varphi = \pi/2) &= 3S_H - S_h - \Delta p \equiv \sigma_{\varphi\varphi}^{\max}. \\ \sigma_{\varphi\varphi}^{\max} - \sigma_{\varphi\varphi}^{\min} &= 4(S_H - S_h) \end{aligned} \tag{7.10}$$

In Fig. 7.9, the variation of tangential (=circumferential) stress at the borehole wall from Eq. (7.9) is shown versus azimuth. The minimum value of tangential stress occurs at $\varphi = 0$ (Fig. 7.9, $\sigma_{\varphi\varphi}^{\min}$) and is aligned parallel to the S_H direction

Fig. 7.9 Circumferential stress versus azimuth at the rim of the circular hole representing the wall of the borehole

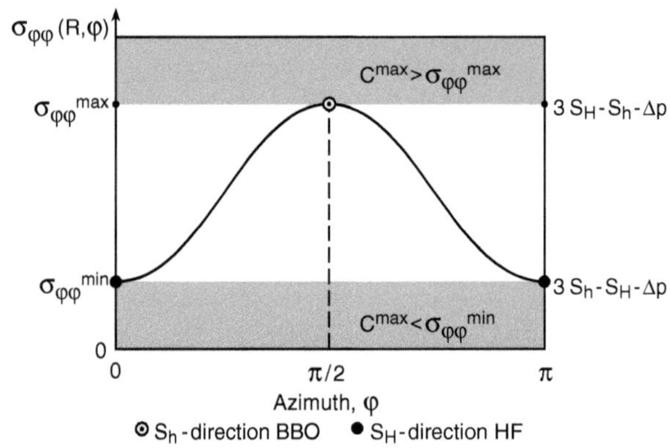

(Fig. 7.8). As we know from Sect. 7.2, this is the simplest criterion for hydraulic fracturing. The borehole wall fails in tension if $p_{crit} - T_0 > 3S_h - S_H$ (Eq. (7.3)), where T_0 is the tensile strength of rock (Fig. 7.9, HF with S_H direction). The maximum value of tangential stress occurs at $\varphi = 90°$ (Fig. 7.9, $\sigma_{\varphi\varphi}^{\max}$) and is aligned parallel to the S_h direction (Fig. 7.8). This is the simplest criterion for the initiation of borehole breakouts. The borehole wall fails if $p_{crit} + C_0 < 3S_H - S_h$, where C_0 is the uniaxial compressive strength of rock (Fig. 7.9, BBO with S_h direction). The onset of spalling (falling out rock material), however, may occur at about half the value of the measured mean peak laboratory uniaxial compressive strength (Martin and Christiansson 2009). We conclude that, in the failure criterion of borehole breakouts, the compressive strength of rock is required in the plane biaxial Kirsch solution, while in the failure criterion of hydraulic fracturing, the tensile strength of rock is required.

Based on the Coulomb fracture criterion (Chap. 3), Zoback et al. (1985) calculated the maximum value of cohesive strength, $C^{\max}$ at which material will fail in terms of radial stress, σ_{rr} circumferential stress, $\sigma_{\varphi\varphi}$ and tangential shear stress, $\sigma_{r\varphi}$ around the borehole wall

$$C^{\max} = \sqrt{1+\mu^2}\sqrt{\left(\frac{\sigma_{\varphi\varphi} - \sigma_{rr}}{2}\right)^2 + \sigma_{r\varphi}^2} - \mu\left(\frac{\sigma_{\varphi\varphi} - \sigma_{rr}}{2}\right), \qquad (7.11)$$

where μ is the coefficient of friction. According to Fig. 7.9, they distinguish three cases. If $C^{\max} > \sigma_{\varphi\varphi}^{\max}$ (Fig. 7.9, upper band) no breakout is observed in the hole. If $C^{\max} < \sigma_{\varphi\varphi}^{\min}$ (Fig. 7.9, lower band), breakouts become as large as to extend nearly around the whole borehole (Fig. 7.3e, f). For the case $\sigma_{\varphi\varphi}^{\min} < C^{\max} < \sigma_{\varphi\varphi}^{\max}$, the size of the region in which rock shear strength is actually mobilized (breakout zone) can be calculated. Assuming $S_H \leq 3S_h$ which is almost always the case in the Earth's crust, Zoback et al. (1985) substituted Eq. (7.8) into Eq. (7.11) and computed the half width of breakouts (Fig. 7.10a, Θ_{BBO}) for given sets of stress ratios (S_H/S_h) and μ values. Figure 7.10b shows the stress ratio (S_H/S_h), which is independent of $C^{\max}$ as

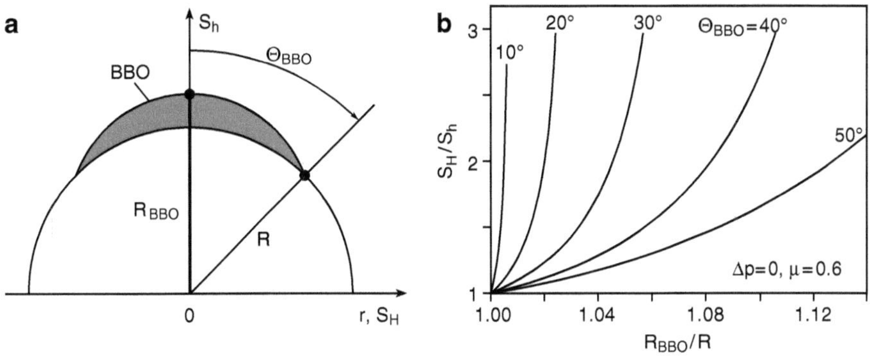

Fig. 7.10 Relationship between horizontal principal stresses, maximum depth, R_{BBO} and half width, Θ_{BBO} of borehole breakouts. **a** Geometrical shape of breakout with respect to principal stresses. **b** Horizontal stress ratios versus breakout radius normalized to borehole radius (modified after Zoback et al. 1985)

a function of breakout depth versus the breakout radius normalized to the borehole radius (R_{BBO}/R). The breakout width (in degrees) is indicated as the curve parameter for a scenario with friction coefficient $\mu=0.6$ and $\Delta p=0$. Little spalling occurs when $S_H \approx S_h$. Breakouts become deeper and wider as the ratio (S_H/S_h) increases but, even for large stress ratios $S_H/S_h \approx 3$, the borehole radius increases by only 10% when Θ_{BBO} is as large as 40°.

Shen (2008) estimated the magnitude of in-situ horizontal stresses from borehole breakout data in a deep geothermal well in Australia. He used the FRACOD numerical tool together with the F criterion (Chap. 3) to simulate breakouts in Habanero No.1 granites. From the width of the breakout, Θ_{BBO} and the depth of the breakout, R_{BBO} he computed both horizontal compressive stress magnitudes at depth of the geothermal reservoir

$$S_H = \frac{1 + (1 - 2\cos\Theta_{BBO})\,\mathrm{X}}{4\,(1 - \cos\Theta_{BBO})} C_0$$
$$S_h = 3S_H - \mathrm{X}C_0; \mathrm{X} = 1 + A\left(\frac{R_{BBO}}{R}\right)^{B},$$ (7.12)

where C_0 is the uniaxial compressive strength of the rock in the borehole wall and A, B are regression parameters fitted to numerical curves quantifying the normalized breakout depth (R_{BBO}/R) versus the stress-strength-ratio of rocks.

Like in the case of tensile failure of the borehole (Fig. 7.3), also in the case of shear failure of the borehole, different geometries of fracture planes in the host rock can be distinguished (Fig. 7.11). Stress magnitudes (radial stress σ_{rr}, tangential stress $\sigma_{\varphi\varphi}$ and axial stress σ_{zz}) along the borehole wall can be arranged in six different ways (Maury and Sauzay 1987; McLean and Addis 1990; Bowes and Procter 1997) from which three cases are only of academic interest (Fjaer et al. 2008). The three shear failure modes in boreholes of practical interest are shown schematically from the inner hole perspective (Fig. 7.11, left) and outer hole perspective (Fig. 7.11, right). Assuming the elastic model to be valid, the orientation of the failure plane (tensile for hydrofracs, shear for breakouts) is given by the orientation of the least and greatest principal stress and the angle of the failure plane which is related to the coefficient of internal friction (Chap. 3). Shear failures occurring parallel to the axial stress $\sigma_{zz}=S_V$, are responsible for the classical shape of a *wide breakout* (Fig. 7.11a, b). The horizontal failure plane is spanned out by the maximum ($\sigma_{\varphi\varphi}$) and minimum (σ_{rr}) horizontal stress. Shear failure occurs in fragments of toroidal shape producing a *shallow knockout* (Fig. 7.11c, d) because the maximum (σ_{zz}) and minimum (σ_{rr}) stress define a vertical plane. Multi-shear fractures occur in the last case, intersecting parallel to the radius producing *high angle en-echelon cracks* in

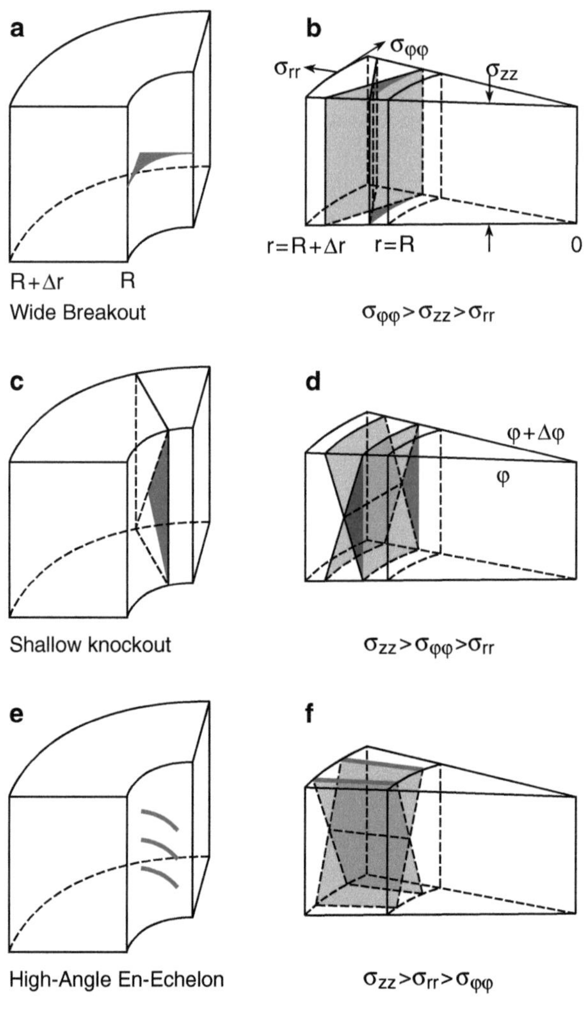

Fig. 7.11 Shear failure process in the borehole wall from inner hole perspective (*left column*) and outer hole perspective (*right column*). Depending on the relationship between the radial, tangential and axial stress we can distinguish three practical cases: **a, b** wide breakouts, **c, d** shallow knockouts and **e, f** high-angle en-echelon cracks (**a, c, e**) modified after Pasic et al. 2007, (**b, d, f**) modified after Fjaer et al. 2008

the borehole wall (Fig. 7.11e, f). The failure in this case occurs in an axial/tangential arc which is defined by the maximum (σ_{zz}) and minimum ($\sigma_{\varphi\varphi}$) stress.

The shear failure of the borehole also involves tensile cracks oriented parallel to the major principal stress (oriented with their normal parallel to the minor principal stress). In the situation corresponding to the case of wide breakouts (Fig. 7.11a, b), these cracks will be oriented parallel to the borehole wall. The initiation and growth of these tensile cracks (mode I) is considered by some authors as the fundamental micro-mechanism for borehole breakouts (Zheng et al. 1989; Ewy and Cook 1990a, b). Rock anisotropy also affects the orientation of borehole breakouts. For isotropic horizontal stresses, the breakouts are oriented normal to the joints of bedding in the rock material (Kaiser et al. 1985). For anisotropic stresses in anisotropic rock we might be lucky to detect a rotation of a stress indicator (e.g., velocity azimuth of drill cores under pressure) in order to separate rock anisotropy from stress anisotropy (Zang et al. 1996a). Otherwise, it is difficult to "extract" anisotropic stresses from anisotropic rock cores.

The specific shape of borehole breakouts beyond failure initiation (Fig. 7.11) is influenced by three parameters: the pore fluid pressure in the well, the magnitudes of principal stresses and the rock strength. Excess pore pressure in the well (Eqs. (7.8)–(7.10), $\Delta p > 0$) leads to smaller size breakouts. A decrease in pore fluid pressure in the well (Eqs. (7.8)–(7.10), $\Delta p < 0$) promotes the development of breakouts. Using thixotrope, dense drilling mud, a borehole can be stabilized when drilled through low-cohesion rocks like shale and siltstone. Anisotropy in far-field stresses around the borehole reduces the region where the borehole is stable. The influence of rock strength and porosity on the final shape of borehole breakouts is shown schematically in Fig. 7.12 for quartz-rich sandstone (Haimson 2007). In high porosity ($\Phi > 20\%$) sandstones where fluids are circulating during drilling we find tabular, slot-shaped breakouts, so-called "*wormholes*" (Fjaer et al. 2008) aligned with the S_h azimuth (Fig. 7.12a). For medium porosity sandstones ($5\% < \Phi \leq 20\%$) we observe "*dog-ear*"-shaped breakouts (Fig. 7.12b), and for low porosity sedimentary and hard crystalline rocks, the micro-mechanism of "*flake-spalling*" breakouts

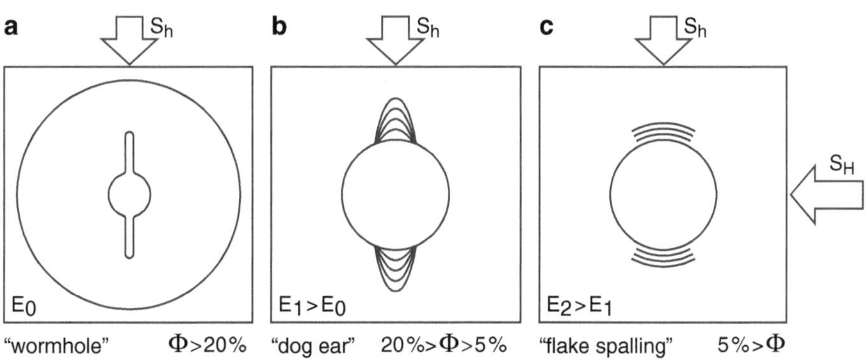

Fig. 7.12 Schematic final shape of borehole breakouts in **a** sand production with wormholes, **b** dog-ear breakouts in sandstone, and **c** flake-spalling process in dolerite or granite

is favoured in the literature and small-size, relatively shallow depth breakouts are observed (Fig. 7.12c).

Zoback et al. (1985) did not consider inelastic deformations occurring as rock around the borehole fails and the corresponding stress redistribution. The boundary element modelling result from Shen et al. (2002) indicate that, in both breakout mechanisms (tensile or shear), shear failures are involved in the final formation of the breakout. Tensile failure alone does not form breakouts. The models predict that the shape (depth) and extent (width) of breakout depend on the stress magnitudes and the fluid pressure in the well. The discrete element results from a joint experimental-numerical study by Cook et al. (2004), show qualitative reproduction of the gross failure mechanisms associated with both hydraulic fractures and borehole breakouts. Besides rock strength, porosity and fluid pressure, the stress regime (Sect. 5.3) also affects the occurrence of breakouts.

Borehole breakouts are reliable indicators for the S_H azimuth at depth which is also confirmed by laboratory experiments (Haimson 2007). For high porosity sandstones like Aztec ($\Phi=26\%$), Mansfield ($\Phi=26\%$) and Berea ($\Phi=25\%$), Haimson (2007) can reproduce wormholes (Fig. 7.12a) in vertical borehole drilling experiments of pre-stressed rocks. He claims the formation of wormholes to be associated with zones of localized grain debonding and crushing, so-called *compaction bands* (Haimson and Kovacich 2003). However, the importance of this mechanism in the field is not clear (Fjaer et al. 2008). All other rocks tested by Haimson (2007) in the laboratory (limestone, granite) develop dog-eared breakouts (Fig. 7.12b) along the minimum horizontal far-field stress springline. Haimson (2007) points out that even if the incipient failure of a breakout is the dilatant microcracking in the zones of highest compressive stress concentration around the borehole, the grain-scale failure mechanism leading to the final shape of the breakouts is rock-type sensitive and can be tensile (mode I) or shear (mode II) crack dominated. More important, Haimson (1997) found that the Coulomb criterion, which is based on the premise that the intermediate principal stress has no effect on compressive strength (Chap. 3), is inappropriate to explain triaxial laboratory breakout data. On the other hand, plotting the same data in the form of octahedral shear stress versus mean stress (so-called *Mogi polyaxial strength criterion*) on striking planes in the σ_2 direction, suggests that Mogi's criterion is the appropriate borehole failure criterion to be used in relating breakout dimensions to far-field stress, at least for the Westerly granites tested.

As in the case of hydraulic fracturing theory (Eq. (7.4)), there are also sources of poroelastic time effects at the borehole wall. We do not intend to present the modified poroelastic biaxial Kirsch solution (e.g., Fjaer et al. 2008), but name at least two main physical reasons for pore-pressure changes that have an influence on BBO: (1) the invasion of borehole fluid and (2) the redistribution of formation stresses. The first mechanism requires a non-sealing wall of the borehole, the second will occur, even if the wall is completely sealed. Solving for the decoupled diffusion equation (Detournay and Cheng 1993; Charlez 1997) and considering production in the well, the tangential stress at the borehole wall is reduced relative to the elastic solution (Eq. (7.10)) for the situation in case (1). For short times, however, an increase in tangential stress in the near-borehole region is observed. This region of high stress propagates outwards with time. In the situation of case (2) when the far-field horizontal stresses are unequal, the drill-out leads to a change in pore pressure, which will dissipate with time to fluid flow. Since the diffusion distance is shortest close to the hole, pore pressure will reach equilibrium fastest here. As a result, the near-well bore behaviour will be drained, whereas the response deeper into the formation will be undrained for short times. Since drained moduli are smaller than undrained (Sect. 5.4), the stress concentration in the near-well bore region will initially be shielded relative to its full elastic value. As the pore pressure imbalance dissipates, the shielding (from borehole failure) disappears and the elastic stress concentration is re-established. This can lead to delayed failure for boreholes in tight formations. For more details and in-depth treatment of poroelasticity with respect to borehole breakouts, we recommend Detournay and Cheng (1988).

In inclined wells (Peng and Zhang 2007; Fjaer et al. 2008), maximum tangential stress values need to be modified to lower values still showing the sinusoidal variation of stress with azimuth (Fig. 7.9). In the end-member case of horizontal wells, tangential stress is constant along the borehole wall (Zoback et al. 2003).

7.3.2 Application of Breakout Analysis

Borehole breakouts in a well can be visualized using optical (camera), mechanical (caliper), acoustic (televiewer) or electrical resistivity (*formation microscanner*, FMS) tools. Due to water-filled wells, optical tools are of limited value. Mechanical caliper logs provide the least-reliable data since they do not provide the detailed shape of the hole but measure the diameter at 3, 4 or 8 points depending on the number of caliper arms used. Borehole televiewers (BHTV) provide continuous, oriented, ultrasonic images of the borehole wall (Zemanek et al. 1970). A narrow pulsed acoustic beam scans the borehole wall in a tight helix as the tool move up the borehole. Advanced tools like FMS (Schlumberger) produce high-resolution resistivity images of the borehole wall that can be used for imaging hydraulic fractures, drilling induced tensile fractures and borehole breakout orientations. For further details, we refer to the two guidelines of the World Stress Map enclosed on the DVD (WSM): one about breakout analysis from four-arm calipers (Reinecker et al. 2008b), and the other about breakout and drilling-induced fracture analysis from image logs (Tingay et al. 2008).

We choose two examples illustrating stress analysis with borehole breakouts. The first one is a classical analysis taken from Zoback et al. (1985) who observed borehole breakouts in a well drilled through granitic rock at Monticello, South Carolina (USA). Figure 7.13a shows a typical reflectivity borehole televiewer record of the section at 795–800 m depth in the well Monticello 2. The vertical dark bands centred about 180° apart correspond to low-amplitude reflections coming from breakout. Figure 7.13b shows a photograph of the oscilloscope record showing how travel time as a function of azimuth is determined for a single rotation of the acoustic transducer at a depth of 797 m in the well. The sinusoidal variation in

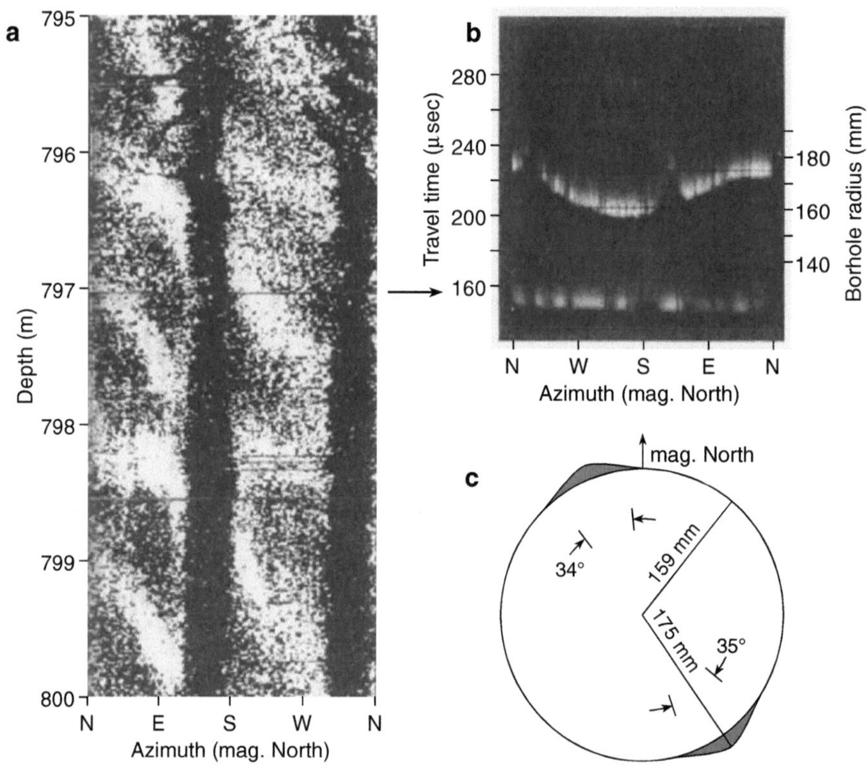

Fig. 7.13 **a** Borehole televiewer record from the well Monticello 2, South Carolina at a depth ranging from 795 to 800 m. *Vertical dark bands* centred π apart correspond to low amplitude reflections coming from BBO. **b** Photograph of oscilloscope record showing travel time versus azimuth of the acoustic transducer rotating at a depth of 797 m. The borehole radius is shown at the right ordinate. **c** Borehole shape corresponding to data shown in **b**. BBO are 35° wide and enlarge the radius of the hole by 16 mm (10%) (after Zoback et al. 1985)

travel time versus azimuth corresponds to the televiewer not perfectly centred in the hole. The two sharp travel time delays west of North (N) and east of South (S) are associated with breakouts. The borehole radius is displayed on the right axis of the diagram shown in Fig. 7.13b. From this radius, the complete shape of the borehole is reconstructed (Fig. 7.13c). The breakout is about $2\Theta_{BBO}=35°$ wide and enlarges the radius of the hole by 16 mm ($r_{BBO}/R=175/159=1.10$). In the example, the S_h azimuth lies in the direction of NNW (N145°E).

A second example is taken from Mastin et al. (1991) who visualized borehole breakouts in the KTB pilot hole at Windischeschenbach, Germany. Beside the 2D televiewer logs of the well in the depth range of 1723–1727 m (Fig. 7.14a), a joint co-operation between Karlsruhe University and Stanford University reveals a pseudo-3D wire-frame diagram of the borehole wall (Fig. 7.14b) based on travel time data of acoustic televiewer data. Reconstructed cross-sections of the borehole indicate an S_h azimuth at the KTB drill site lying in the NE direction at a depth of 1725 m (Fig. 7.14c, open arrows).

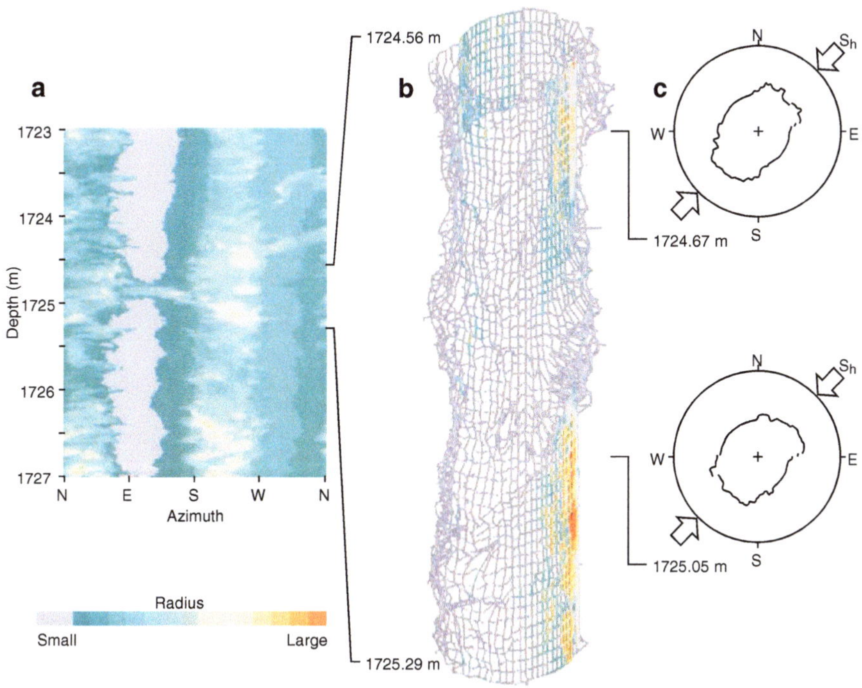

Fig. 7.14 Borehole televiewer data from KTB pilot hole. **a** Travel time log of the hole at a depth from 1723 to 1727 m. *Two vertical stripes in azure color* indicate borehole breakouts. **b** 3D wire-frame diagram shows borehole wall geometry. *Colors* indicate amplitude of reflected pulse (warmer color stand for higher amplitude). **c** Cross-sectional shape of the borehole at a depth of 1724.67 and 1725.05 m. Breakouts are elongated parallel NE-SW (after Mastin et al. 1991)

Note-Box Borehole breakouts are enlargements of the borehole wall caused by stress-induced failure of wells occurring 180° apart. In vertical wells, the diametrically faced zones of broken and fall-out rock material occur at the azimuth of minimum horizontal compressive stress and have a consistent orientation in a given well or field. Breakouts can have a length of between centimetres up to several hundred metres. The geometry of breakouts stores information on in-situ stress orientation and magnitudes. The method reveals stress information valid at very great depths (11.6 km). If no breakouts occur at a shallower depth, the method is inapplicable. Inventing techniques for determining anisotropic stresses in anisotropic rock mass is a future key challenge in rock mechanics.

Chapter 8
Measuring Crustal Stress: Core-Based Methods

In-situ stresses are important for many processes in tight gas reservoirs, ranging from siting the well, to drilling, completion and long-term production. The orientation of the stress field, which controls the azimuth of the hydraulic fracture (Sect. 7.2), is important for field development as optimum drainage in a tight reservoir will depend on the drainage from a hydraulic fracture. If wells are not sited properly, drainage patterns will overlap and production may be uneconomic. Stress magnitudes are important for wellbore stability and hydraulic fracturing. For fracturing, in particular, the stress difference between lithologies in a geological sequence is the major control on fracture-height growth. The focus of this chapter is shifted from borehole methods to core-based methods for in-situ stress estimates. As pointed out in Sect. 7.1, the crucial assumption in core-based stress estimates is that the dominating portion of the stress-relief cracks observed in deep drill cores is caused by the relief of present-day stresses. Only with this assumption, the azimuth of relief-crack populations can be related to in-situ stress azimuth, and the closing pressure of relief cracks to in-situ stress magnitudes.

Among the core-based methods distinguished in the previous chapter (Table 7.1, category (3) mechanism), were the strain based methods (ASR, DRA, DSA), the wave velocity analysis (DWVA, WVA), and the fracture pattern inspection methods (CD, DIF). When appropriately applied, these techniques can yield valuable data on the stress field at depth. We select the anelastic strain recovery method (Sect. 8.1) applied to *freshly recovered cores* and the Kaiser Effect methodology for recalling previous maximum stress from *relaxed cores* (Sect. 8.2). Subsequent sections provide information on the technical background, procedures, equipment, analyses, as well as a case study for the core-based techniques selected.

8.1 Anelastic Strain Recovery

Anelastic strain recovery (ASR) is a method which uses the anisotropic strain retardation of a "freshly" drilled, oriented core in order to obtain estimates about the azimuth and magnitude of the in-situ stress field (Teufel 1983). Retarded strains are

A. Zang, O. Stephansson, *Stress Field of the Earth's Crust*,
DOI 10.1007/978-1-4020-8444-7_8, © Springer Science+Business Media B.V. 2010

measured with inductive displacement transducers about 3–4 h after the core has been cut in-situ. The azimuth of maximum retarded strain correlates with the direction of maximum compressive in-situ stress in the borehole. Assuming homogeneous, linear viscoelastic rock, the ratio of principal in-situ stress magnitudes can be computed from measured retarded strains.

Voight (1968) suggested that there is empirical justification for considering the recovered anelastic strain to be proportional to the total recoverable strain (both elastic and anelastic) and, hence, to the pre-existing state of stress in rhelogically-isotropic core material. An approximate estimate of the in-situ stress state at depth can then be made by instrumenting oriented cores immediately upon removal from deep boreholes. If the rock material behaves viscoelastic, then the strain relief along principal axes will be uniform with time and the azimuths of principal strain relief over a given time interval will correspond to the initial in-situ strain conditions. Thus, principal strain azimuths from ASR will correspond to principal in-situ stress azimuths at depth.

Teufel (1983) realized Voight's (1968) ideas by presenting a reliable and practical device for determining the directions and magnitudes of principal horizontal in-situ stresses in deep formations. Teufel assumed the nature of the recovery process to be a consequence of crack formation. Note that the "direct" proof of crack genesis during the recovery process, was discovered at a later stage when, independently, Teufel (1989) and Wolter and Berckhemer (1989) attached a piezoceramic sensor to the surface of relaxing cores from deep wells. They found that a cracking noise and clicking noise, so-called acoustic emissions (AE), accompanied the strain recovery process. More specific, they demonstrated that the curves of cumulative AE counts versus recovery time are very similar to those of the measured anelastic recovery strain versus time. The integrated approach to relate anelastic strains, seismic velocities and crack closure pressures measured on the same core specimen was presented by Teufel (1989) and Zang et al. (1989).

Following the work by Simmons et al. (1975), Teufel (1983) stated that the nature of the ASR process for isotropic, homogeneous rock at depth prior to coring can be physically explained if the strain recovery is considered as being a product of crack genesis. Cracks are induced in the rock matrix as the core expands in response to the release of the in-situ stress. Cracks are aligned by the direction of principal stresses. The total crack volume due to strain recovery is proportional volumetrically to the corresponding in-situ stress magnitudes (Sect. 4.3). A fully relaxed core exhibits a distinct microcrack fabric at atmospheric pressure (Earth's surface) which can be quantified by elastic or (ultra-) sonic velocity anisotropy. We discussed the sensitivity of cracks to stress in Sect. 4.4.1.

The determination of in-situ stress magnitudes from ASR measurements is considerably more difficult than the determination of stress azimuths and requires a model of the relaxation process. The zero-order model was developed by Blanton (1983) for isotropic and transversely isotropic core material that allows the stresses to be calculated from the relative magnitudes of the principal strains at any time during the relaxation process. In his 2D isotropic model, Blanton postulated global creep compliance and assumed a constant Poisson ratio throughout the recovery process. To calculate S_H and S_h, the model requires S_V from overburden, the pore pressure, Poisson ratio, and a poroelastic constant. Various rock types were investigated using ASR data interpreted in terms of Blanton's 2D creep model: volcanic tuff (Teufel 1982), sandstone, siltstone and mudstone (Teufel 1983), granite (Engelder 1984), sandstone, siltstone and shale (Warpinski and Teufel 1986), chalk (El Rabaa 1989), gneiss and metabasite (Wolter and Berckhemer 1989), oceanic basalt (Brereton et al. 1990). In all studies, vertical stress is assumed to be one of the principal stresses.

In their refined, first-order creep model, Warpinski and Teufel (1986) used two independent creep compliances, one for dilatation and one for distortion. A linear three-parameter model was adopted for both compliances by using the viscoelastic correspondence principle. Warpinski and Teufel identified the relaxation process as an intergranular mechanism within the rock matrix, more specific as crack formation, grain boundary sliding and intergranular fracturing. A poroelastic constant in this model needs not be specified.

The second-order creep model of Matsuki (1991) extends the 2D approach to the 3D case by using six independent ASR normal strains measured. However, few practical applications of the 3D model are reported (Matsuki and Takeuchi 1993; Ito et al. 1997; Lin et al. 2006b, c). The Matsuki model states that the ASR depends on the in-situ stress tensor components, pore pressure, temperature change, thermal expansion and the two creep compliances introduced by Warpinski and Teufel (1986). Matsuki's 3D creep model overcomes the shortcoming of previous 2D models that the vertical stress is the principal stress and is calculated from density logs. In addition, Matsuki (1991) stated that in-situ stress determined by previous works using 2D creep models are inaccurate because the dependence of ASR compliances on applied stress was ignored. For Ogino tuff, Matsuki calibrated the creep compliances by freezing-in controlled applied stress magnitudes over 12 h in uniaxial, triaxial and hydrostatic laboratory tests with subsequent determination of recovery strains. Matsuki replaced two-element rheological models by four-element substances and compensated anelastic mean normal strains for changes in pore pressure and temperature. Brereton et al. (1995) pointed out that complex recovery curves (e.g. multi-directional core contraction in basaltic rock) can be explained by the effect of pore-pressure decay with recovery time. For determining correct principal stress magnitudes, any creep model used for fitting measured recovery strain curves would require a pore-pressure-decay curve for reference. However, other reasons for abnormal core contractions are indicated like thermal expansion and rock anisotropy in-situ.

Laboratory studies on the physical mechanism involved in ASR are rare (El Rabaa and Meadows 1986; Matsuki 1991; Wang et al. 1997a; Barr and Hunt 1999). While most of the authors focussed on studies with natural rock material, Wang et al. (1997) did a systematic study on strain recovery response of synthetic sandstones cemented in the laboratory with different applied stress scenarios. They created their synthetic samples from a mixture of sand, silica flour and sodium silicate solution, which underwent a one-step cementation process (Sect. 4.3) where silicates reacts with CO_2 gas to form the cement. Isotropic, anisotropic, single and multiple stress fields were frozen-in into the synthetic sandstones and the characteristics of the recovery strains were determined. Only for the scenario that a single stress field was applied to the synthetic sandstone, in-situ stresses could be obtained from measured recovery strains. The recipe for forming synthetic sandstones using CO_2 technique was actually developed by Holt and Kenter (1992).

Simple rheological models are used to understand recovery strain curves. Note that these models do not represent any explanation of the microphysical mechanism behind, but rather are a convenient tool to fit measured recovery strain data. While Blanton (1983) started off with a two-element model, Matsuki (1991) already used a four-element substance. Barr and Hunt (1999) applied more complex rheological models like the Generalized Kelvin substance or the Burgers model in order to explain strain-recovery curves and the Kaiser Effect retention spans (Sect. 8.2) in Carnmenellis Granite, UK. They correlate the cumulative number of measured AE with microstrains from laboratory and geologically pre-stressed cores. The loss of recalled maximum stress at low stress levels in terms of primary relaxation, is best fitted with the Burgers model. Longest relaxation times are obtained for cores with geological applied pre-stress, while shortest relaxation constants are obtained for laboratory pre-stressed samples. Also, Gunzburger et al. (2006) used a Burger substance and a Generalized Kelvin-Voigt substance to reproduce the immediate elastic response, as well as the primary and secondary creep of a hard-clay layer in the "Bure argillite" formation in the eastern part of the Paris sedimentary basin, France.

The most recent studies on ASR (Lin et al. 2006b; Lin. et al. 2007a) used Matsuki's (1991) creep model to determine the 3D in-situ stress in deep scientific drill holes. Drilling the Niitsu well in the Niigata Basin, Japan in a submarine seismogenic zone, Lin et al. (2006b) found that ASR is well-suited for determination of principal stress magnitudes and azimuths in 3D for the case of isotropic rocks. While mudstone (measured depth 2374 m) showed expansion recovery microstrains in the order of 10^{-3} μm m^{-1}, other rock types investigated (andesite (4542 m), basalt (3980 m) and dolerite (3175 m)) showed recovery strains of one order of magnitude less, 10^{-4} μm m^{-1}. These values fall well within the range of ASR strains published by Zang and Berckhemer (1993) for 32 drill cores of the KTB pilot hole taken from a depth of 1488–3858 m. Within the Taiwan Chelungpu-fault drilling project, Lin et al. (2007b) applied ASR to sand and siltstone cores. In this case, the core-based method helped to resolve the change in stress pattern beneath the Chelungpu fault after the 1999 Chi-Chi Mw=7.6 earthquake. ASR instrumented sandstone cores resolved a change in principal stress orientation above and below the fault zone inspected. Bloch et al. (2006) developed a fully coupled poroviscoelastic model for determining stress from ASR measurements in the oil industry.

Even if today ASR apparatus are highly sophisticated devices in terms of (1) measuring recovery deformations with resolutions smaller than 1 microstrain in six independent directions, (2) using dummy core specimens to minimize temperature as well as humidity effects and (3) measuring two test cores simultaneously, Lin et al. (2006b) state that the problem of obtaining in-situ stress data from ASR on anisotropic rock cores at depth is still unsolved. The majority of rocks and sedimentary rock in particular, however, cannot be regarded as fully isotropic material.

8.1.1 Rheological Basis

The study of time-dependent rock stress-strain behaviour is known as rheology, from the Greek *rhei*, meaning flow, and *logos*, meaning study. Rheological models are analogues of material behaviour, formed from assemblages of mechanical components like spring (*Hooke* substance), dashpot (*Newton* substance) and slider (*St. Venant* substance). The *Kelvin-Voigt* model consisting of a Hookean elastic spring (ideal elastic, $\sigma=E\varepsilon$) and a Newtonian damper (ideal viscous, $\sigma = \eta\dot{\varepsilon}$) connected in parallel (Fig. 8.1a), can be used to mathematically describe the stress relaxation of a freshly recovered drill core.

For the case of a Kelvin-Voigt model, also called *firmo-viscous* substance, the total stress is the sum of the stress carried by the spring and the stress carried by the dashpot:

$$\sigma = E\varepsilon + \eta\dot{\varepsilon}. \tag{8.1}$$

This differential equation can be solved using the *Laplace transform* formalism supposing that the stress $\sigma=\sigma_0$ is suddenly applied at time $t=0$ to the system that is initially unstrained. The governing equation takes the form

$$\sigma_0 = E\varepsilon + \eta\dot{\varepsilon}, \tag{8.2}$$

with the following solution for the time-dependent deformation

$$\varepsilon(t) = \varepsilon_\infty \left(1 - \exp\left(-\frac{t}{\tau}\right)\right). \tag{8.3}$$

When stress σ_0 is applied, the strain increases asymptotically from zero to its final, steady state elastic value, $\varepsilon_\infty=\sigma_0/E$, with the time constant $\tau=\eta/E$ that is characteristic of the Kelvin-Voigt substance (Fig. 8.1b).

Voight (1968) was the first to propose that the state of stress in the vicinity of a borehole can be determined by measuring anelastic strains of drill cores taken from

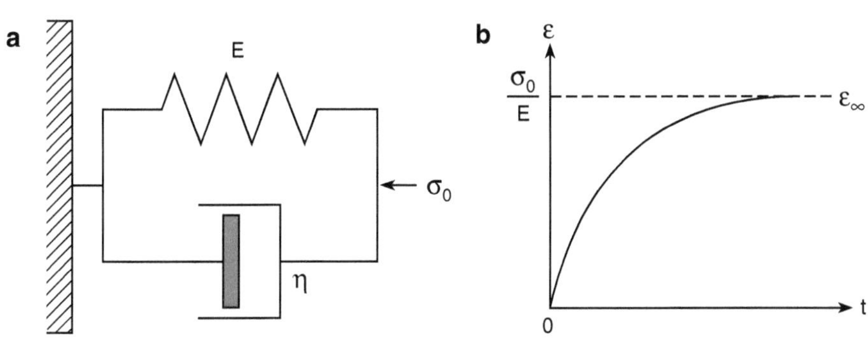

Fig. 8.1 Mechanical model of a Kelvin-Voigt substance (**a**) and response of the substance to applied stress (**b**)

the hole. The critical assumption in his relaxation theory is that of an isotropic rheology. The empirical function to describe the recovery deformation of a "relaxing" drill core (Eq. (8.2)) is dependent on time, the elastic and viscous material properties, as well as the frozen-in residual stresses of the core.

In a hypothetical strain-time curve (Fig. 8.2b) from core drilling, the in-situ coring process ($t=t_0$) with elastic ($\overline{AB}$) plus anelastic, time-dependent behaviour ($\overline{BC}$) of the rock core ($t>t_0$) is indicated. In the time interval for ASR measurements, t_2-t_1 only part of the anelastic strain recovery is registered, from which ε_∞ is extrapolated. The simplest model that exhibits both an instantaneous strain ($\overline{AB}$) and a late-time strain ($\overline{BC}$) is the *generalized Kelvin model* which consists of a Kelvin-Voigt element arranged in series with an extra spring. For interpreting measured ASR curves, however, the Kelvin model (Fig. 8.1a) is sufficient, since only the tail of the exponential core recovery (Fig. 8.2b, $t_2 - t_1$) is captured.

Blanton (1983) showed that from measured differential strains (Fig. 8.2b, $\Delta\varepsilon=\varepsilon(t_2)-\varepsilon(t_1)$) in a different direction from the core, the azimuth of S_H (Fig. 8.2a, assumed to be parallel to $\Delta\varepsilon_H$) as well as the ratios in principal stress magnitudes, S_H/S_V and S_h/S_V, can be calculated. For homogeneous, linear viscoelastic core material, characterized by relaxation modulus and a constant Poisson ratio v throughout time, it follows for an instantaneous removal of in-situ stress, that the stress ratios can be calculated from deformation differences $\Delta\varepsilon_h$, $\Delta\varepsilon_H$ (horizontal principal) and $\Delta\varepsilon_V$ (vertical) measured on deep drill cores:

$$\begin{aligned}\frac{S_H}{S_V} &= \frac{(1-\nu)\Delta\varepsilon_H + \nu(\Delta\varepsilon_h + \Delta\varepsilon_V)}{(1-\nu)\Delta\varepsilon_V + \nu(\Delta\varepsilon_H + \Delta\varepsilon_h)}\\ \frac{S_h}{S_V} &= \frac{(1-\nu)\Delta\varepsilon_h + \nu(\Delta\varepsilon_H + \Delta\varepsilon_V)}{(1-\nu)\Delta\varepsilon_V + \nu(\Delta\varepsilon_h + \Delta\varepsilon_H)}\end{aligned}. \tag{8.4}$$

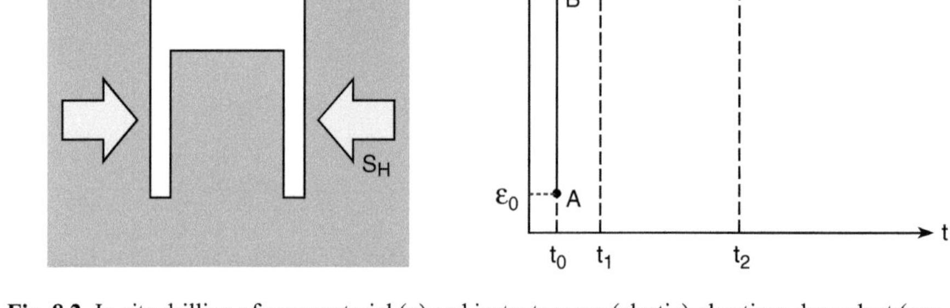

Fig. 8.2 In-situ drilling of core material (**a**) and instantaneous (elastic) plus time-dependent (anelastic) behaviour of the rock core (**b**) (modified after Teufel 1983). Time of in-situ coring is t_0, while $t_2 - t_1$ is the time interval for ASR measurements

Equation (8.4) is only valid if (1) one of the principal stresses is vertical and its magnitude is equal to the weight of the overburden and (2) the vertical axis of the core is parallel to the vertical principal stress. Each ASR stress estimate requires oriented core material.

Figure 8.3 schematically illustrates the microphysical cracking process behind the measured viscoelastic ASR curves. In the case of in-situ isotropic core material (Fig. 8.3a, b), upon relief from an anisotropic in-situ stress field, core samples tend to expand most in the direction of maximum stress relief (S_H direction) and least in the direction of minimum stress relief (S_h direction). Core samples acquire a new

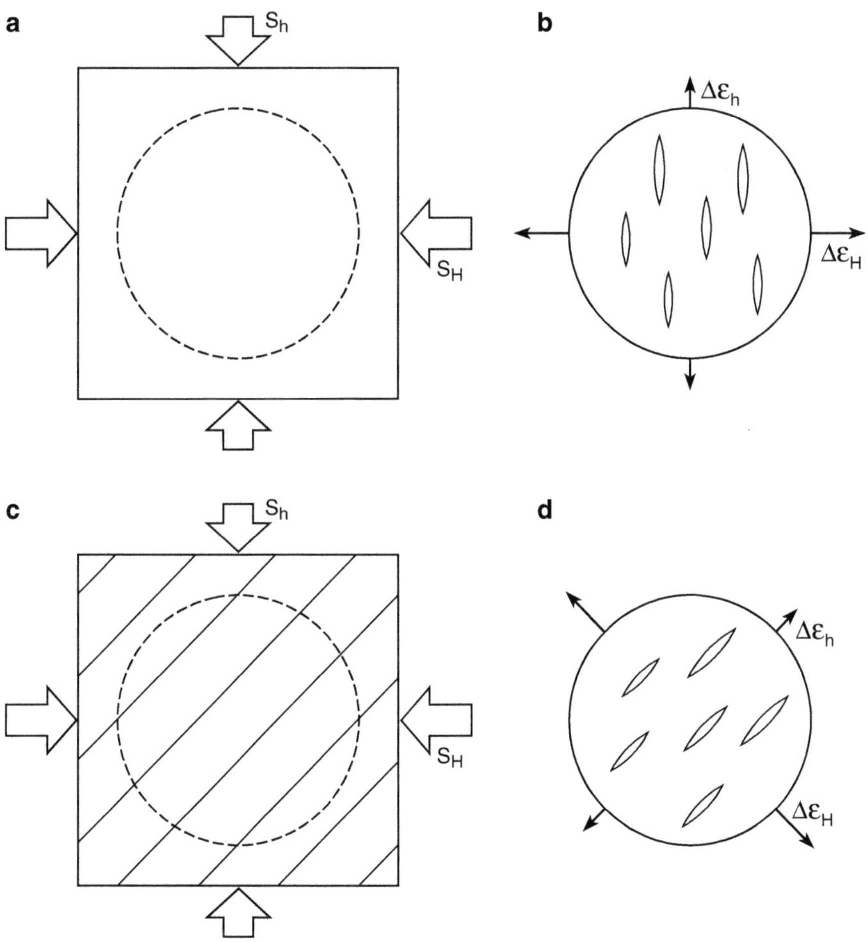

Fig. 8.3 Drilling of in-situ isotropic core material (**a**) generates stress-relief cracks aligned with the anisotropic in-situ stress field, i.e. with maximum strain recovery occurring parallel to the former maximum horizontal principal in-situ stress (**b**). Drilling of in-situ anisotropic core material (**c**) in the worst case scenario, generates stress-relief cracks which are aligned with the plane of rock anisotropy, and do not reflect in-situ stresses at depth (**d**)

microfabric due to the opening and propagation of cracks that are aligned in the directions of pre-existing in-situ principal stresses (Fig. 8.3b). In the case of in-situ anisotropic core material (Fig. 8.3c, d), however, the crack population developing upon stress relief may not reflect either-in-situ stress azimuths (by crack orientation) nor in-situ stress magnitudes (by crack closure pressure). For anisotropic core material, sophisticated methods like the rotation of the pole of *P*-wave slowness under pressure (Zang et al. 1996a) in combination with ASR data (Zang and Berckhemer 1993) have to be applied in order to obtain information on in-situ stresses. In the worst case scenario (Fig. 8.3d), textural anisotropy of rock at depth is responsible for stress-relief cracking and for the memory loss of rock to the pre-existing in-situ stress state during the recovery process.

8.1.2 Relaxation Apparatus

Teufel (1983) and Teufel and Blanton (1986) used clip-on disk gauges placed on cores to measure displacements across the entire core diameter (Fig. 8.4). Cores were wax sealed to minimize strain errors due to dehydration. Three disk gauges mounted 45° to each other in the horizontal plane, and one vertical, parallel to the core axis were able to record rock deformation with a sensitivity of about 10 $\mu m\,m^{-1}$ (microstrains) for a drill core 102 mm in diameter. Cores selected for ASR are always massive and homogeneous. Measurements are made in a constant temperature environment. Strain relief began 4 h after the core was cut in-situ and ended within about 40 h after the first data point was collected. A weighted average of the principal strains was calculated for each core, that is, the directions and magnitudes initially measured are more significant than those at the end of the recovery process. Teufel (1983) stated that core-based methods (ASR, DSA, WVA) cannot be used if a microcrack fabric exists at depth due to tectonic or other reasons.

Teufel (1993) carefully analyzed limiting factors of ASR stress data and concluded that nine parameters are significant: change in temperature, dehydration, decay of pore fluid pressure, inhomogeneous recovery strain, rock anisotropy, drilling-mud rock interaction, residual strains, insufficient core recovery time and the accuracy of core orientation. If cores are obtained from deviated wells, the method requires at least six independent strain gauges. In this case, the Matsuki (1991) method needs to be applied to determine the 3D stress tensor in-situ.

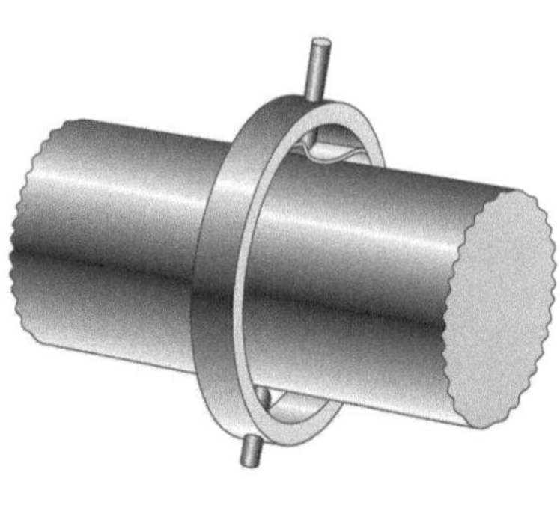

Fig. 8.4 Underlying physical principle of the "*Teufel-Apparatus*" for measuring anelastic recovery strains: support ring with Linear Velocity Displacement Transducer (LVDT) mounted on a wax-sealed drill core (schematically, after Engelder 1993)

Fig. 8.5 The "*Aulbach-Berckhemer-Wolter Apparatus*" for measuring recovery strains: six horizontal and one vertical displacement transducers (*LVDT*) as well as one Acoustic Emission (*AE*) sensor are attached to a freshly recovered drill core from the KTB pilot-hole. The apparatus is placed in a climate box to minimize temperature effects (courtesy of Klaus Wolter)

8 ASR

As a second device, we describe the *multi-component dilatometer* built at the University of Frankfurt/Main and used by Wolter and Berckhemer (1989) to determine in-situ stresses from drill cores of the KTB pilot hole (Zang et al. 1989). Figure 8.5 shows the dilatometer (six horizontal and one vertical inductive displacement transducers) with a resolution of 0.1 microstrains and one piezoceramic transducer for detecting acoustic emissions (Fig. 8.5, AE) during the strain recovery process. The apparatus is placed in a climate chamber where temperature is held constant at 25±0.1°C (Fig. 8.5, metallic box).

After removal from the wire line core barrel, a massive, cylindrical specimen 94 mm in diameter and free of visible fractures, is cut to 100 mm length from the bottom part of the rock column marked with a reference line for reconstructing absolute North orientation. As in the case of the "*Teufel-Apparatus*", the core is coated with wax to keep the moisture content constant. The hermetically sealed core is positioned in the centre of the multi-component dilatometer and stands on three steel pins. The tips of the six horizontal and one vertical displacement transducers are pushed through the wax coating to directly touch the rock surface. After reaching thermal equilibrium, attaching the AE-sensor and zero adjustment, the ASR measurement starts. In summary, we miss 5 h of strain relaxation: 1 h when bringing the core to the Earth's surface, 1 h while cutting and coating the core, and 3 h to reach thermal equilibrium in the climate box.

Measured ASR curves for a lamprophyre (core #526G2u) from a depth of 2231 m in the KTB pilot hole are shown in Fig. 8.6. For reference, $\varepsilon_{1,2}$ are the principal

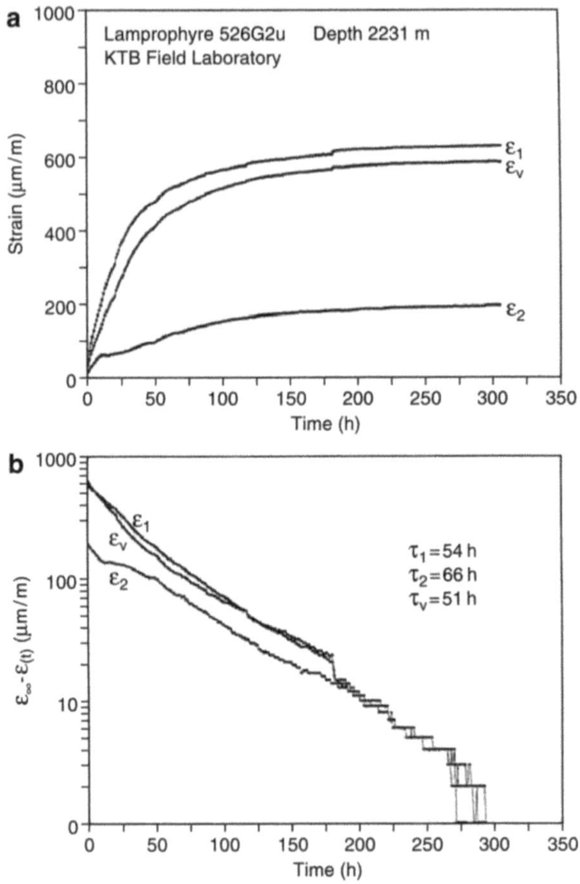

Fig. 8.6 **a** Principal recovery strains and **b** semi-logarithmic, normalized recovery strains, both versus time for quasi-isotropic dyke rock (core #526G2u, depth of 2231 m) from the KTB pilot borehole (modified after Wolter and Berckhemer 1989)

horizontal strains calculated from six measured horizontal strains, and ε_V is the vertical strain recorded parallel to the core axis (Fig. 8.6a). The total duration time of the measurement on the lamprophyre core specimen was 305 h, i.e. about 13 days. In Fig. 8.6b we fitted the ASR data with a Kelvin-Voigt substance using Eq. (8.3), and determined in a semi-logarithmic plot the corresponding relaxation times of the process (τ_1=54, τ_2=66, τ_V=51 h) with average value $\bar{\tau}$=57 h. The procedure for determining asymptotic strains (ε_∞), relaxation times for strain recovery (τ) and relaxation times for the cracking process (τ_{AE}), as well as parameters of the complete 3D strain ellipsoid in the drill cores, are provided by Zang and Berckhemer (1993). Here, recovery strain anisotropy is discussed together with ultrasonic velocity anisotropy and tensile strength anisotropy (Brazilian test) of 52 crystalline cores (amphibolite, gneiss, metagabbro, lamprophyre) from the KTB-VB well with depths ranging from 127 to 3888 m.

Applying Blanton's (1983) creep model using Eq. (8.4), we determined in-situ stresses from the quasi-isotropic lamprophyre core (Fig. 8.7). At the end of the recovery process, stress magnitudes calculated are (S_H=64, S_h=47 MPa) with ν=0.3 and $S_V=\rho gz$=61 MPa (Fig. 8.7a). The peak in stress magnitudes, however, is reached

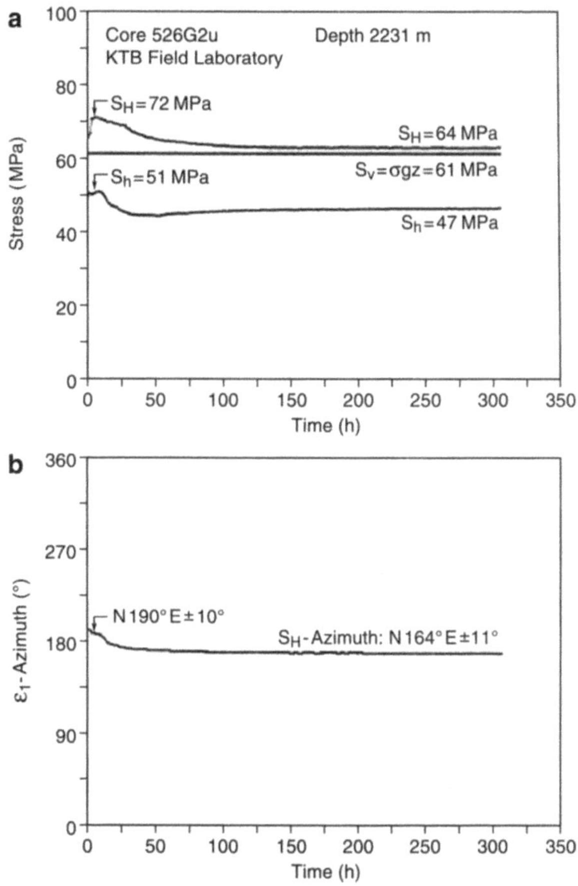

Fig. 8.7 **a** Principal in-situ stress magnitudes and **b** azimuth of maximum horizontal principal stress inferred from maximum recovery strain versus time for the same dyke rock as shown in the previous figure (modified after Wolter and Berckhemer 1989)

at recovery times of about 10 h (S_H=72, S_h=51 MPa), and should be, according to the weighted average approach of Teufel (1983), more reliable for stress estimates. The S_H azimuth inferred from our orientation of the final value of the ε_1 direction is N164°E ± 11° for the KTB site (Fig. 8.7).

Figure 8.8 shows the cumulative AE number versus recovery time for the same dyke specimen. Apart from AE bursts (Fig. 8.8, arrows), the cumulative AE-curve indicates a Kelvin-Voigt recovery behaviour. The characteristic time, τ_{AE}=56 h compares well with the average value of relaxation times from the principal recovery strains, $\overline{\tau}$=57 h. The intimate relationship between the time constant of acoustic relaxation and the time constant of recovery strain is discussed in detail by Zang and Berckhemer (1993) for different rock types. All 1723 AE detected have frequencies between 100 and 500 kHz. Optical microscope and scanning electron microscopy studies on the same, totally relaxed, core materials, reveal two types of cracks. One population of relatively larger cracks (length scale in the order of millimetre to centimetre) occur in carbonate veins running through the dyke rock material. A second class of relatively smaller cracks with low aspect ratios (pore-shape) have length scales (pore radii) in the order of micrometer and are located inside quartz

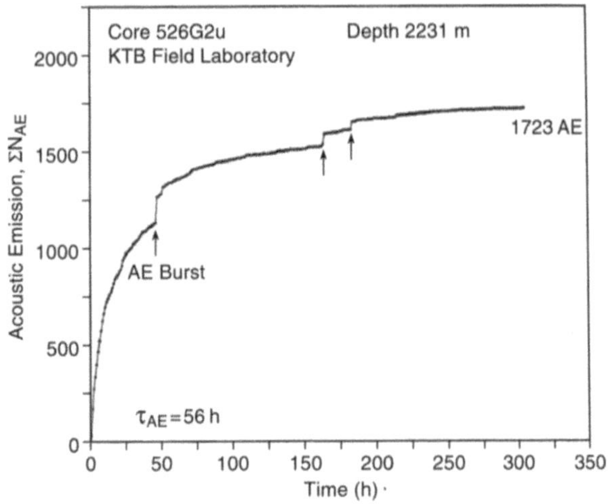

Fig. 8.8 Cumulative number of acoustic emissions (*AE*) versus recovery time for the same core material, as shown in the previous figure. Note cascades of *AE* (*bursts*) producing sudden steps in the overall Kelvin-Voigt recovery behaviour (*arrows*) (modified after Wolter and Berckhemer 1989)

and feldspar grains. It is not clear which population of cracks is responsible for the AE bursts observed, and which population is causing the overall firmo-viscous behaviour.

Note-Box Anelastic strain recovery is a core-based method to estimate in-situ stress magnitudes and orientations from instrumenting a freshly recovered drill core obtained from deep wells. For in-situ isotropic core material, the direction of maximum strain recovery is parallel to the azimuth of maximum horizontal stress in the borehole. Assuming linear viscoelastic material behaviour, principal in-situ stress ratios can be calculated. Increasing the numbers of independent recovery strain data to six and calibrating creep compliances to applied stress, the complete stress tensor can be determined. Like in borehole techniques, the problem of in-situ stress estimates from in-situ anisotropic drill cores has not yet been properly solved.

8.2 Kaiser Effect

The *Kaiser effect* (KE) is the phenomenon that a material under stress emits acoustic waves, so-called acoustic emissions (AE), only after the previously applied stress level is excited. In material sciences, Joseph Kaiser (1950) was the first to describe this memory effect on metals, rock and wood in tension. In the rock mechanic community, Goodman (1963) introduced the Kaiser effect by testing sandstone and Kurita and Fujii (1979) by testing crystalline rocks in compression. In Fig. 8.9, the Kaiser effect is shown schematically for two re-loading cycles of a laboratory

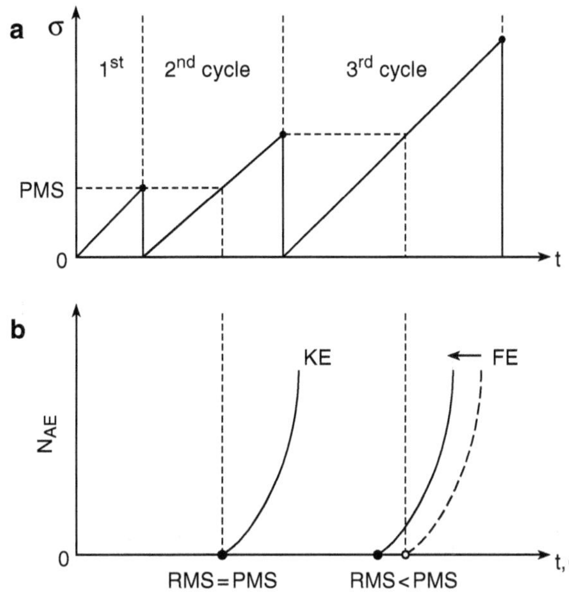

Fig. 8.9 Cyclic loading of a virgin rock core in the laboratory indicated by applied stress versus time curve (**a**), and a measured number of acoustic emissions versus time or applied stress in the second and third loading cycles (**b**)

compressed rock core. In Fig. 8.9a, a fictitious maximum stress is applied to the rock in the first (virgin) loading ramp with subsequent (abrupt) unloading. The fictitious stress value, the so-called *previous maximum stress* (PMS, Stuart et al. 1993), is increased for each subsequent cycle. The ideal, laboratory KE is demonstrated in Fig. 8.9b. If the point of significant AE onset, the so-called *recalled maximum stress* (RMS, Hughson and Crawford 1986) is equal to PMS (Fig. 8.9b, dot RMS = PMS), the KE with laboratory-frozen-in stresses is verified par excellence. Approaching the fracture strength of rock (Fig. 8.9a, third cycle), however, significant AE appear at a lower stress level compared to the previous applied maximum stress (Fig. 8.9b, RMS < PMS). This phenomenon is called *Felicity effect* (*FE*, in this section, otherwise Finite Elements) and can be quantified in terms of the *Felicity ratio* (*FR*)

$$FR = \frac{RMS}{PMS} \begin{cases} = 1 \rightarrow KE \\ < 1 \rightarrow FE \end{cases}. \tag{8.5}$$

The Felicity effect is demonstrated by a shift in the AE curve expected for the ideal KE behaviour (Fig. 8.9b, dashed line) towards lower stress values, as indicated by the actually measured AE curve (Fig. 8.9b, solid line). An alternative plot for discriminating KE and FE in rock is shown in Fig. 8.10. Pollock (1989) plotted cumulative AE numbers for different loading cycles versus load and separated KE behaviour (Fig. 8.10a, AB-CBD) from FE behaviour (CBD-EFG). Note that the "Felicity nose" obtained in a modified plot by removing constant AE numbers during unloading phases (Fig. 8.10b), can only result from comparing cumulative AE numbers of two neighbouring cycles (Fig. 8.10, second and third cycles).

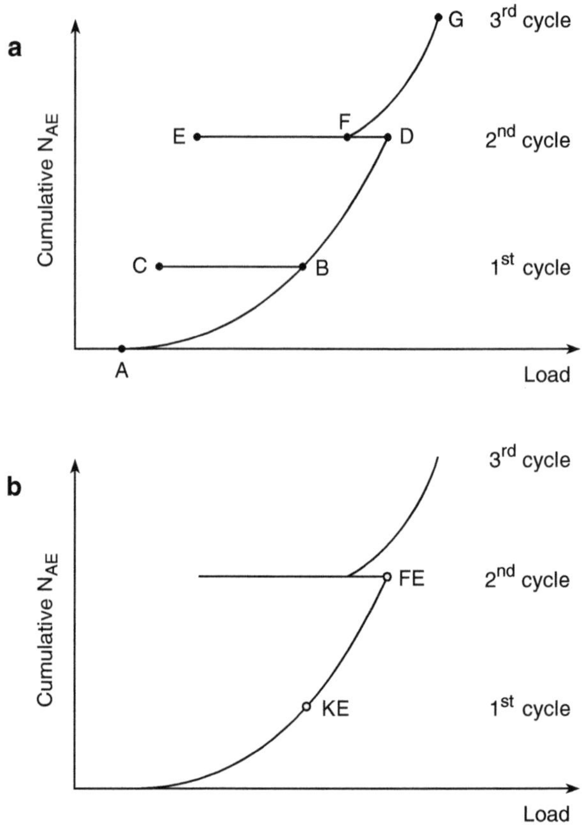

Fig. 8.10 Cumulative number of acoustic emissions versus load for three loading cycles (**a**) after Pollock (1989) and the same curve without unloading events for demonstrating the ideal Kaiser effect (*KE*) and the Felicity effect (*FE*) in rock (**b**)

Determining RMS is not a simple task. Like in the interpretation of highly non-linear pressure-time-records of hydraulic fracturing data (Sect. 7.2), the point of onset of AE activity is also difficult to fix in rock. In general, the AE onset or take-off point is assumed to be equal to RMS and can be determined from cumulative AE curves versus stress, $N(\sigma)$ as shown in Fig. 8.9b, from the first derivative of that curve, $dN(\sigma)/dt$ and the second derivative, $d^2N(\sigma)/dt^2$. Using primary data $N(\sigma)$, RMS can be recognized as an inflexion point indicating the change in slope by bilinear regression or the pivot point method (Shen 1995). The determination of the inflexion point can be sharpened using $dN(\sigma)/dt$ with bilinear regression or, even more precisely, by using the AE hit rate square $(dN(\sigma)/dt)^2$ as demonstrated by Hardy et al. (1989). Wenzel (1990) was the first to, besides $N(\sigma)$ and $dN(\sigma)/dt$, also use the second derivative $d^2N(\sigma)/dt^2$ for determining RMS in deep drill cores from KTB. More recent techniques include the method of maximum curvature (Momayez et al. 1992) and the technique of using normalized slopes of the $N(\sigma)$ curve (Villaescusa et al. 2002). In addition, the relative changes of the AE curves, i.e. the difference in AE number or AE rate for two subsequent loading cycles, are also used to further sharpen the picking procedure of the AE onset point (Yoshikawa and Mogi 1981,

1989). A good review in processing AE curves for determining precise RMS values in detail and of KE in rock in general is given by Lavrov (2003).

In brittle rock, KE is best-pronounced if the PMS does not exceed the stress value at which dilatancy begins. This is demonstrated by plotting the Felicity ratio of rock versus PMS normalized to the rock uniaxial compressive strength, C_0 (Fig. 8.11). The closer PMS to the ultimate rock strength, the less-pronounced is KE. This has been demonstrated by Li and Nordlund (1993) using FR in the second cycle uniaxial compression test versus peak stress of the first cycle normalized to the compressive strength. Among the rock tested by Li and Nordlund (1993), most of them exhibit good KE (marble, gneiss, granite, gabbro, greenstone, porphyry, chalcopyrite ore), while the Kiruna magnetite had low FR values indicating poor KE. The range of FR values resulting in well-pronounced KE was $0.3 \leq FR \leq 0.8$ for most of the rock types investigated. The lower boundary, FR=0.3 is due to the fact that below this value only little AE are detected anyway, even without preloading, and the quiescence can be misinterpreted as KE. The upper boundary, FR=0.8 is due to the effect of dilatancy. Note, however, that for rocks with plastic behaviour at room temperature (e.g. rock salt), the KE can be also observed in the post-failure region (Filimonov et al. 2002). The effect of confining stress on AE onset in ductile rock has been investigated by Filimonov et al. (2005).

In general, it is not possible to state that rock memorizes PMS during recent stress history, since the rock memory depends on (1) the duration time between *pre*loading and *re*loading and (2) the duration time of *pre*load. The larger the duration time of PMS, the clearer is KE, which was verified by Michihiro et al. (1989) on different rock types (tuff, mudstone, sandstone, schist). However, this is not true for andesite (Yoshikawa and Mogi 1989). Increasing the time delay between

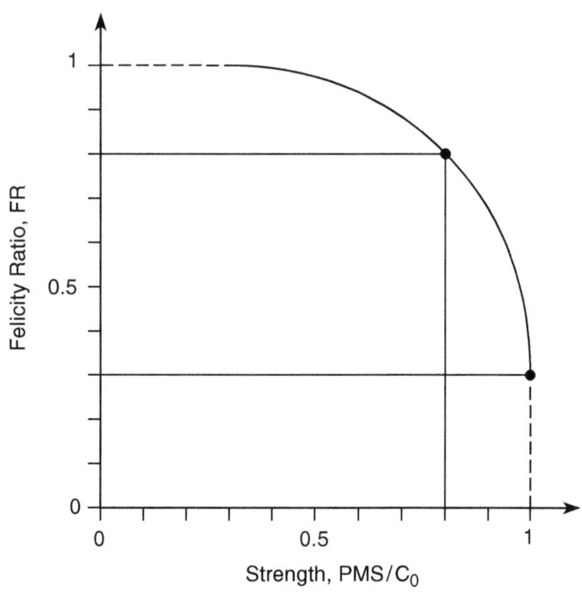

Fig. 8.11 Felicity ratio (*FR*) versus previous maximum stress *PMS* normalized to uniaxial rock compressive strength (modified after Li and Nordlund 1993)

successive loading cycles results in decreasing KE quality. Koerner and Lord (1989) found well-pronounced KE when the delay was less than 10 h, but for delays over 40 days, KE disappeared in different rock types (gneiss, schist, carbonate, mud-, lime-, sandstone). In rock salt, however, the laboratory KE could be observed with time delays over 2 weeks (Filimonov et al. 2001), and in granite up to delays of 300 days (Shin and Kanagawa 1995). The KE loss over time can be quantified by the *Kaiser effect retention span* (KERS) as introduced by Barr and Hunt (1999) while investigating Carnmenellis granite from UK. The KE degradation in the course of time is attributed to ASR (Michihiro et al. 1992; Barr and Hunt 1999). This "*Alzheimer*" in KE of rock is sensitive to PMS. The closer PMS is to ultimate rock strength (Fig. 8.11, $PMS/C_0 \to 1$), the faster the decay in KE memory (the more-pronounced is KE Alzheimer). KE memory loss in rock, besides increasing time delay between successive loading cycles, can also be supported by water saturation and heating of rock which was shown by Yoshikawa and Mogi (1981). A *thermal Kaiser*, the recall of previous maximum temperature exists (Yong and Wang 1980; Zogala et al. 1992; Shkuratnik et al. 2007), while an *aqua-Kaiser* (recall of previous maximum humidity conditions) needs to be proven. No significant influence of loading rate on KE has been found (Lavrov 2001).

Early applications of KE in rock (Kurita and Fujii 1979) assumed that the normal stress component in each direction of in-situ rock mass can simply be retrieved by uniaxial reloading of a rock core oriented in that direction. Holcomb in several publications (1983–1993) showed, however, that this is not true. Holcomb (1993a, b) concluded that understanding the role of triaxial stresses is the main prerequisite for correct application of KE in reliable rock stress estimates. In the following, for simplicity, we refer to studies using the KE observed on uniaxially stressed rock cores as *uniaxial stress approach*, and studies using the KE observed in triaxially stressed cores as *triaxial stress approach*.

8.2.1 Triaxial Kaiser Stress Approach

Physically, the triaxial stress approach is based on the detection of the damage surface in brittle materials using AE (Holcomb and Costin 1986a). By definition, the *damage surface* is the locus of points in stress space (Chap. 2) where the criterion of crack growth (Chap. 3) is just satisfied. A 6D damage surface in rock is schematically visualized in S_1-S_2-space (2D cut perpendicular S_3) in Fig. 8.12. Any stress state lying on the damage surface is on the verge of causing cracking in the rock (Fig. 8.12, area of cracking). Assuming the cracking progress to be associated with AE, KE is observed when the laboratory-applied stress path (Fig. 8.12, line with arrow) touches the damage surface (Fig. 8.12, black dot). In concept, the damage surface for elastic-brittle material is similar to the yield surface for elastic-plastic material (Sect. 5.5).

Fully screening the damage surface in rock involves (1) a crack growth model and (2) a true-triaxial experimental rock testing system. In 1993, Holcomb used,

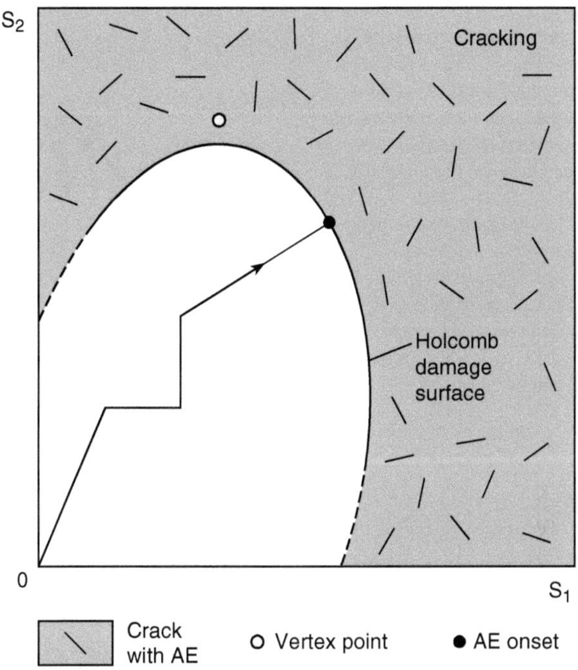

Fig. 8.12 Damage surface in rock visualized in S_1-S_2 stress space. Any stress state on the Holcomb damage surface is on the verge of causing cracking in rock (*shaded area*). The cusp in the Holcomb line corrected for the Felicity effect is called vertex point V

for example, a penny-shaped crack-growth model (Costin 1983, 1987) to map the damage surface in parallelepipeds 54 × 54 × 25 mm in size of Tennessee marble with a servo-controlled true-triaxial system (Wawersik et al. 1990) located at Sandia National Laboratories (Alberquerque, NM). In 1996, Pestman and van Munster used a wing-crack growth model (Sect. 3.3) to map the damage surface on 50 mm cubes of oven-dry Darley Dale sandstone with a true-triaxial apparatus located at Shell Research, Rijswijk, NL (Sayers et al. 1990). The premise in investigating stress memory effects in rock under polyaxial stress states is that measured AE are well described by the anticipated crack model. For different rock types, it may be convenient to pick micromechanically different crack models (Sect. 3.3.1) in predicting damage surfaces. For sandstone, a pore crack (Eq. (3.18)) or Hertzian fracture model (Eq. (3.19)) may be more appropriate than for granite where a wing-crack model (Eq. (3.14)) is more suitable.

In 2001, Pestman et al. presented a novel technique to estimate crustal stress magnitudes using AE-controlled polyaxial reloading of cores in the laboratory. The so-called *Pestman-Kenter-van Munster* (PKM) method is based on scanning Holcomb surfaces in rock with the premise that one point of the damage surface corresponds to PMS experienced by the rock core. Probing the damage surface of sandstone outcrop samples with PKM in S_H-S_h space (Fig. 8.13) involves increasing $S_{H,h}$ values while keeping the ratio S_H/S_h constant for each cycle (Fig 8.13, loading path with arrow) until the onset of AE signalled that a point on the damage surface is reached (Fig. 8.13, point 1). Then, $S_{H,h}$ values are decreased to prevent further damage.

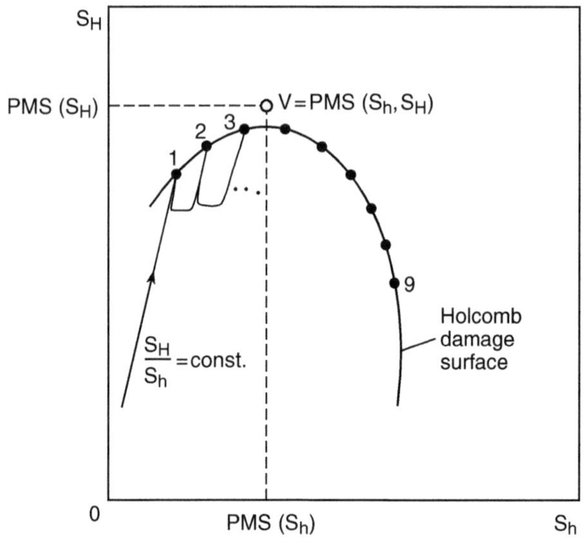

Fig. 8.13 Probing the damage surface in rock with the PKM method involves increasing stress values while keeping the stress ratio constant (*stress path with arrow*) until the onset of acoustic emissions signalled that a point on the damage surface is reached (*numbered dots*). The vertex point, *V* in PKM corresponds to PMS components experienced by the rock (modified after Pestman et al. 2001)

To probe a second point, the ratio S_H/S_h is decreased and stresses increased again until the onset of AE (Fig. 8.13, point 2). This procedure in PKM is repeated until the whole Holcomb surface of the sandstone had been mapped out (Fig. 8.13, Holcomb surface). The point of maximum curvature of the Holcomb surface corrected for the Felicity effect is called the *vertex* (*V* in Figs. 8.12 and 8.13). The vertex point according to PKM, corresponds to PMS experienced by the rock core under consideration. Assumptions of PKM for obtaining Holcomb surfaces are (1) that the rock core must be reloaded along principal axes known from other sources (e.g., ASR, DSA), and (2) that the rock strength is large enough not to fail during coring but low enough that PMS takes the sample into the inelastic regime of deformation necessary for RMS. The PKM method for in-situ stress estimates was applied to cores taken at a depth of 3997 m in the Munnekezijl field (Pestman et al. 2002).

An alternative method to determine the vertex point in rock is layed out in Lavrov (2003), who used a 2D cut of the damage surface in (S_1-S_3)-S_3-space (Fig. 8.14). Triaxial axisymmetric loading with principal stresses $S_1=S_V$ vertical load and $S_3=p_c$ horizontal confining pressure shows that the damage forming loading path is represented by two straight lines, the intersection of which is the vertex point to infer PMS components, as indicated by Pestman and van Munster (1996). Vervoort and Govaerts (2006) used Lavrov's (2003) bilinear method in conventional triaxial compression tests of crinoidal limestone to determine both slopes, $k_1>k_2$ of the damage surface projection onto the (S_1-S_3) plane versus S_3 for the same material (Fig. 8.14, slopes $k_1>k_2$). Earlier studies revealed either k_1 (e.g. for rock salt $0.5<k_1<0.6$, Filimonov et al. 2001) or k_2 (e.g. for Westerly granite $k_2=2.6$, Holcomb and Rudnicki (1986), or for Kuru granite $k_2=2.8$, Li and Nordlund (1993); Li (1998)). Vervoort and Govaerts (2006) found $k_1=1$ and $k_2=0.5$ for a lower Carboniferous limestone. Note that it is critical that the rock core is oriented precisely towards the principal

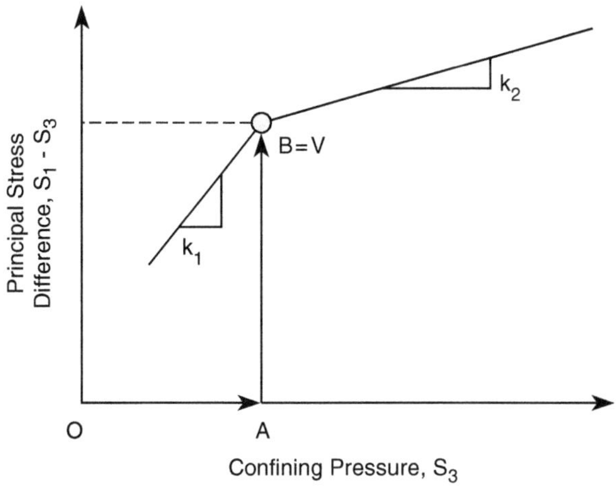

Fig. 8.14 Damage surface formed by triaxial axisymmetric loading with principal stresses, S_1-S_3. The damage-forming loading path is represented by two straight lines, *OA* and *AB*. The vertex point, $B = V$ corresponds to the previous maximum stress state (modified after Lavrov 2003)

axes in the laboratory experiments. In case of small tilt angles (>10°) between the loading direction and the principal stress direction, KE is blurred (Lavrov 2003). This is due to the fact that the AE onset is a strong function of the angle between the PMS axis and the axis of the core (Holcomb and Costin 1986a, b).

8.2.2 Uniaxial Kaiser Stress Approach

The more simplistic (experimental wise) uniaxial stress approach was used by several authors to determine the in-situ stress from KE observed in the laboratory before the publication of Holcomb's (1993a) generalized theory of KE (Holcomb and Martin 1985; Hughson and Crawford 1986; Momayez and Hassani 1992) and after (Seto et al. 1999; Villaescusa et al. 2002; Lehtonen and Särkkä 2006). As stated by Holcomb (1993a), if indeed in-situ stress can be determined via uniaxial testing, then there must be another phenomenon responsible for the relationship between the onset of AE and the stress history of the rock taken from the Earth's crust. One way to solve this controversy is to postulate a different mechanism for the laboratory KE and the in-situ KE. It has never been shown that the mechanism producing KE for core retrieved from the Earth's crustal stress field (*in-situ KE*) is the same as that seen after the sample is pre-stressed in the laboratory (*laboratory KE*). Holcomb (1993a) hypothesized that when relaxation cracks (Sect. 8.1) are forced to close (e.g. by DSA, WVA) the interference leads to AE via a crush-up of mismatched asperities and that is the AE activity observed in uniaxial tests.

Even if it is impossible to determine the full in-situ stress tensor by uniaxial compression tests of core specimens, convincing studies appeared showing KE stresses inferred from drill cores in close agreement to in-situ stresses determined using more-accepted methods such as overcoring, borehole breakouts and hydraulic fracturing. In this context, we highlight the study of Momayez and Hassani (1992) who determined principal stresses and their orientation at the Atomic Energy of Canada Underground Research Laboratory (URL). Second, we draw the attention of the reader to the Western Australian School of Mines (WASM) where experimental verification of AE in-situ stress measurements has been demonstrated using the uniaxial tests (Villaescusa et al. 2003).

In 2002, Villaescusa et al. presented the WASM-AE technique in combination with *deformation rate analysis* (DRA) to estimate in-situ stresses in four Australian mines. In the AE technique, the second or higher-order load cycle is used to determine KE, since most noise from asperity breakage during crack closure is reduced and PMS can be estimated by a clear onset. In addition, the DRA method (Yamamoto et al. 1990) was applied in the laboratory to the same core material, whereby PMS is obtained from the change in the gradient of the stress-strain curve under cyclic uniaxial compression of rock. The bending point in the strain difference function, defined as the difference in inelastic strain between the two cycles is identified with the normal component of in-situ stress along the loading direction of the core. WASM methodology consists of isolating a master core (diameter ≥ 50 mm) from the rock mass where in-situ stress data are wanted. Then the master core is undercored (Sect. 6.2) by six smaller sub-cores (diameter 20 mm) with different directions. Uniaxial compression tests are carried out on all sub-cores and AE inflexion points are determined in bilinear fits. Assuming uniaxial RMS to work, the full stress tensor in the master core can be determined from the six independent inflexion points (Villaescusa et al. 2002). Both KE inflexion points from AE curves and bending points from inelastic strain curves are used to determine PMS values in comparison to standard stress estimation techniques. The WASM-AE method has been used for in-situ stress estimates at more than 50 mine sites in Western Australia (Villaescusa et al. 2006).

In Fig. 8.15, in-situ stress data from different methods are compared for the Perseverance nickel mine located 500 km north from Kalgoorie in Western Australia. Data are taken from Villaescusa et al. (2006). In Fig. 8.15a, the three principal stress magnitudes are derived from (1) overcoring (OC) with the CSIRO hollow inclusion-cell method above 800 m depth, (2) hydraulic fracture tests (HF) in depths from 770 to 1120 m, and (3) WASM-AE method based on KE in uniaxial testing for depths greater than 1120 m. Note that the four KE stress-data points measured well fall on the regression lines of the minor (σ_3), intermediate (σ_2) and major principal stress (σ_1) calculated from all stress data collected. In Fig. 8.15b, the orientation of principal stress axes is shown for the four KE stress data from Fig. 8.15a. The *Schmidt stereographic projection* of the stress orientation indicates σ_3 as vertical stress, since data points are close to the centre of the circle (Fig. 8.15b, cross with dip 90°), while σ_1 is oriented at about N45°E or N225°E, respectively, with zero or small dip angles (Fig. 8.15b, small solid circles at the outer rim of the large circle with dip < 10°). The consistent orientation of the stress axes from KE are in good agreement with ovularity ratios from borehole breakouts obtained in the depth range from 1100 to

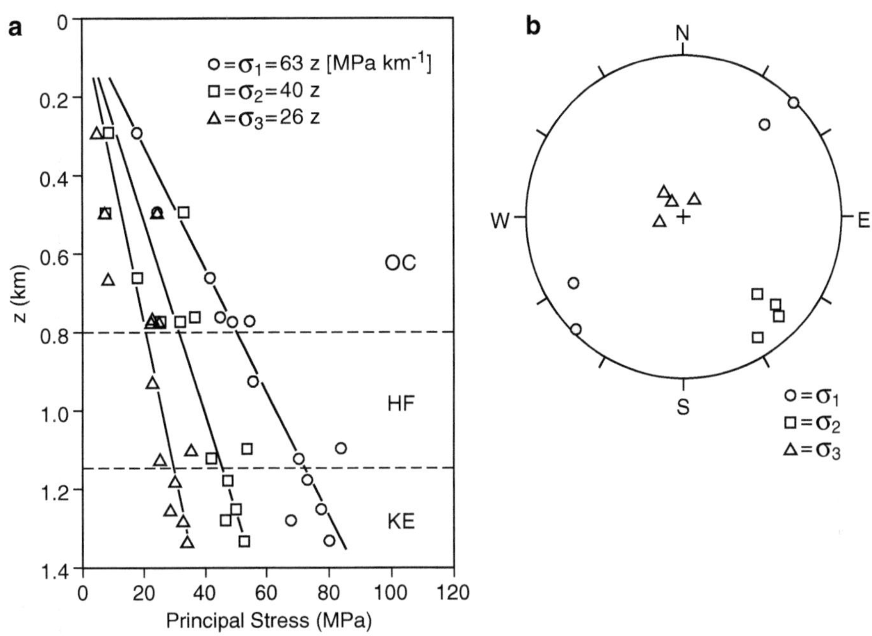

Fig. 8.15 In-situ stress estimates in the Perseverance nickel mine, Western Australia. **a** Stress magnitudes from overcoring (*OC*), hydraulic fracturing (*HF*) and AE onset (*KE*) versus depth. **b** Orientations of in-situ stresses for the four KE stress data from **a** at a depth ranging from 1175 to 1330 m (after Villaescusa et al. 2006)

1400 m, N230°E±30° in the same mine (Villaescusa et al. 2006). With the help of in-situ stress data from KE in the Preserverance mine feasibility study, it was possible (1) to demostrate a clear rotation of the in-situ stress below a depth of 1100 m (11 level), and (2), more important, to extend mine life past 11 level and beyond the year 2013. Dight (2006) emphasized that Villaescusa's combined Kaiser effect and deformation rate analysis uniaxial stress approach is valid.

In summary, the uniaxial Kaiser stress approach is a method where the stress tensor of a master drill core taken from great depth is determined by reloading six sub-cores with different orientations in laboratory uniaxial compression tests. The AE onset point in uniaxial Kaiser stress can produce credible results (Lehtonen and Särkkä 2006) the method, however, is still in development and far from fully reliable. This is also demonstrated by the fact that KE stress data are usually linked to DRA (Lin et al. 2006a) or other core-related stress determination techniques (ASR, DSA, WVA).

8.2.3 Core Damage and Kaiser Stress

For stress determination it is essential to know if the in-situ KE, i.e. nailing PMS on drill cores retrieved from deep inside the Earth's crust, is due to the present-day stress, paleo-stress or the peak stress imposed during the coring process. One phenomenon affecting both, the uniaxial and triaxial Kaiser stress approach, is the problem of rock alteration induced by the in-situ coring process (Santarelli and Dusseault 1991). During in-situ drill-out (1 m core per 10 min depending on rock type), the rock is subjected to an extreme stress state (Fig. 8.16). The vertical stress S_V is gradually removed as the core bit approaches from above (Fig. 8.16a). The horizontal stress S_H is removed only after point P gets inside the core barrel (Fig. 8.16b). Large deviatoric stress, $S_V - S_H$ can build up due to the time variation of stress components (Fig. 8.16d) effective during drill-out of the drill core with stress-free surfaces (Fig. 8.16c). In the case of high deviatoric stress, the rock can fail in-situ during the drill-out and core disks (CD) can result (Maury et al. 1988). Deviatoric stresses below failure can still be high enough to cause permanent damage documented in the form of microcracks in a fully relaxed core at surface (Fig. 8.16c). Note that stress changes in Fig. 8.16d take into account the effect of a perfect mud

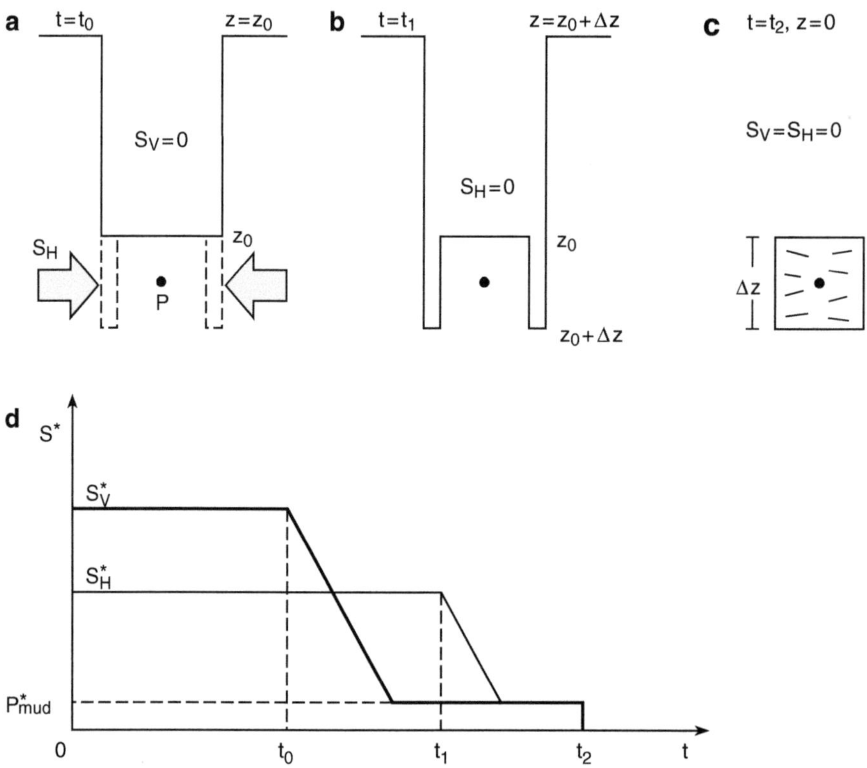

Fig. 8.16 Rock alteration due to the in-situ drill-out of rock material (modified after Holt et al. 2000a). In a sequence of three pictures, the redistribution of stresses inside a drilled-out core is demonstrated: **a** removal of vertical stress by drilling the borehole, **b** removal of horizontal stress by in-situ overcoring and **c** final drill core with stress-free surface condition and stress-relief cracks inside. Time variation of effective stress components **d** during the in-situ drill-out can produce large deviatoric stresses inside the rock core

cake, resulting in effective vertical stresses equal to the mud overbalance p^*_{Mud}. Using three synthetic sandstones manufactured under stress, Holt et al. (2000a) quantified core damage in terms of stiffness ratio, ultrasonic velocity and KE onset for virgin, pre-drilled cores (in-situ) compared to altered, drilled cores (at surface). Stiffness ratios of pre-drilled cores compared to drilled cores were 2–3, the stress sensitivity of velocities were more pronounced for drilled, cracked material and the AE onset point for KE was much broader, obscured in the drilled compared to the virgin, compacted core material.

The effect of core damage can be visualized with the concept of damage surfaces as used by the PKM method (Pestman et al. 2002). In Fig. 8.17, the Mohr-Coulomb polygon indicates the initial damage surface of rock before the core is taken (cf. Fig. 3.3). During drill-out (Fig. 8.16b), the initial damage surface is puffed up according to the precise combination of minor and major horizontal effective stress (Fig. 8.17, Holcomb line). Drill-out-induced core damage can be quantified by the difference in initial (Mohr-Coulomb envelope) and the Holcomb damage surface (Fig. 8.17, shaded area). The cusp in the Holcomb line, corrected for the Felicity effect, indicates the magnitudes of PMS used for Kaiser-stress determination (Fig. 8.17, *V*=vertex point). The PMS deduced from a Holcomb damage surface, however, is a superposition of virgin damage of the core in-situ (Coulomb-Mohr line) and the drill-out-induced part of core damage. In order to understand the effect of core damage, discrete particle models were used (Holt et al. 2000b, 2001, 2003). In 2006, based on Particle Flow Code modelling, Gorodkov et al. concluded that the

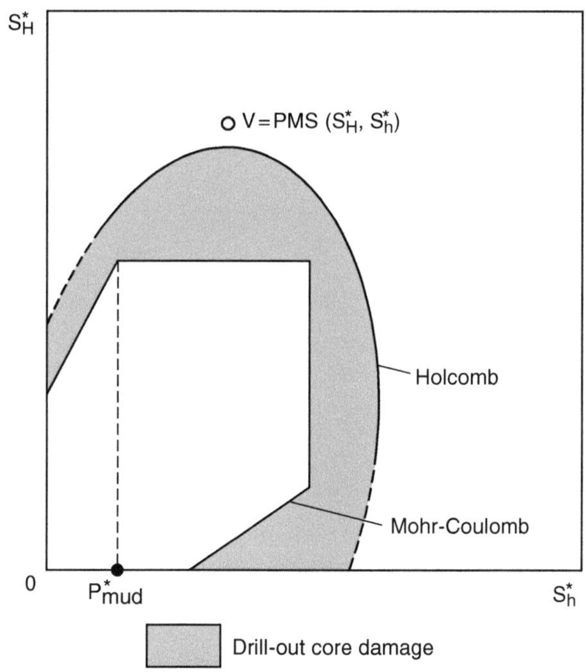

Fig. 8.17 In effective in-situ stress space, the previous maximum stress (*PMS*) deduced from Holcomb damage surface by the PKM method is a superposition of the virgin damage of the core in-situ (Coulomb-Mohr polygon) and the core damage induced by the drill-out process (*shaded area*) (Based on figure 2 of Pestman BJ, Holt RM, Kenter CJ, van Munster JG (2002). Field Application of a Novel Core-Based In-Situ Stress Estimation Technique. Paper SPE 78158 presented at the SPE/IRSM Rock Mechanics Conference, Irvin Texas, 20-23 October)

stress path during coring seemed to be less damaging than anticipated. To account for the shaded area in Fig. 8.17, they used numerical rock models consisting of up to 305,000 particles with different radii distributions. They found that horizontal stresses in part of the assembly which is to become a core, grow by 20% during the drill-out operation before they decrease. An alternative approach to separate in-situ stress magnitudes from PMS in cores at the time of drilling was anticipated by Dight (2006). Based on DRA curves, he could separate two inflection points, the first, lower stress value for in-situ stress and the second, higher value for the maximum stress experienced by the core during drill-out.

8.2.4 Microphysical Models

Application of KE in rock-stress estimates requires a firm understanding of the physical mechanism behind the phenomenon. We close this section by reviewing important micro-mechanisms in rock (grain aggregates) responsible for KE, as well as some theoretical models used to reproduce KE results obtained from rock. Generally speaking, KE and other stress memory effects in rock can be regarded as fracture-induced physical phenomena (Lavrov 2005). We separate KE mechanisms into crack and damage models. While in the latter, local failure of rock, producing AE is simulated by an abstract damage variable (or damage tensor), in physical crack models, the actual mechanism of producing single AE events at grain scale (e.g. sliding crack ligaments, extending new crack surfaces) is anticipated.

The *sliding crack model* (Stevens and Holcomb 1980) accounts for KE only if the dynamic friction between crack faces equals the static friction coefficient. If the dynamic friction is smaller than the static coefficient, the stress memory is lost (Alzheimer effect) since stick-slip events occur on the sliding crack. The *reversible Griffith crack model* (Holcomb and Stevens 1980) gives rise to a hysteresis since the opening stress is higher than the closing stress of the crack. Considering a population of such hysteresis-producing cracks allows the simulation of stress-memory effects (Holcomb 1981; Guyer et al. 1995; Guyer and Johnson 1999). The *penny-shaped crack-grow model* (Costin 1983) was used (as we know) to calculate the damage surface expansion under true-triaxial stresses. The local tensile stress in rock is relieved as a penny-shaped tensile crack propagates. The final crack length, after completing loading, therefore corresponds to PMS. The *wing-crack model*, consisting of one shear mother crack and two tensile daughter cracks at the tips of the mother crack (Chap. 3), was used for simulating KE by Pestman and van Munster (1996) and Lavrov (1997). AE activity for stresses smaller than RMS is due to friction along existing cracks, while for stresses larger than RMS, AE activity is caused by crack extension (Yoshikawa and Mogi 1989). In this way, a wing crack can produce AE below RMS (multiple reactivated mother shear crack) and above RMS (extension of tensile wing cracks). An alternative approach implies that KE is caused by *void closure* (Holcomb 1993a, b) instead of crack growth. Crack models, thought to be responsible for KE, are summarized in Fig. 8.18.

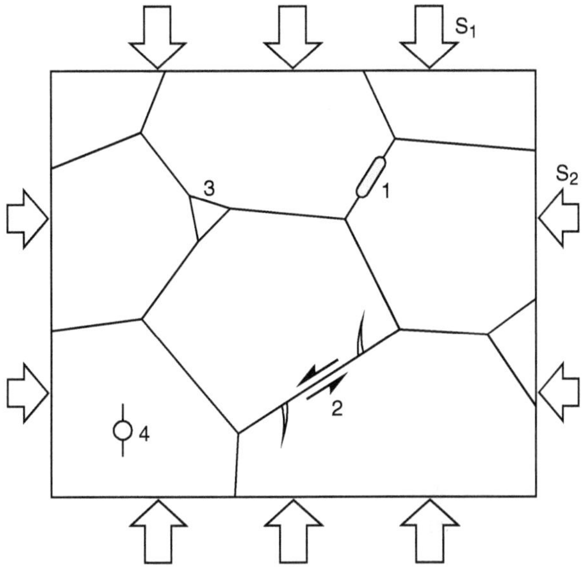

Fig. 8.18 Crack models responsible for acoustic emissions (*AE*) and Kaiser effect (*KE*) in rock: (*1*) Griffith crack, (*2*) sliding and wing crack, (*3*) void closure and (*4*) pore crack

Enhancing rock-stress understanding goes hand in hand with numerical analysis (Cundall 2001; Hart 2003). A powerful feature of explicit, dynamic discrete codes is the ability of simulating AE (and also KE) by the breakage of contact bonds in discrete assemblies of spheres simulating the rock material (Hazzard and Young 2000). Studies by Holt et al. used the commercial Particle Flow Code (PFC) to model core mechanics (2000b), understand stress-relief effects during drill-out (2003), and assess the feasibility of core-based stress determination techniques (2001). In addition, the PFC code (Itasca Consulting Group Inc. 2004) was used to investigate stress-memory effects in rock (Louchnikov et al. 2004) as well as screening critical parameters in determining KE stresses, e.g. the confining pressure as is the case in the triaxial stress approach (Louchnikov et al. 2006).

Tang et al. (1997) used a scalar-damage parameter to model the number of AE observed in laboratory-compressed rock material. Their theoretical model of KE is based on the assumption that local damage in rock sub-elements is the cause for detecting AE. Implementing a *Weibull* distribution for the local elemental strength of rock, they reproduced AE activity in extremely heterogeneous rock (shape parameter $m=1.5$, lognormal distribution) and nearly homogeneous rock ($m=4$, normal distribution). Their damage model was used by Qin et al. (1999) to determine geo-stresses from laboratory KE measurements. In 2003, Fu et al. simulated multiple rock-cycling tests in uniaxial compression with Tang's damage model and found good agreement with experimentally observed phenomena. In 2004, Hamiel et al. presented the first viscoelastic damage model and verified KE by a power-law relation between the damage variable and elastic moduli leading to a non-linear coupling between the rate of damage evolution and the damage variable itself. Their damage variable is roughly zero for stresses smaller than PMS. The onset of damage in every cycle depends on the damage level reached in previous cycles in consistency with KE. In 2007, Yuan et al. verified KE by applying their damage model to experimental data obtained from uniaxial tests on concrete. From our viewpoint, reproducing KE (i.e. AE onset, stress memory) does not mean that the actual physical mechanism used is correct for the rock specimen under consideration.

Summarizing our KE knowledge gained in this section, we can follow the synoptic picture in Fig. 8.19. A drilled-out core specimen at surface (Fig. 8.19a) contains microcracks due to the superposition of the in-situ damage (pre-existing Griffith cracks) and drill-out induced cracks. Assuming isotropic core damage in Fig. 8.19a, omni-directional measurements of the *P*-wave slowness would result in a constant value of a drill core at surface (Fig. 8.19b). Virgin loading of the drill core (Fig. 8.19c, first cycle) produces, according to the crack model used, a re-opening and in part extension of the pre-exsisting cracks. Acoustic emissions (AE) measured in this cycle (Fig. 8.19d) can be used to infer the in-situ Kaiser effect of the rock sample. Removing the first uniaxial re-loading stress (Fig. 8.19e) cracks in part close, but in comparison to the virgin drill core (Fig. 8.19a), the residual length of closed cracks is increased. Due to the removal of anisotropic stress, the core after the first reloading cycle is characterized by a maximum *P*-wave slowness parallel to the loading direction (Fig. 8.19f). The slowness maximum is caused by an additional set of horizontal cracks resulting from vertical unloading of the core (Fig. 8.19c, d). The second reloading of the core (Fig. 8.19g) opens both pre-existing cracks and freshly-induced unloading cracks of the first cycle. For simplicity, we avoid drawing wing

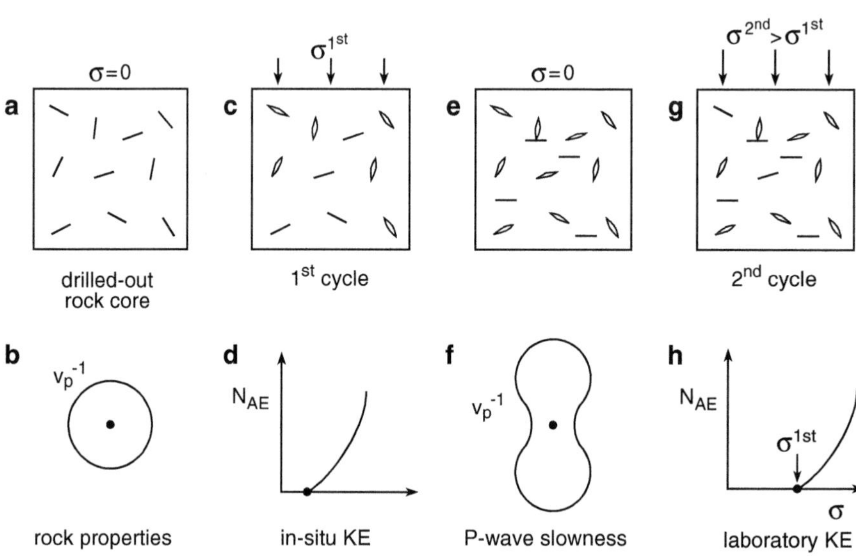

Fig. 8.19 Synoptic picture of the Kaiser effect (*KE*) in rock split into crack-related phenomena in physical space (*above*) during laboratory cyclic loading of rock, and indirect indicators of cracks in rocks (*below*) by ultrasonic velocity (**b**, **f**) and acoustic emissions measurements (**d**, **h**). In **d**, the in-situ Kaiser effect and in **h**, the laboratory Kaiser effect is demonstrated

cracks (Chap. 3) which can also develop in uniaxial compression and show single separated (Fig. 8.19e) or merged single cracks only (Fig. 8.19g). AE measured in the second cycle (Fig. 8.19h) can be used to determine the laboratory-induced PMS from the first cycle (AE onset at σ^{1at}), whereby the laboratory KE is demonstrated.

To physically solve the problem of in-situ stress estimates from KE, we have to understand (1) the freezing-in mechanism of rock stress in-situ closely related to the residual stress phenomenon (Sect. 4.4), (2) the actual microphysical mechanism producing AE during relaxation of deep drill cores related to anelastic strain recovery (Sect. 8.1), and (3) the actual reason for AE activity during re-pressurizing the relaxed cores with stress-relief cracks inside (core damage) related to DSA, DRA, WVA. In this sense, KE is the most ambitious leading-edge technology for stress estimates from cores and worthy to close Chap. 8.

Note-Box Kaiser effect is defined as the phenomenon that a material under stress emits acoustic emissions only after the previous maximum stress is reached. If the recalled maximum stress in rock cores under cyclic load equals the previous maximum stress, the Kaiser effect par excellence is verified. In all other cases, the Felicity ratio (recalled maximum divided by previous maximum stress) is used to limit the application range of this effect in rock. Memory loss (Alzheimer) is due to chemical effects and heating. Experimental techniques to determine the Kaiser stress use either uniaxial or triaxial rock testing. Six sub-cores, drilled in different directions of a master core, are required to obtain the full in-situ stress tensor from uniaxial tests. The scanning of the complete damage surface of rock is required with polyaxial compression tests in the triaxial approach. Both approaches lead to credible Kaiser stresses. The method, however, is still under development and far from fully reliable as a stand-alone technique for stress determination in rocks.

Part III
Interpreting Stress Data

Chapter 9
Local Stress Data

Local stress data are usually tied to underground excavations (mines, tunnels) or boreholes which are drilled into the Earth's crust. The most important reasons for designing excavations and drilling holes are withdrawal of natural materials like hydrocarbon or mineral resources, deposit of human waste materials like nuclear waste or CO_2, and geothermal energy as well as pure scientific purpose. We choose three sites to show different applications of stresses in the Earth's crust and facets in stress determination involved.

9.1 Continental Deep Drilling Site KTB, Germany

M9.1 Zoback

The German Continental Deep Drilling Program (KTB) is an example of two scientific drill holes. KTB was designed to study the properties and processes of the deeper continental crust by means of a superdeep borehole. Among major research themes like the nature of seismic reflectors, the thermal structure of the crust, the transport of crustal fluids and the evolution of the central European Variscan bedrock, the Earth's crustal stress field played a key role (Emmerman and Lauterjung 1997). The project was conducted in three phases (Table 9.1). Phase (1) was a wise decision and involved drilling a pilot hole into the steeply dipping metasediments of NE Bavaria. The pilot hole, KTB-VB (*VorBohrung*) was drilled from 1987 to 1990 to a final depth of 4000 m. Phase (2) involved drilling the main hole, KTB-HB (*HauptBohrung*) at 200 m distance from the pilot hole from 1990 to 1994 to a final depth of 9101 m. Due to the bottom hole temperature of 265°C, the planned 10–12 km depth of KTB (Fuchs et al. 1991) could not be realized. The deep crustal laboratory phase (3), starting in 1995, was designed to operate long-term fluid-injection experiments in the deeper Earth's crust using cross-correlation data of both bottom hole sections at KTB-VB and KTB-HB. Hydraulic tests during the deep laboratory phase (3) included a massive fluid production (2002–2003) and fluid injection test (2004–2005) in the KTB-VB well (Kümpel et al. 2006). Even today, scientific experiments are underway and a visitor centre at the drill site provides information about milestones of the project.

A. Zang, O. Stephansson, *Stress Field of the Earth's Crust*,
DOI 10.1007/978-1-4020-8444-7_9, © Springer Science+Business Media B.V. 2010

Table 9.1 Fact sheet of KTB separating the project into three distinct phases

Phase (1) *KTB-VB*—Drilling the pilot hole

Start	End	Total Depth	Cores, Core Diameter	Reference
1987, 22 Sep.	1989, 4 Apr.	4000 m 13,124 ft	3594 m, Ø = 94 mm	Emmerman and Wohlenberg (1989)

Drill rig GH 1400 E total height 49 m (162 ft); open hole section 3850–4000 m, bottom hole temperature $T = 120°C$; 14 hydraulic fracturing tests between 805 and 3011 m

Phase (2) *KTB-HB*—Drilling the main hole

Start	End	Total Depth	Cores, Core Diameter	Reference
1990, 6 Oct.	1994, 12 Oct.	9101 m 29,859 ft	84 m, up to Ø = 234 mm	Emmerman and Lauterjung (1997)

Drill rig UTB-1 total height 83 m (272 ft); open hole section 9030–9101 m, bottom hole temperature $T = 265°C$; two hydraulic fracturing tests at 6018 and 9066 m

Phase (3) Deep Crustal Laboratory—Cross Hole Hydraulic Tests

Year	Test	Depth	Fluid, Rate	Seismicity	Reference
1994	Injection *KTB-HB* 24 h	9.1 km	210 m^3 $CaBr_2/KCl_2$ at rate 550 l min^{-1} with 50 MPa well head peak	400 FIS*	Zoback and Harjes (1997) Jost et al. (1998)
2000	Injection *KTB-HB* 60 days	9.1 km leak at 5.4 km	4000 m^3 fresh water at rate 30–90 l min^{-1} with 30 MPa well head peak	2799 FIS	Baisch et al. (2002) Baisch and Harjes (2003) Rothert et al. (2003)
2002–2003	Produc. *KTB-VB* 1 year	4.0 km	22,300 m^3 saline fluid at flow rates −29# to −57# l min^{-1}	–	Kümpel et al. (2006) Gräsle et al. (2006)
2004–2005	Injection *KTB-VB* 10 months	4.0 km	84,600 m^3 fresh water at rates 185–196 l min^{-1} with 20 MPa well head peak	3000 FIS	Shapiro et al. (2006)

* *FIS* fluid-induced seismicity (earthquakes due to propagating pore pressure front)
\# negative flow rates = production test

Geographically, KTB is located near the village of Windischeschenbach in NE Bavaria about 40 km SE from the "Wagner-City" Bayreuth. Geologically, the KTB drill site is located at the boundary between the Saxothuringian (ST) and the Moldanubian (MN), two major tectonostratigraphic units of the Hercynian fold belt in Central Europe (Fig. 9.1). Basement rocks are separated from Permo-Mesozoic foreland sediments by the Franconian Lineament (Fig. 9.1a, b), a NW-SE trending system of reverse faults. The KTB site is about 4 km east of FL and south of the boundary between ST and MN units. The ST-MN boundary is regarded as a suture zone formed by the closure of an early Paleozoic oceanic basin during the Variscan collision in Carboniferous times.

The drill site is located in a small, isolated tectonometamorphic unit called the Zone of Erbendorf-Vohenstrauß (Fig. 9.1a, ZEV) which is characterized by medium-pressure metamorphism dating 410 and 380 Ma and relics of an earlier high-pressure metamorphism (O'Brian et al. 1997). The area was intruded by late to post-Variscan granites between 326 and 310 Ma (Wendt et al. 1992). From late-Variscan to subrecent times, the Bohemian terrane was affected by several episodes of brittle deformation (Zulauf 1993). A *terrane* is a fragment of crustal material broken-off from a tectonic plate and sutured to another plate. The latest major tectonic event indicated by extensional structures of Tertiary age is the so-called Eger graben (Malkovsky 1987). The ZEV nappe complex consists of gneisses and metabasic rocks with minor metapegmatites. The KTB-HB encountered steeply inclined units belonging to ZEV over the entire drilled section. Figure 9.1b shows a schematic SW-NE profile through ZEV down to 10 km depth (Emmerman and Lauterjung 1997). The drilled crustal segment consists of an alternating sequence of three lithologic units: (1) paragneisses, (2) metabasites and (3) layers of gneisses and amphibolites. Most rocks show a penetrative foliation which dips steeply between 50 and 80° to the SW or NE. A prominent fault system at approximately 4 km depth (Fig. 9.1b, seismic reflector SE2) was penetrated by both drill holes. Two more fault systems were penetrated by KTB-HB, one between 6850 and 7260 m (Fig. 9.1b seismic reflector SE1), and the other between 7820 and 7950 m. Both systems consist of a number of individual fault planes, dipping steeply to the NE and can be correlated with the trace of FL at surface (Fig 9.1a, Franconian Line).

In Fig. 9.1c, the crustal profile of Fig. 9.1b is put into the context of the continental crust as a whole. In drilling the top third of the crust, the temperature at bottom hole at KTB-HB reached about 265°C. Scientific results of KTB are published in KTB-Reports and in a special volume of the *Journal of Geophysical Research*. Here, Emmerman and Lauterjung (1997) gave an overview of major results. Scientific challenges of KTB from a drilling engineer perspective are summarized by Bram et al. (1995). The overriding goal of KTB was to provide scientists with two permanent, accessible deep holes for research in the Earth's crust. The asymmetric telescope (one 4 and one 9.1 km deep hole 200 m apart) into the crust had a budget of 500 million DM (~270 million euros).

Figure 9.2 shows an aerial view of the KTB site (1997) with the drill rig of HTB-HB in the foreground, and the cleared platform of the KTB-VB in the background (left). The drill rig of KTB-VB completed drilling in 1994 (not seen on the photograph) was a 49 m high Gulliver mast construction. At that time the world largest land rig UTB1 for drilling KTB-HB had a total height of 83 m (272 ft). In the meantime, the well site of KTB-VB was cleared, the rig of KTB-HB was dismantled. The derrick of KTB-HB, however, remains as a monument to KTB's achievement and serves as an observation deck accessible from the visitor centre.

The integrated stress measurement strategy in the KTB project is illustrated in Fig. 9.3. At field scale, stress information is inferred from both natural seismicity (NS) in terms of swarm earthquakes from the Eger rift 50 km NE of KTB (Ibs-von Seht et al. 2006) and fluid-induced seismicity (FIS) focal mechanisms from stimulation tests at the KTB site (see references Table 9.1, KTB Phase (3)). At borehole

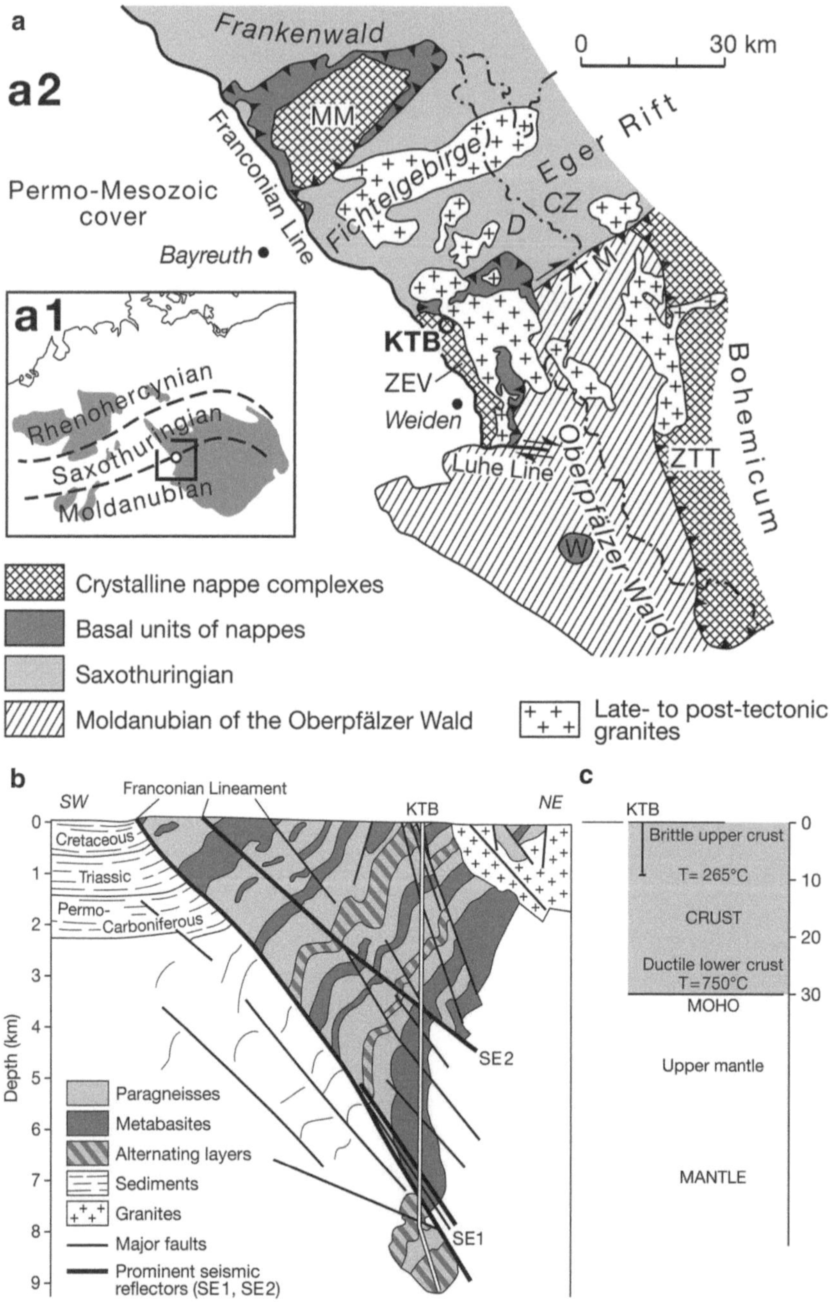
a
a2
Frankenwald
0 30 km
MM
Franconian Line
Fichtelgebirge
Eger Rift
Permo-Mesozoic cover
CZ
D
Bayreuth
ZTM
a1
Rhenohercynian
Saxothuringian
Moldanubian
KTB
ZEV
Weiden
Bohemicum
Luhe Line
Oberpfälzer Wald
ZTT
W
Crystalline nappe complexes
Basal units of nappes
Saxothuringian
Moldanubian of the Oberpfälzer Wald
Late- to post-tectonic granites
b
Franconian Lineament
SW
KTB
NE
Cretaceous
Triassic
Permo-Carboniferous
Depth (km)
SE2
SE1
Paragneisses
Metabasites
Alternating layers
Sediments
Granites
Major faults
Prominent seismic reflectors (SE1, SE2)
c
KTB
Brittle upper crust
T= 265°C
CRUST
Ductile lower crust
T=750°C
MOHO
Upper mantle
MANTLE
Depth (km)

Fig. 9.2 Aerial view of the KTB Drill Site in Windischeschenbach with drill rig of the KTB Main Hole in the *foreground* (total height 83 m) and the cleared spot of the KTB Pilot Hole at 200 m distance in the background (*left*). The building of the KTB field laboratory is located next to the two drill holes (© ICDP, GFZ German Research Centre for Geosciences)

scale, stress data are obtained from hydraulic fracturing in cased and perforated intervals in the KTB-VB (Rummel 1991; Baumgärtner et al. 1993) and the KTB-HB (te Kamp et al. 1995a, b; Brudy et al. 1997), from borehole breakout analysis (Mastin et al. 1991; Brudy et al. 1993) and from the analysis of drilling induced fractures (Brudy and Zoback 1999). At laboratory scale, core-based stress measurements started with an original 94 mm diameter core cut to 100 mm length for anelastic strain recovery (Wolter and Berckhemer 1989). Subsequently, the core was split

---

Fig. 9.1 Location of KTB drill site. **a** Geologic map of the western rim of the Bohemian Massif in NE Bavaria. (*a1*) Variscan basement outcrops in Middle Europe with zones according to Kossmat (1927) now interpreted in terms of terranes: Rhenoherzynian (*RH*), Saxothuringian (*ST*), Moldanubian (*MN*). (*a2*) Tectonometamorphic units: crystalline nappe complexes, basal units of nappes: *MM* Münchberg, *ZEV* Zone of Erbendorf Vohenstrauß, *ZTT* Zone of Tepla-Taus forming the western part of Bohemian terrane, *ST*, *MN* of the Oberpfälzer Wald, late- to post-tectonic granites, overthrust: *ZTM* Zone of Tischenreuth-Mähring, *W* Winklarn (after Weber 1990). **b** Depth section SW-NE through the ZEV nappe complex with KTB-HB. The metamorphic sequence of ZEV has early high pressure metamorphism at 475 Ma, and later, Barrovian-type metamorphism at 375 Ma. The Falkenberg granite to the NE intruded at 310 Ma. Foreland sediments to the SW have Upper Cretaceous to Permo-Carboniferous ages (after Emmerman and Lauterjung 1997). **c** Depth profile at *KTB* in context to the continental crust and upper mantle (after Yardley 1997)

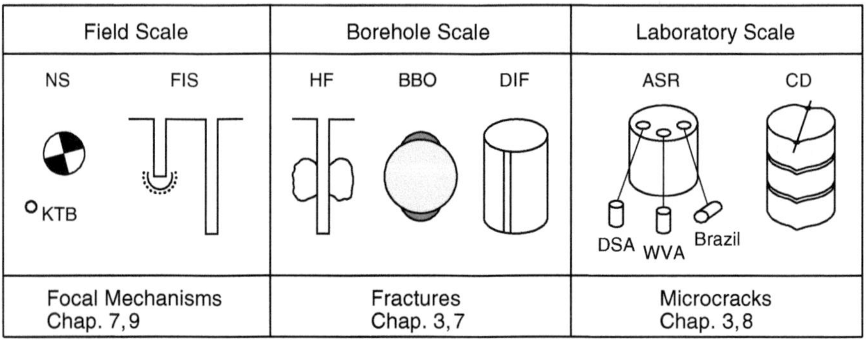

Fig. 9.3 Integrated stress measurement strategy at KTB

into three rock disks 30 mm in thickness (Zang and Berckhemer 1993), one for stress estimates from differential strain analysis (Baumann 1993), one for stress estimates from wave velocity analysis (Zang et al. 1996a) and one for determining the anisotropic rock strength through Brazilian tests. Core disking has been analyzed in terms of stress directions (Wolter et al. 1990). Stress data were also provided by Haimson and Chang (2002) from true triaxial strength tests of KTB amphibolite cores.

Benchmark stress data from hydraulic fracturing in the pilot hole are visualized in Fig. 9.4. Stress magnitudes determined by HF (Fig. 9.4a) indicate a strike

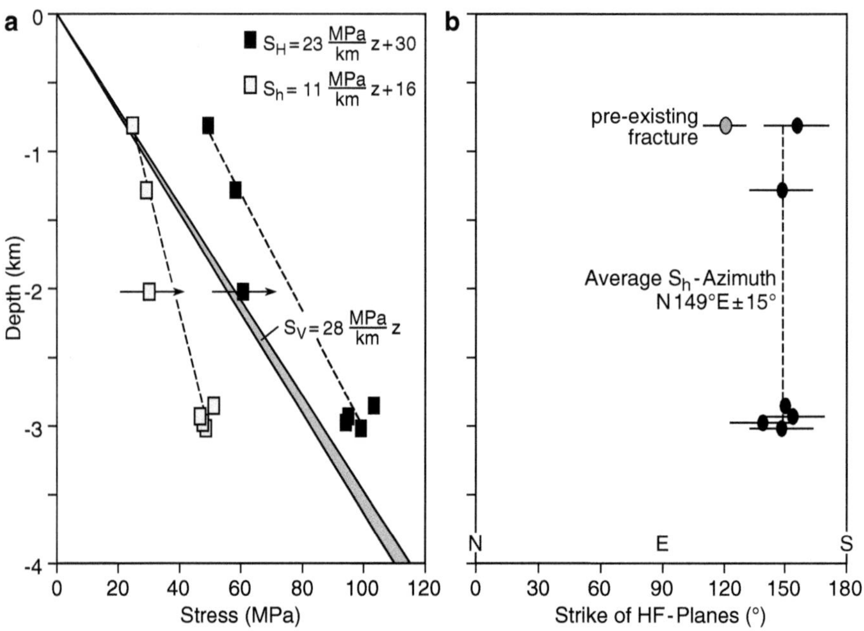

Fig. 9.4 Stress data from hydraulic fracturing in the KTB pilot hole, KTB-VB **a** stress magnitudes and **b** direction of maximum horizontal compressive stress (after Rummel 1991)

slip faulting regime ($S_h<S_v<S_H$) over the depth range analyzed (805–3011 m) with horizontal gradients, dS_h/dz of 11 MPa km^{-1} and dS_H/dz of 23 MPa km^{-1} and a vertical stress gradient, dS_v/dz of 28 MPa km^{-1} (Rummel 1991). Except for one rock joint which was opened at 805 m in the KTB-VB, only axial fractures were induced or extended (Baumgärtner et al. 1993). The strike of the hydraulic fracture planes, assumed to be parallel to the S_H direction in quasi-isotropic rock (Sect. 7.2), is shown in Fig. 9.4b. Excluding the pre-existing joint data (Fig. 9.4b, open symbol), the average S_H azimuth at the KTB drill site from hydraulic fracturing planes was determined to be N149°E ± 15°.

In Fig. 9.5, stress data from borehole-based methods (BBO, HF) and from core-based methods (ASR, CD, DSA) are summarized for KTB-VB. In Fig. 9.5a, the hydrofrac stress profile from Fig. 9.4a is compared to stress magnitudes obtained from differential strain analysis (DSA, Baumann 1993) and wave-velocity analysis (WVA, Zang et al. 1996a). All three methods applied (DSA, HF, WVA) indicate a strike-slip faulting stress regime. The slope dS_H/dz is 23 MPa km^{-1} as determined from WVA and HF, whereas the value from DSA is 12 MPa km^{-1} larger. The slope of dS_h/dz is 21 MPa km^{-1} as determined from DSA and WVA, while that from HF is 10 MPa km^{-1} smaller. Vertical stress calculated from average rock density varied between 2750 and 2840 kg m^{-3} for WVA and HF, and is determined to be ~28.7z (mega pascal) from re-pressurized DSA rock cubes.

Each method indicated in Fig. 9.3 can provide information on stress direction. The direction of S_H is assumed to be parallel to the axis defined by the core disking low saddle points (Fig. 9.3, CD), the direction of measured maximum differential strain (Table 7.2, DSA), the direction of maximum strain recovery (Fig. 8.3a, ASR), the direction of minimum P-wave velocity (Fig. 4.9d, WVA), the azimuth of the hydraulic fracture in-situ (Fig. 7.2, HF), the orientation of drilling-induced vertical fractures (Fig. 9.3, DIF) and perpendicular to the diametrically opposed borehole breakouts (Fig. 7.8, BBO). In Fig. 9.5b, the S_H azimuth averaged over the depth range of investigation is indicated for different methods as vertical lines for the KTB-VB. Results are N163°E ± 22° from 1.2 to 3.6 km for CD analysis (Wolter et al. 1990), N177°E without an error bar from 855 to 3858 m for DSA (Baumann 1993), N168°E ± 25° for two lamprophyres investigated by ASR from 2051 to 3573 m depth (Zang et al. 1990), N161°E ± 14° from 3 to 4 km depth obtained from BBO (Mastin et al. 1991) compared to the HF data set from Fig. 9.4b. The average value of five average values from different methods revealed a direction of S_H at the drill site of N162°E (Fig. 9.5b, stress rosette).

For reference in Fig. 9.5c, the S_h-azimuth inferred from different methods in the pilot hole is compared to the azimuth and dip of the foliation of the rock mass (Mastin et al. 1991). The quality of stress data is influenced by the degree of rock anisotropy in-situ (Chaps. 4, 7, 8). The influence of rock anisotropy on in-situ stress estimates has been addressed by Amadei and Goodman (1982), Borsetto et al. (1984) and Amadei (1996). The problem of gaining reliable stress data in highly anisotropic rock formations, however, is not yet solved. Most formulas for stress estimates presented in Chaps. 7 and 8 assume elastic, isotropic, homogeneous rock material. There is uncertainty, however, when the analysis breaks down for anisotropic rocks in terms of the modulus ratio (Worotnicki 1993; Amadei 1996; Hakala et al. 2007).

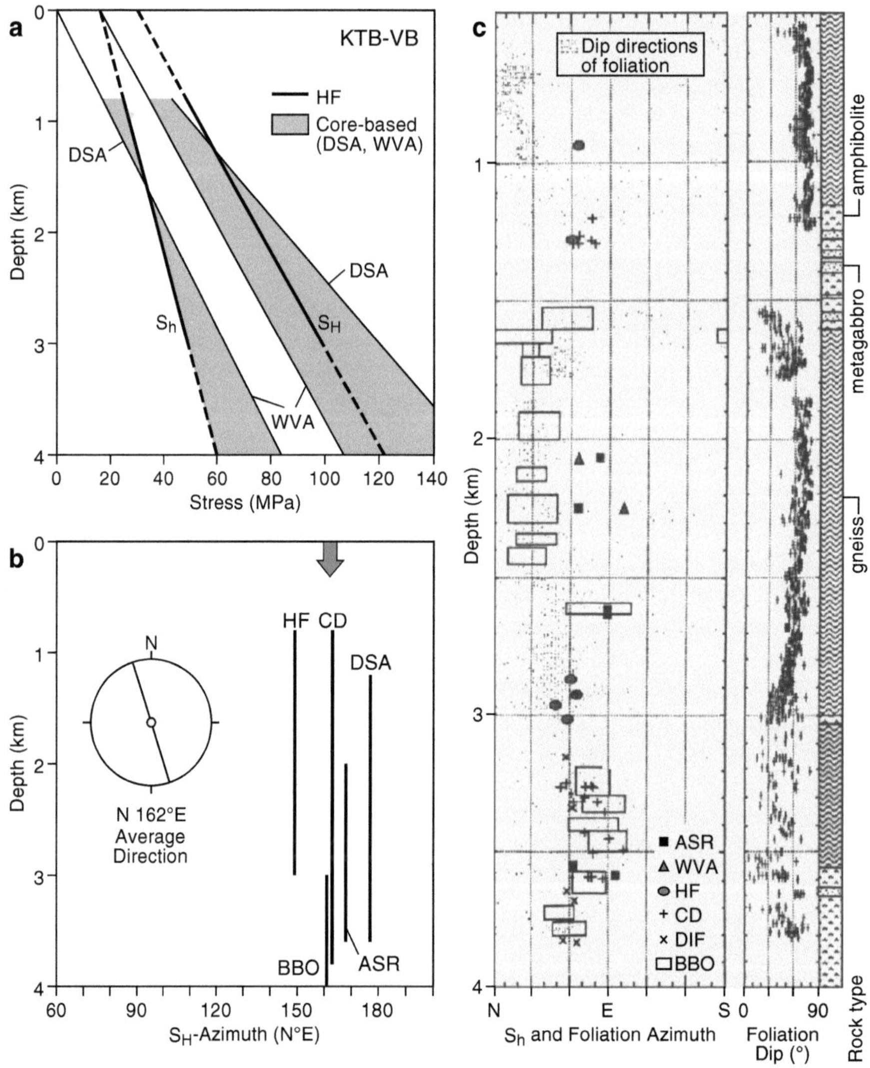

Fig. 9.5 Stress data for KTB-VB from the integrated stress approach in comparison to the HF benchmark data **a** stress magnitudes and **b** stress orientation. **c** High resolution stress orientation plot together with the azimuth and dip of foliation (anisotropy plane) of KTB rocks (modified after Mastin et al. 1991)

Dight and Dyskin (2008) determined anisotropic rock moduli by testing small sub-cores drilled in different directions (cf. WASM-AE method in Sect. 8.2). Without solving the problem of reliable stress data in anisotropic rock, the plot in Fig. 9.5c is a prerequisite to rate the quality of stress data in anisotropic rock masses.

Stress data from KTB-HB together with the results from KTB-VB are seen in Fig. 9.6. Stress magnitudes in Fig. 9.6a are provided from HF (Engeser et al. 1993; Rummel and Zoback 1993; Zoback et al. 1993; te Kamp et al. 1995a, b), from

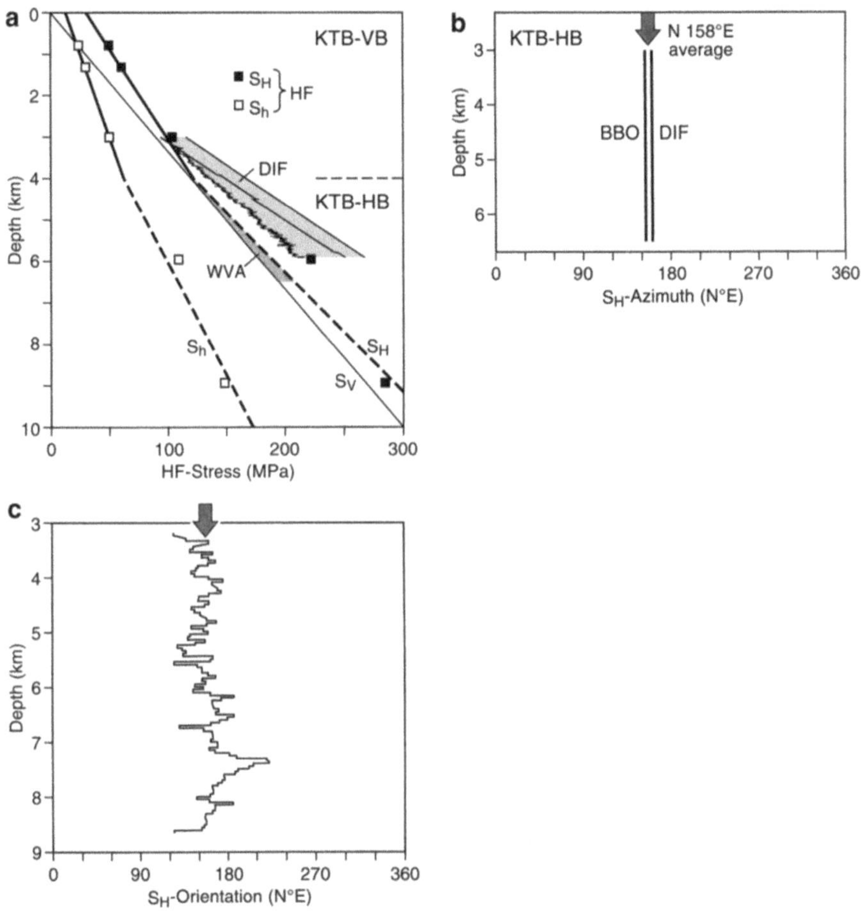

Fig. 9.6 Stress data for KTB-HB together with stress data from KTB-VB from the integrated stress approach **a** stress magnitudes obtained from *HF*, *DIF*, *WVA* and **b** stress orientation from *BBO* and *DIF*. **c** High resolution stress orientation plot from BBO in the depth range from 3.2 to 8.6 km in the KTB-HB (after Brudy et al. 1997)

DIF and DITF (Brudy et al. 1997; Brudy and Zoback 1999) and WVA (Zang et al. 1996a). The integrated stress measurement strategy indicates a linear increase of stress magnitudes with a depth down to 9 km in the continental crust at KTB. As pointed out by Zoback et al. (1993) and later by Brudy et al. (1997), the KTB stress results support a high-strength upper crust, in which the state of stress is in equilibrium with its frictional strength (Chap. 5). In addition, Zoback et al. (1993) suggest that plate-driving forces in the continental lithosphere in this part of Western Europe are transmitted principally through the brittle upper crust.

In Fig. 9.6b, the average value of the S_H azimuth inferred from breakouts (Brudy et al. 1993) and drilling-induced fractures (Natau and Röckel 1993) for the depth range from 3 to 6 km in the KTB-HB are visualized. Data were obtained from four-arm caliper measurements with the Borehole Geometry Tool (BGT), the acoustic borehole televiewer (BHTV) and the electrical imaging tool Formation MicroImager (FMI), see Chap. 7. The average value of the two average values from BBO and DIF is N158°E, which compares well to the average value for the S_H azimuth from the pilot hole N162°E (Fig. 9.5b). For reference, the most reliable stress orientation data from BBO in the depth range from 3.2 to 8.6 km are shown in Fig. 9.6c (Brudy et al. 1995). For this profile, BBO orientations from BHTV measurements between 3.2 and 6.8 km depth were used; below this depth the BBO of four-arm caliper data recorded by BGT and FMS were used (Brudy et al. 1997). Stress orientations are averaged over depth sections of 50 m. Brudy et al. (1997) conclude that except an abrupt change in stress orientation at about 7.2 km (Fig. 9.1b, fault system), the orientation of S_H is quite constant over the entire depth interval (Fig. 9.6c, variation from 135° to 180°).

In Fig. 9.7, the S_H-azimuth at KTB (average value from KTB-VB (N162°E) and average value of KTB-HB (N158°E)) is placed into context to the paleostress directions near the drill site (Fig. 9.7a), the World Stress Map data entries (Fig. 9.7b) and the results from swam-quake analysis in the nearby Eger rift (Fig. 9.7c). From Fig. 9.7a, it is evident that the in-situ stress field at KTB (N160°E) deviates by about 30° from the paleostress field (N120°E) as determined from open cracks and fluid inclusions (Vollbrecht et al. 1994) and secondary overcoring of KTB-VB cores (Zang et al. 1996b). In Fig. 9.7b the S_H azimuth at KTB is seen in comparison to stress data from the World Stress Map (Mastin et al. 1991). This data set includes HF test results in the vicinity of the KTB drill site (te Kamp et al. 1995), overcoring (OC), breakout (BBO), as well as stress data from geologic indicators and fault plane solutions of natural seismicity. In Fig. 9.7c, focal mechanisms of swarm quakes (Ibs-von Seht et al. 2006) are placed into a tectonic map (Peterek and Schröder 1997, thick lines) and into structural features from a geology map (thin lines). Three stress orientation rosettes indicate the directions of maximum principal stress, S_1, and minimum principal stress, S_3, in three sub-regions of the area investigated: (1) Vogtland, (2) NE-Bavaria and (3) KTB. Directions shown are average values of each individual region after Ibs-von Seht et al. (2006) findings. The NW-SE strike of maximum principal stress in the Vogtland (N145°E) seems to rotate clockwise by 30° into the NNW-SSE orientation of S_1 at KTB deduced from focal mechanisms of fluid-induced seismicity at KTB. Note that relevant data for in-situ stress at KTB also came from phase (3) of the project (Table 9.1) where massive hydraulic tests in the two open wells produced fluid-induced microearthquakes.

Fig. 9.7 Maximum horizontal in-situ stress azimuth at KTB in comparison to **a** the corresponding paleostress directions near the drill site (modified after Vollbrecht et al. 1994; Zang et al. 1996b), **b** an extract from the European Stress Map showing trajectories of the most compressive horizontal stress as determined by various indicators (modified after Müller et al. 1992) *OC* Overcoring, *HF* Hydraulic Fracturing, *BBO* Borehole Breakouts, *NS* Natural Sesmicity and **c** results from swarm-quake analysis in the nearby Eger rift (modified after Ibs-von Seht et al. 2006) *FL* Franconian Line, *MLF* Marianske Lazne Fault, *MB* Marktredwitz Basin (after Wolter 1987)

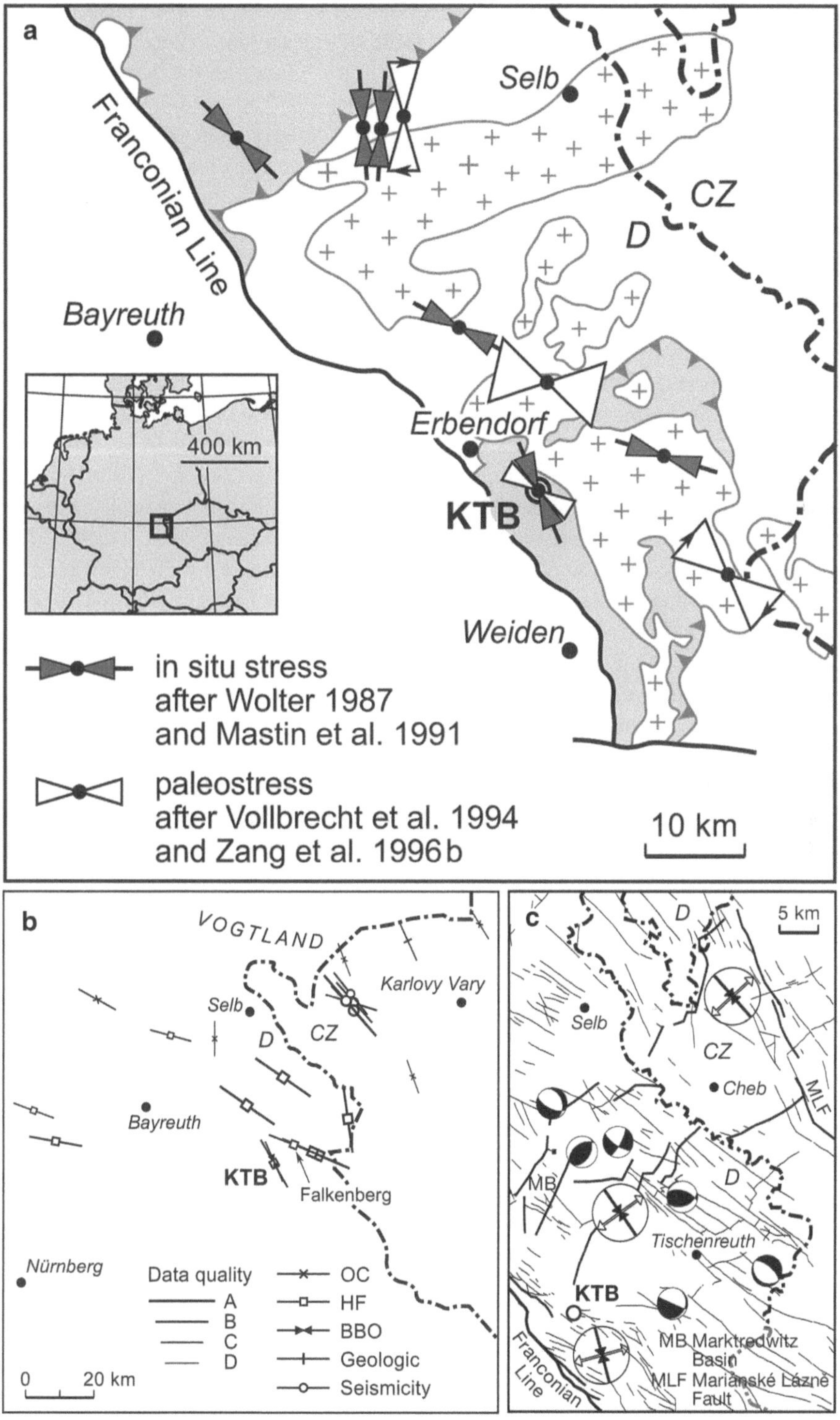
a
Selb
CZ
D
Franconian Line
Bayreuth
400 km
Erbendorf
KTB
Weiden
in situ stress
after Wolter 1987
and Mastin et al. 1991
paleostress
after Vollbrecht et al. 1994
and Zang et al. 1996b
10 km
b
VOGTLAND
Karlovy Vary
Selb
D
CZ
Bayreuth
KTB
Falkenberg
Nürnberg
Data quality
A
B
C
D
OC
HF
BBO
Geologic
Seismicity
0
20 km
c
D
5 km
Selb
CZ
Cheb
MLF
MB
D
Tischenreuth
KTB
Franconian Line
MB Marktredwitz Basin
MLF Mariánské Lázně Fault

The 1994 short-term (24 h) fluid-injection test at 9.1 km depth in the KTB-HB produced about 400 microearthquakes (Table 9.1, phase (3)). A total of 200 m^3 of heavy brine was injected into the ~70 m open hole section near the bottom of KTB-HB at rates of up to 600 l min^{-1} (Zoback and Harjes 1997; Jost et al. 1998). Seismicity during the experiment was monitored by one borehole sonde in the KTB-VB at about 4 km depth and 73 surface, short-period seismometers. Fluid-induced seismicity (FIS) was located maximum 100 m away from the hole, and only above 9.1 km depth with magnitudes smaller than $M=1.2$. Most FIS events had a strike-slip faulting mechanism with NNW trending P axis (Zoback and Harjes 1997) assumed to align parallel to the S_H azimuth at the drill site. Since very small (≤1 MPa) pore pressure perturbations were found to trigger FIS, Zoback and Harjes (1997) concluded that Byerlee's law ruling the coefficient of sliding friction (Chaps. 3, 5) is valid to 9.1 km depth and that the continental crust at KTB is critically stressed. Ito and Zoback (2000) examined fracture permeability and in-situ stress in the KTB pilot hole using core and logging data. In the entire depth range studied (3–7 km), they found rock with permeable fractures which lie close to the Coulomb failure line for a coefficient of friction of about 0.6. Contrarily, this implies that critically-stressed faults in the Earth's crust are also the most permeable faults.

The 2000 long-term (60 days) fluid-injection test in the well KTB-HB was hampered by a casing leakage at about 5.4 km depth (Baisch et al. 2002). A surface seismic network consisting of 39 three-component seismometers and one borehole seimometer at 3827 m depth in the well KTB-VB (yielding a low magnitude threshold $M_W=-2.5$ for events, Bohnhoff et al. (2004)) was used to map 2800 FIS events over 3 months duration including the injection period. During the injection phase, 4000 m^3 fresh water were pumped over a period of 60 days at flow rates between 30 and 70 l min^{-1} in the KTB-HB. In terms of fluid-induced seismicity, the variation of well head pressure with time is important, whereas in Table 9.1 only peak values of the pressure are listed. Bohnhoff et al. (2004) found dominating strike-slip faulting mechanisms with inferred maximum principal stress direction oriented N-S ± 15° by analyzing 125 fault plane solutions from the 2000 FIS events.

Triggering mechanisms of fluid-induced seismicity at KTB are discussed by Baisch and Harjes (2003) and Rothert et al. (2003) based on the model suggested by Healy et al. (1968). In the *Healy model*, increasing fluid pressure along pre-existing faults causes a reduction of the effective normal stress on the fault plane resulting in the release of tectonic stress. Fluid-induced seismicity occurs at fractures where elevated pore pressures (Chap. 5) exceed the critical value to cause Coulomb failure (Chap. 3). Baisch and Harjes (2003) assume the Healy model valid at KTB because (1) FIS moment tensor orientations are consistent with the orientation of the local in-situ stress field from the integrated stress approach, and (2) repeating FIS events (seismic multiplets) show hypocenter locations generated from the same fault plane. FIS hypocenters of the 2000 fluid injection test were found to group along the outer rim of FIS events located after the 1994 fluid injection test (Baisch et al. 2002, Baisch and Harjes 2003). Most of the hydraulic energy induced (10^{11} J) is converted into potential energy to rise the ground-water level at KTB; only a small fraction is converted into seismic energy (10^8 J), as pointed out by Baisch and Harjes (2003). They therefore argue that the fracture network is able to memorize the previous maximum fluid pressure it has experienced during the field test (see Kaiser effect, Sect. 8.2). Crustal memory indicates that shear stress relaxations caused by fluid injection in a tectonically passive region such as KTB persisted for at least 6 years (1994–2000). Shapiro and Dinske (2007) were more sceptical about the applicability of the Kaiser effect in hydraulic-fracture-related seismicity. By analyzing microseismicity data from Carthage Cotton Valley gas field in East Texas (USA), the authors found that the Kaiser effect is absent on reopening the hydraulic fracture, but is clearly observable after the termination of the active fluid injections.

According to Ito and Zoback (2000), the most permeable faults are those which are critically stressed. Therefore, detecting zones of high permeability or high hydraulic diffusivity allows locating high stressed regions in the Earth's crust. Rothert et al. (2003) estimated the hydraulic diffusivity at KTB drill site based on a method published by Shapiro et al. (1997, 1999, 2000). Shapiro et al. (1997) obtained the spatial position, r of the FIS triggering front in an effective, isotropic, homogeneous, poroelastic medium at time t

$$r = \sqrt{4\pi D t}, \tag{9.1}$$

where D is the scalar hydraulic diffusivity [$m^2 s^{-1}$]. Note that also in the Perkins-Kern-Nordgren (PKN) model known from the theory of hydraulic fracturing (Economides and Nolte 2000), the half-length of the hydraulic fracture is given by the square root of the injection time, t. The FIS front is regarded as spatial surface which separates the regions of the relaxed (earthquakes) and unrelaxed (no earthquakes) pore-pressure perturbations. Calculated D-values at KTB do not show an inverse relationship with depth, as expected for a crust with progressively closing faults with depth. In contrast, D-values increase from $0.004\,m^2 s^{-1}$ at 5.4 km depth to $0.05–0.3\,m^2 s^{-1}$ at 9.1 km depth (Rothert et al. 2003). A complete analysis of the D-tensor by Rindschwentner (2001) showed that (1) the principal components of D_{ij} are oriented parallel to the Franconian Line (Fig. 9.7) and (2) the horizontal projection of the principal axis is parallel to the S_H azimuth at the drill site, N160°E. FIS events are strongly correlated with the dip of the seismic reflector SE1 (Fig. 9.1b). This was inferred from the shape of FIS hypocenter distributions being spherical at the 5.4 km injection test where no fault system was crossed by the KTB-HB, and being planar elongated when the hole crossed the high reflectivity zone SE1 (Rothert et al. 2003).

The result of the 1-year production test (2002–2003) and the 10-months injection test (2004–2005) are summarized in Kümpel et al. (2006). During the 10-months injection test, a total of $84{,}600\,m^3$ of fresh water was pumped into the open hole section of KTB-VB. About 3000 FIS events were detected by the borehole seismometer and 150 FIS events were located by the local seismic surface network (Shapiro et al. 2006). Seismicity started only after injecting a fluid volume approximately equivalent to the amount previously extracted (crustal memory, Table 9.1). The crystalline crust at KTB was mechanically stable when pore pressure was reduced below its natural level by the production test, but became instable (producing earthquakes) when small (~0.1–1 MPa) positive pore-pressure perturbations were created by the fluid injection test.

Exercise 9.1 Triggering front of fluid-induced seismicity at KTB.

(a) Visualize the triggering front, r (km) of fluid-induced events at KTB in the time interval $0 \leq t \leq 1000\,\text{h}$ for two different scalar hydraulic diffusivities, $D = 0.005\,\text{m}^2\,\text{s}^{-1}$ and $D = 0.05\,\text{m}^2\,\text{s}^{-1}$. Use Eq. (9.1).
(b) What is the physical meaning of the area under the curves plotted in (a)?
(c) At what distance from the hole, r FIS events occur 3 weeks after injecting fluids at 9.1 km depth assuming a rock diffusivity of $D = 0.05\,\text{m}^2\,\text{s}^{-1}$?

Note-Box In drilling the top third of the continental crust at KTB we can conclude that (1) Byerlee law is valid from surface to 9.1 km depth and (2) the crust is in brittle failure equilibrium at temperatures of 265°C. The integrated stress measurement strategy revealed a strike-slip faulting regime from 805 to 9101 m, while below 4 km the maximum horizontal stress gradient is close to the vertical stress gradient. The best estimated in-situ S_H direction from borehole and core methods coming from both wells is N160°E. The S_H azimuth obtained from stress inversions of fluid-induced earthquakes at KTB is N180°E, the average value of S_H azimuth of Central Europe from WSM is N145°E, and the paleostress direction at KTB is N120°E. Crustal memory, a type of larger scale Kaiser effect, indicates that shear-stress relaxations caused by fluid injection in a tectonically passive region persist at least for 6 years after pumping has stopped. The fact that the hydraulic diffusivity at KTB aligns with in-situ stress and that fluid-induced seismicity aligns with the prominent seismic reflector (fault system), indicates that the rock-mass anisotropy, also at borehole scale, controls in-situ stress orientations.

9.2 Nuclear Waste Site Olkiluoto, Finland

There are very few substances on Earth which receive the same care and attention or generate such strong emotions as do radioactive waste. An exceptional and increasing amount of resources have been directed towards finding a permanent solution to their disposal and to closing the nuclear fuel cycle. This effort has largely been focussed on conditioning, packaging and burying the waste and spent nuclear fuel in a suitable geological environment at an appropriate depth. Rock stress magnitudes and orientation are of utmost importance for location, construction and long-term safety of a waste repository (Stephansson 1997).

M9.2 Sjöberg

Sweden and Finland are two countries where ca. 50% of the electricity is produced by nuclear power. Sweden has three power plants with nine units in operation and Finland has four nuclear power plant units in operation and a fifth unit under construction. The total amount of spent fuel from the lifetime of the reactors is about 6000 t of uranium for each of the countries. The final disposal plans for the two countries are based on the so-called KBS-3 concept, which was developed by the Swedish waste handling organisation SKB. The basis of the concept is the multi-barrier principle in which the spent nuclear fuel is implemented using several engineer barriers that support one another. The idea is that any deterioration of the performance of one barrier does not jeopardize the total safety of long-term isolation of the spent fuel. In the Swedish and Finnish safety concept, the long-term isolation and the disposal system is planned to retain radioactive releases into the geosphere and biosphere for at least 100,000 years.

The underground repository will be built in the bedrock of granites and gneisses at a depth of 400–700 m. The principal of KBS-3 disposal concept on multiple engineered barriers by the copper canister, bentonite buffer and tunnel backfill in a fractured rock mass is shown in Fig. 9.8. The spent nuclear fuel assemblies are placed in copper canisters with a cast iron insert. Before the canister is placed in the about 8 m deep deposition hole, the space between the canister and the bedrock is filled with hard compacted bentonite blocks. Bentonite is a volcanic ash and when it is compacted, it has a very low permeability and thereby prevents groundwater from circulating around the canister and causing corrosion. Bentonite is also plastic and therefore protects the canister from failure due to minor displacements in the surrounding rock mass. The deposition tunnels and the surface connections via tunnels and shafts are backfilled and sealed with low-permeable natural material to ensure that the repository and the surrounding bedrock are restored to a natural

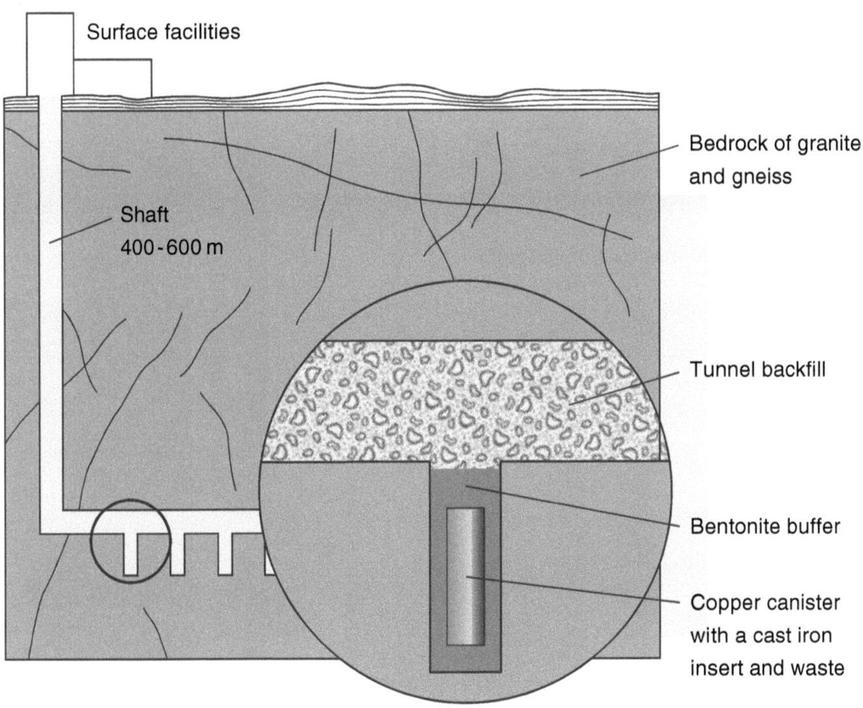

Fig. 9.8 Spent nuclear fuel disposal concept according to the Swedish KBS-3 system with multiple engineered barriers. The repository is located at a depth of 400–600 m in hard rock

state. Backfilling also prevents unauthorized access to the repository. The depth of the repository and the geometry and orientation of the deposition tunnels are governed by the state of stress at the site.

Prior to selection of a site for the repository spent nuclear fuel and radioactive waste in Finland, geological and geophysical investigations, including stress measurements with overcoring and hydraulic methods, were conducted in four different communities in Finland. Based on a screening procedure, one site—Olkiluoto—was selected by the nuclear waste managing organization Posiva Oy.

Olkiluoto is a large island about 10 km^2 and separated from the main land in the east by a narrow strait (Fig. 9.9). The Olkiluoto nuclear plant with two reactors in operation and the third under construction is located in the western part of the island together with an underground repository for low and intermediate radioactive waste. The suitability of Olkiluoto as a location for spent nuclear fuel has been investigated over a period of more than 15 years by means of ground and air-based methods and from shallow and deep (300–800 m) boreholes and trenches. The central part of the island has been selected as a site area for the rock characterization with the aim of finding the most suitable location for the deposition tunnels for spent fuel canisters. Inside the site area, Posiva has started the tunnelling work to reach the rock characterization facility and underground laboratory called ONKALO to be located at 420 and 500 m level.

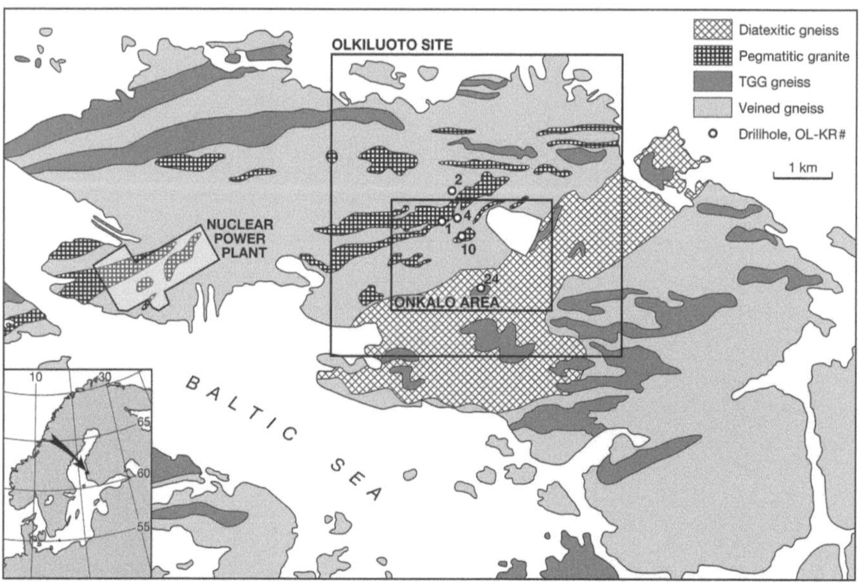

Fig. 9.9 The Olkiluoto nuclear power plant and the site area for location of a repository of nuclear waste in Finland. Also shown is the location of the rock characterization facility area *ONKALO* and the major rock types on the island. More than 40 deep diamond boreholes have been drilled. Overcoring stress measurements was used in boreholes *OL-KR10* and *OL-KR24* and hydraulic fracturing in boreholes *OL-KR1, 2, 4* and *10* (modified after Andersson et al. 2007)

The rocks on the island of Olkiluoto are Precambrian in age and can be divided into high-grade metamorphic rocks and igneous rocks. The metamorphic rocks include various migmatite gneisses, veined gneisses, diatexitic gneisses and tonalite-granodiorite-granitic gneisses (TGG-type). Igneous rocks comprise pegmatitic granites and sporadic narrow diabase dykes on the island, Fig. 9.9. The veined gneisses are the main rock volume of the area and they are highly anisotropic. Composite foliation is fairly constant over large distances with the dip direction being ca. 30° to the southeast.

The stress measurements with overcoring and hydrofracturing technique in the deep boreholes at Olkiluoto were done over a long period of time from the late 1980s to 2003 (Klasson and Leijon 1990; Ljunggren and Klasson 1996; Sjöberg 2003). A recent compilation of the results from the two methods is presented by J.A. Hudson in Andersson et al. (2007) and depicted in Fig. 9.10. When using hydraulic fracturing stress measurements, the stresses in the plane perpendicular to the borehole axis can be determined (Chap. 7). The vertical stress is estimated from the weight of the overburden. From the overcoring stress measurements, the full, three-dimensional stress tensor can be determined and from the tensor, the stresses in the horizontal and vertical direction have been calculated.

All stresses increase with depth and the least horizontal and vertical stresses are about the same. The upper regression line for the vertical stress corresponds to the weight of the overburden. The maximum horizontal stress is distinctly larger than the other principal stresses, which is typical for glaciated terrains of Precambrian rocks (notice the difference in stress scale of the diagrams in Fig. 9.10). The orientation of the maximum horizontal stress shows a large variability between different methods and between different boreholes. The average orientation of the maximum horizontal stress at Olkiluoto is about E-W, which differs from the general NW-SE orientation of the maximum horizontal stress for northern Europe, (cf. European stress map in Chap. 10). Since the magnitude of the vertical and least-horizontal stress is close, the stress regime could change at great depth from thrust-faulting to strike-slip. Stress data after applying the Kaiser effect (Sect. 8.2) of the gneiss from borehole OL-KR14, are presented in Fig. 9.10 (Lehtonen 2005). The method seems to over-predict the vertical stress, but remaining results are within the regression lines.

The nuclear waste management companies Posiva in Finland and SKB in Sweden have jointly supported further developments of the overcoring and hydraulic methods and data interpretation for rock stress determination. Some of the results from this effort have been presented in a special issue of the *International Journal of Rock Mechanics and Mining Science* (Hudson and Cornet 2003). The improvements are related to transient strain analyses of the recorded strains during the overcoring, another development looks for possible core damage during overcoring and, finally, new stress evaluation computer codes allow for the calculation of the stress tensor for anisotropic material—an option which is important for stress analysis of the gneissic rocks at Olkiluoto. Based on these developments, Hakala and Sjöberg (2006) have developed a system to rank overcoring stress data with respect to strain responses from overcoring and biaxial testing with the following qualitative

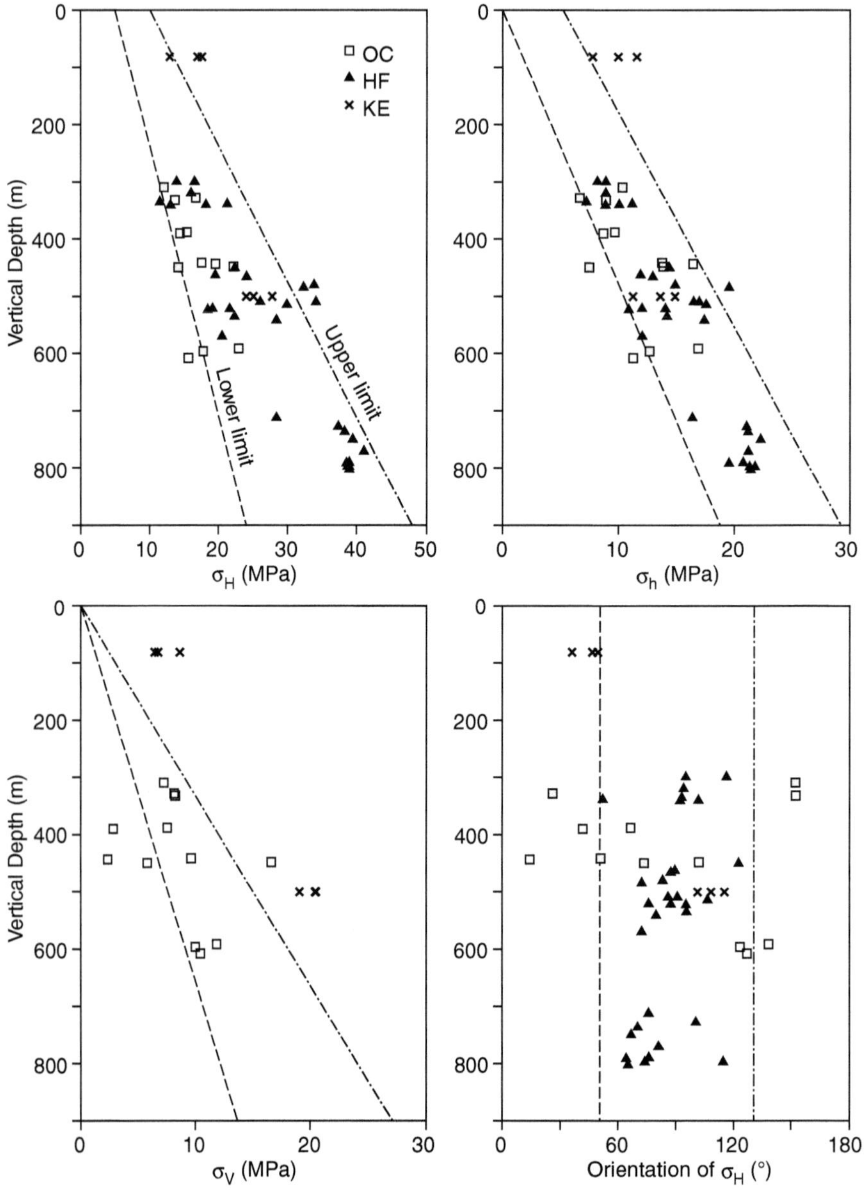

Fig. 9.10 Magnitude of horizontal and vertical stress and orientation of maximum horizontal stress from overcoring, hydraulic fracturing and Kaiser effect methods applied at Olkiluoto, Finland. Linear regression lines are shown for overcoring and hydraulic fracturing except for vertical stress where only overcoring was used in the regression (after Andersson et al. 2007)

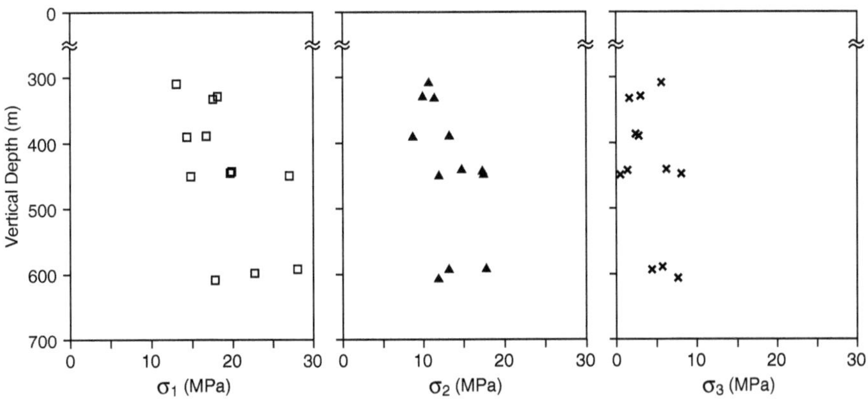

Fig. 9.11 Re-analyzed overcoring principal stress data ranked good and/or moderate from deep boreholes at Olkiluoto (modified after Andersson et al. 2007)

rankings: good, moderate, poor and rejected. When applying this method to all the data presented in Fig. 9.10, only 12 out of 18 measurements were ranked as good and/or moderate. The re-analyzed normal stresses are, in general, somewhat lower in magnitude than the original values (Fig. 9.11).

The re-analyzed magnitude of the three principal stresses σ_1, σ_2, and σ_3 versus depth are shown in Fig. 9.11. The scatter in the data makes it difficult to determine a single, representative stress profile of each of the three principal stresses versus depth. The full stress tensor has been calculated for the re-analyzed data and the orientation of the three principal stresses for each of the measuring points are shown in Fig. 9.12.

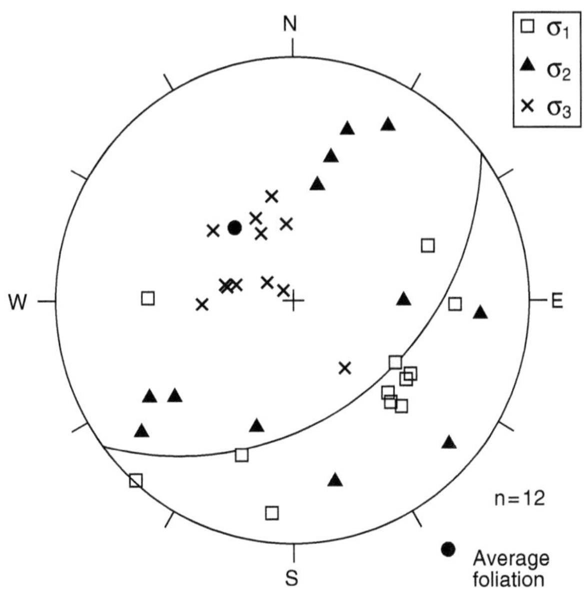

Fig. 9.12 Stereographic projection of the principal stresses from the re-analyzed overcoring stress data from the deep boreholes at Olkiluoto (modified after Andersson et al. 2007)

Several interesting interpretations can be done from analyzing the data in Fig. 9.12. Firstly, we notice that all the data for the minimum principal stress σ_3 coincide with the orientation of the average pole of the foliation, meaning that the minimum stress is oriented perpendicular to the foliation (planar structure in the gneiss). Further, the majority of the maximum and intermediate principal stresses are oriented close to the great circle to the mean foliation pole, meaning that the maximum and intermediate principal stresses are located in the foliation plane. At Olkiluoto, the bedrocks are known to have been through five different phases of deformation. The third phase of plastic deformation has generated the major foliation in the area. Later phases of deformation have re-folded the foliation and this can be interpreted from the variation of the orientation of the stresses in the plane of the foliation. Analysis of the laboratory rock mechanics testing of the deformability of the gneisses from Olkiluoto assuming transversely anisotropy shows a deformation anisotropy of veined gneisses of $E/E' = 1.3$ where E is Young's modulus parallel to the foliation and E' perpendicular to the foliation (cf. Sect. 4.4.1). Hence, the direction of the least-principal stress tends to coincide with the direction of minimum deformability. Notice that the re-analyzed data presented in Fig. 9.12 has considered the rock anisotropy in the analysis of the principal stresses.

Note-Box The present knowledge of the state of stress for the Finnish nuclear waste repository at Olkiluoto shows that the orientation of the maximum horizontal stress is nearly E-W and differs from the general trend of a NW-SE orientation for most of Central and North Europe. The data supports the interpretation of Olkiluoto being located in a thrust faulting stress regime where $S_H > S_h > S_v$ and where the maximum horizontal stress at about repository level at 500 m depth varies between 15 and 31 MPa. The state of stress at the site seems to be governed by the orientation of the foliation in the gneisses so that the least-principal stress is oriented perpendicular to the foliation and the maximum and intermediate stress is lying in the foliation plane. Additional stress measurements in the tunnel of the underground facility ONKALO and the repository at depth will certainly narrow the variance of the stress data and increase the overall knowledge of the state of stress at the site.

9.3 San Andreas Fault Observatory at Depth, USA

M9.1 Zoback

The San Andreas Fault Observatory at Depth (SAFOD) is a comprehensive project to drill to a depth of about 3 km into the hypocentral zone of repeating micro-earthquakes on the San Andreas Fault. SAFOD is located in Parkfied, California, 1.8 km southwest of the surface trace of the San Andreas Fault (Fig. 9.13), which represents a major transform plate boundary and corresponding stress conditions (Chap. 11). Annual slip of the Pacific Plate relative to the North American Plate is about 48 mm (Fig. 9.13, arrow), resulting in continuous stress accumulation between plates and discontinuous stress relaxation during earthquakes along the fault.

Fig. 9.13 Map of California showing the San Andreas Fault and the location of the deep continental drilling site of *SAFOD*. Portions of the fault that ruptured in major historical earthquakes are shown in *solid line* and the creeping, microseismically active parts with *dots* (after Hickman et al. 2004)

SAFOD is motivated by the need to answer fundamental questions about the physical and chemical processes, including rock stresses, controlling faulting and earthquake generation within a major plate-bounding fault. Despite intense research over many decades, numerous fundamental questions about the physical and chemical processes acting within the San Andreas Fault (SAF) and other plate-bounding faults remain open. For example, it is often proposed that high pore pressure exists within SAF and that the variation of pore pressure will affect the fault behaviour (Rice 1992; Byerlee 1993). These hypotheses are not proven. How large, plate-bounding faults such as SAF accomplish the slip required of them by global kinematics given ambient stresses and the lack of a distinct heat flow anomaly at the trace of the fault are other outstanding questions.

The goals of SAFOD are to establish a multistage geophysical observatory with fault-zone monitoring activities scheduled to continue for at least 20 years (Hickman et al. 2004). Drilling, sampling and downhole measurements and monitoring directly in the fault zone will advance our understanding of the mechanical behaviour of a major active fault zone at hypocentral depth. Fault zone rocks and fluids have been sampled for laboratory analyses and extensive downhole geophysical measurements, including estimation of stresses from geophysical loggings, have been performed in a pilot borehole drilled in 2002 west of the San Andreas Fault. The main hole was drilled vertically to 1.5 km depth (phase 1, 2004) and then steered northeast to a vertical depth of 3.1 km (phase 2, 2005), taking it through the fault zone. In the third phase (2007), four multilateral core holes were drilled off the main hole, each extending 250 m into regions of earthquake repeaters. SAFOD is being outfitted as a long-term fault observatory. Monitoring includes wide-dynamic range seismological observations of earthquake nucleation and rupture together with continuous monitoring of temperature, fluid pressure and strain during the earthquake cycle.

Throughout eastern and southern California and the state of Nevada, a uniform NNE-SSW direction of maximum horizontal compressive stress is observed from focal mechanisms, borehole breakouts, hydraulic fracturing and geological observations (Zoback et al. 1980; Zoback et al. 1987; Zoback et al. 1989; Townend and Zoback 2000), Fig. 9.14. This is consistent with the superposition of stresses arising from lateral variations in lithosphere buoyancy in western United States and

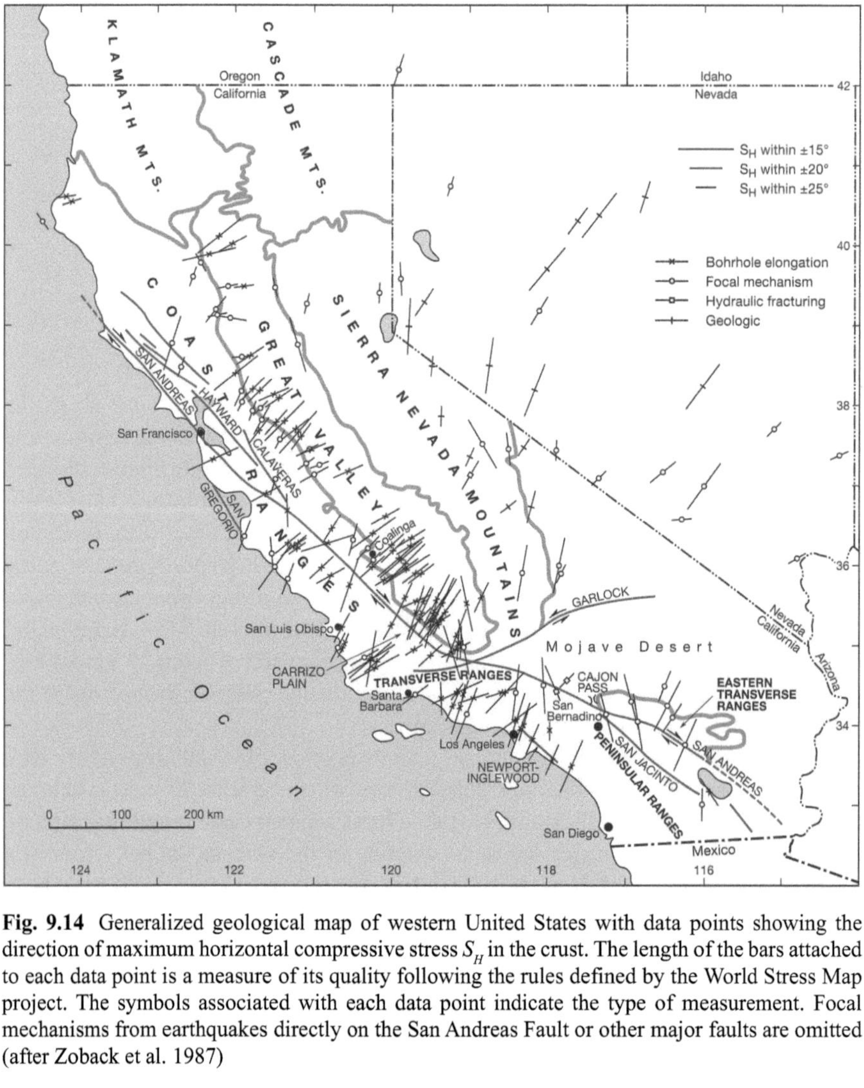

Fig. 9.14 Generalized geological map of western United States with data points showing the direction of maximum horizontal compressive stress S_H in the crust. The length of the bars attached to each data point is a measure of its quality following the rules defined by the World Stress Map project. The symbols associated with each data point indicate the type of measurement. Focal mechanisms from earthquakes directly on the San Andreas Fault or other major faults are omitted (after Zoback et al. 1987)

the far-field interaction between the North American Plate and the Pacific Plate (Flesch et al. 2000) as the latter is displacing relative to North American Plate with a velocity of 48 mm year^{-1} (Fig. 9.13) derived from continuous GPS data (Murray and Segall 2001).

In central California and closer to the San Andreas Fault, the axis of maximum horizontal stress rotates and lies at a high angle to the strike of the fault, cf. Fig. 9.14. The rotation of the maximum horizontal stress is assumed to be due to a low frictional angle of the rock along the fault. The frictional coefficient of common rock types measured in the laboratory is in the range of $\mu = 0.6–1.0$ (Chap. 3). Applying these values to a typical strike-slip fault such as the San Andreas will generate an angle of 30–35° between the axis of maximum horizontal stress and the strike of the fault. The Andreas Fault is not a well-oriented Anderson fault, as observations of S_H azimuth close to the fault support. Hardebeck and Michael (2004) found an angle of 55 ± 22° for the S_H direction with respect to the fault as determined from stress inversion of earthquakes. Townend and Zoback (2004) used 70 earthquakes closer than 10 km distance from the fault trace and found 68 ± 7°. These two azimuths from earthquake stress inversion can be compared with borehole stress data from SAFOD at the interval 1600–1750 m (55 ± 11°), and at borehole interval 2050–2200 m (69 ± 14°). According to Townend (2006), much of the horizontal stress field adjacent to SAF is provided by lateral variations in crustal buoyancy (Chap. 11). SAF in conclusion is oriented at higher angles (~60°) to the S_H azimuth than a typical Anderson fault based on Byerlee friction and hydrostatic pore pressure (~30°). For very high angle, 80°–85°, the ratio between shear and normal stress acting on the fault is very low, leading to high normal stress across the fault and extremely low friction. Townend and Zoback (2000) estimated the shear traction on the plane parallel to the fault at the depth of 7.5 km to be as little as ~10 MPa and the normal traction ~180 MPa, which corresponds to a friction coefficient of 0.06.

The angle between the orientation of the maximum horizontal stress and the strike of the fault in southern California is on average about 65°. This is slightly less compared to central California and shows that the fault slips at a low shear/normal stress ratio corresponding to friction coefficients of 0.2–0.3. Overall we can conclude that the state of stress in southern and central California is consistent over a 600 km-long section of the fault and that the stress field is in agreement with the results of calculations of buoyancy-derived stresses from the thermally uplifted Basin and Range province and far-field shear related to the plate motion of the North American and Pacific plates (Flesch et al. 2000). From this it can be concluded that the San Andreas Fault is a weak fault compared to intraplate faults. On a broader scale, the origin of the North American intraplate stress field is discussed in Humphreys and Coblentz (2007). Stress orientations in most of westernmost North America are consistent with right-lateral transform coupling with the Pacific Plate (Liu and Bird 2002) and oblique subduction of the Juan de Fuca slab beneath Washington and Oregon State (Wang et al. 1997b).

The state of stress in seismogenic depths in the vicinity of the SAF has been the subject of intense investigation and controversy over time. The SAFOD project aims, among other things, to provide data to verify the lack of heat-flow anomaly from the SAF, and how this information can be related to the stress and strength of the fault. One approach to quantify the average resisting stress on active faults has been to examine evidence for heat generation from frictional sliding. If the fault had a high frictional strength, the frictional resistance over geological time should produce a detectable heat flow anomaly centered on the fault. Brune et al. (1969) and later Lachenbruch and Sass (1980) conducted heat flow measurements over the San Andreas Fault but found no anomalies of frictional heat flow and therefore concluded that the average strength of the fault is very low and does not exceed 20 MPa.

The SAFOD drilling site is located 1.8 km southwest of the vertical San Andreas Fault on a segment of the fault that moves through a combination of aseismic creep and repeating micro-earthquakes, Fig. 9.13. *Aseismic creep* of a fault is defined as a time-dependent fault deformation without any detectable earthquake activity. The figure shows a map of California SAF and sections that ruptured in major historical earthquakes. The creeping and micro-seismically active segment of the fault is indicated in dots. This 300 km long, straight section of SAF runs roughly between Cholame Valley in the SE (Parkfield earthquake) to San Juan Bautista in the NE (Loma Prieta earthquake) with a creep rate as high as 28 mm year^{-1} (Moore and Rymer 2007). The drill site lies at the extreme northern end of the rupture zone of the 1966, magnitude 6 Parkfield earthquake, and the most recent in a series of events that have ruptured the fault five times since 1857. Also, the Parkfield region is the most comprehensively instrumented section of a fault worldwide and a large number of geophysical surveys have been conducted in the area before and after the selection of the SAFOD drilling site. A compiled geological and geophysical data set from a section across the Parkfield region around the SAFOD pilot hole is presented in Fig. 9.15. The San Andreas Fault in the Parkfield area juxtaposes relatively undeformed magnetic and non-magnetic Salinian granite basement on the southwest against deformed Franciscan Complex rocks on the northeast (McPhee et al. 2004). Magnetic rocks within the Salinian basement cause the high magnetic anomaly to the southeast. Serpentinite is common in the area, giving rise to a magnetic high and gravity low in the NE part of the profile.

To prepare for the scientific and technical foundation for SAFOD, a 2.2 km deep pilot hole was drilled in the summer of 2002. The deviation of the pilot hole was less than 3° from vertical for most of the length of the borehole. A comprehensive suite of geophysical logs was collected in the pilot hole from a depth of 775–2150 m in the highly fractured Salinian granite west of the San Andreas fault zone (Boness and Zoback 2004). High-quality logs were obtained over the open-hole interval of the pilot hole using acoustic and electrical borehole imaging tools. The geophysical logs showed extensive borehole breakouts and local sections with drilling-induced tensile fractures in response to tectonic stress concentrations at the borehole wall. The azimuths of breakouts and drilling-induced fractures differ by 90° and offer reliable information on the orientation of the horizontal principal stresses in vertical boreholes (Sect. 7.3). Azimuths of borehole breakouts and drilling-induced fractures together with inferred azimuths of maximum horizontal stress (S_H) are shown in Fig. 9.16.

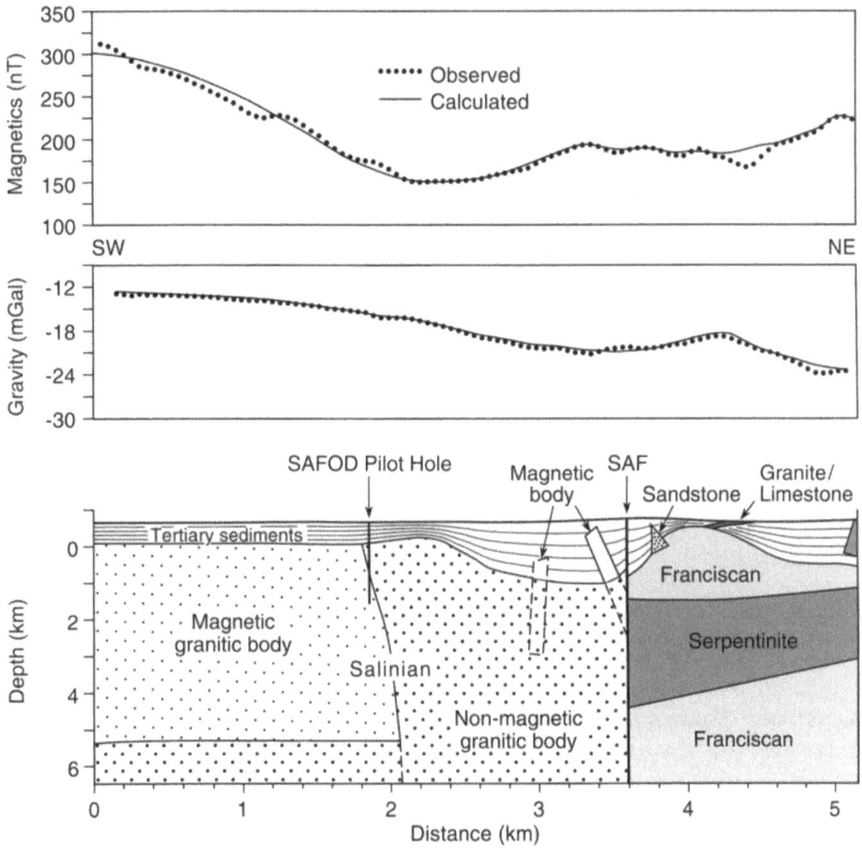

Fig. 9.15 Geological model with fit to ground magnetic and gravity data for a NE-SW profile across the San Andreas Fault and through the *SAFOD Pilot Hole* in the Parkfield area (after McPhee et al. 2004)

Hickman and Zoback (2004) presented the mean orientation of S_H over 10 discrete, 150 m long depth intervals in the borehole and found a local variation of principal stress within each interval due to slip on nearby faults. They also observed a shift in azimuth of S_H with depth relative to the strike of SAF so that the angle changed from ~44° at 850–1000 m depth to ~69° at 2050–2200 m. The angle between the azimuth of S_H and the strike of SAF is in agreement with that observed in Central California at a greater distance from the fault (Townend and Zoback 2004) and with inversion of focal mechanisms from microearthquakes along the creeping section of the fault (Provost and Houston 2001). Three depth intervals in the pilot hole were free from borehole breakouts (Fig. 9.16) and are interpreted as zones of stress relief associated with slip on faults penetrated by the pilot hole. These zones are also characterized by low seismic velocity, high velocity anisotropy and low density (Boness and Zoback 2004). Using data on

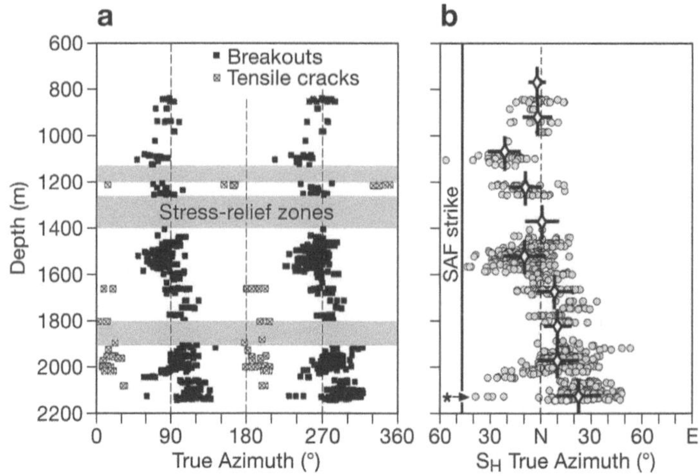

Fig. 9.16 Azimuths of **a** stress-induced borehole breakouts and drilling-induced tensile fractures and **b** maximum horizontal principal stress S_H versus depth in the pilot hole of SAFOD. Stress-relief zones in the borehole correspond to stress-relieved zones where the borehole penetrates major weak zones in the rock mass. The strike of the San Andreas Fault (N46°W) is depicted for reference. Notice the reorientation of breakouts and maximum horizontal stress with depth (modified after Hickman and Zoback 2004)

unconfined compressive strength (C_o), tensile strength (T_o) and elastic constants of the rock and knowing formation fluid pressure (P_p), drilling mud weight, extent of wellbore cooling and breakout width, the stress magnitudes of maximum and minimum horizontal principal stress can be determined (Chap. 5).

Figure 9.17 shows the stress constraint from a depth of 1671 m in the Pilot Hole of SAFOD following the methodology developed by Moos and Zoback (1990) (Sect. 5.4). Contours of unconfined compressive strength (C_o), tensile strengths (T_o) with uncertainties and vertical stress S_V from the weight of the overburden are shown for the SAFOD Pilot Hole together with limits of crustal frictional strength for the regimes normal faulting (NF), strike-slip faulting (SS) and reverse faulting (RF). Result of this analysis indicates that the stress magnitudes at depth 1671 m are $S_H = 113 \pm 14$ MPa and $S_h = 49 \pm 9$ MPa.

Hickman and Zoback (2004) made a similar stress estimate for four additional depth intervals in the pilot hole (Fig. 9.18). The magnitude of the vertical stress S_V was estimated as the weight of the overburden and the pore pressure was taken to be hydrostatic. The stress state shows a gradual transition from strike-slip faulting stress regime in the upper part of the pilot hole to reverse faulting where S_h become the least-principal stress at greater depth.

From the measured orientation of the maximum horizontal stress in Fig. 9.16 and the estimated stress values at different depth in the pilot hole presented in Fig. 9.18, Hickman and Zoback (2004) calculated the ratio of shear to effective normal stress on planes parallel to SAF. The result of the calculation, assuming effective stress state, shows a stress ratio equal to the apparent coefficient of friction of about 0.6

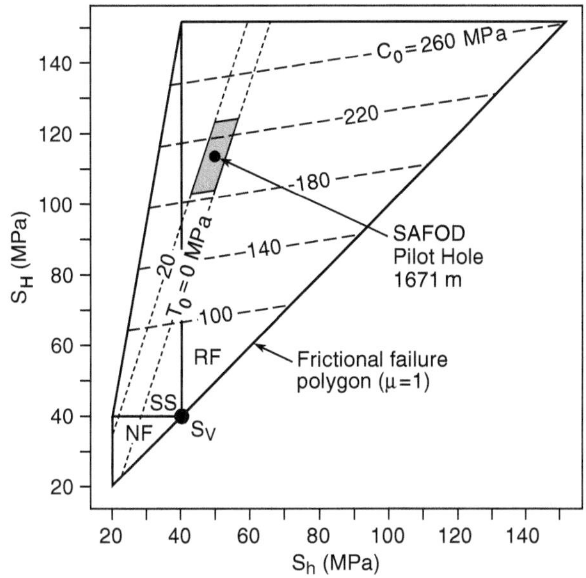

Fig. 9.17 Stress constraint of S_H and S_h from depth 1671 m in the *SAFOD Pilot Hole* using unconfined compressive strength C_o derived from *P*-wave velocity measurements and estimated tensile strength T_o. *NF*, *SS* and *RF* denote normal, strike-slip and reverse faulting regimes (after Hickman and Zoback 2004)

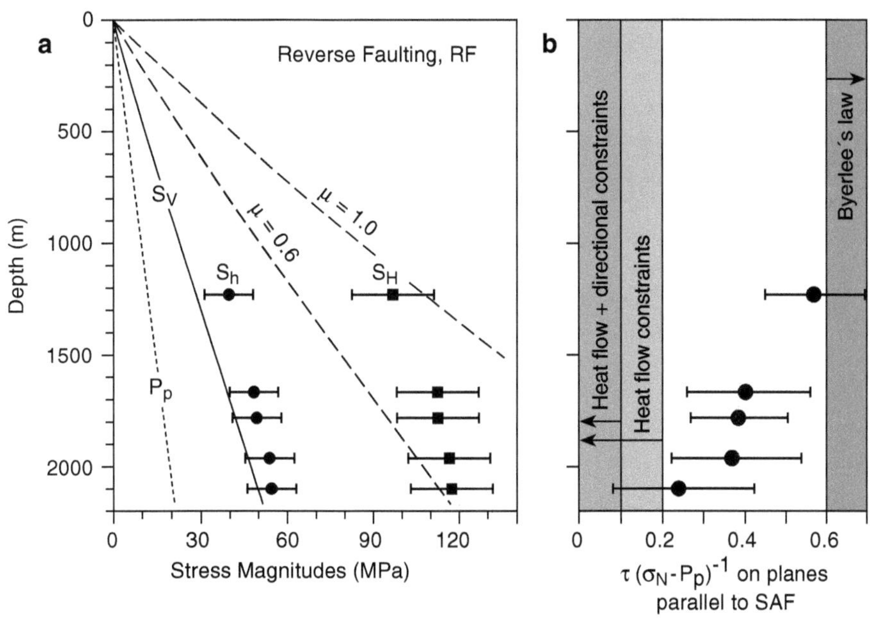

Fig. 9.18 Estimated stress magnitudes and stress ratio versus depth in the Pilot Hole of SAFOD. **a** A reverse faulting stress regime in the upper part of the pilot hole is changed towards a strike-slip faulting regime at depth. **b** Calculated shear stress versus effective normal stress onto planes parallel to the San Andreas Fault together with bounds of Byerlee's Law and heat flow constraint (after Hickman and Zoback 2004)

for the shallowest depth in the borehole. This value is in accordance with the results from application of Byerlee's Law. At greater depth, the value of the ratio diminishes and approaches low values of about 0.2, which indicates a low apparent coefficient of friction and small shear strength.

Drilling and logging in SAFOD Main Hole revealed a 250 m broad fault zone, with several discrete, localized narrow faults (2 m wide) characterized by low P- and S-wave velocities (Zoback et al. 2008). Fault creep at two of these narrow zones has been identified by shearing events of the casing. Stress orientations to within 100 m of the activated fault zone are nearly perpendicular to the strike of the fault. Instead of high pore pressure, abnormally-high (significantly above lithostatic) magnitudes of all three principal stresses are observed inside the core of the fault. All of the observations are consistent with the SAF being a weak fault embedded in an otherwise strong crust. The weak fault hypothesis result from low friction coefficients (Table 3.1, wet talc $\mu=0.16$) and not from super-lithostatic pore pressure conditions or dynamic weakening.

To learn more about the San Andreas Fault being a strong or a weak fault in a high- or low strength crust, the reader is referred to a contribution by Hardebeck and Michael (2004). Here, three end-member models of the strength of the SAF are discussed: the strong fault strong crust approach (SFSC) by Scholz (2000), the weak fault strong crust approach (WFSC) by Brune et al. (1969), Zoback et al. (1987), Mount and Suppe (1987), Zoback (1991), Lachenbruch and Sass (1992), Zoback (2000), Provost and Houston (2001), Sobolev et al. 2003, Hickman and Zoback (2004), Williams et al. (2004), Moore (2005)—the most popular model, and the weak fault weak crust approach (WFWC) by Molnar (1992) and Hardebeck and Hauksson (2001). Apart from the SFWC approach (strong fault in weak crust) which never appeared in literature, all hypotheses above are based on two weak geophysical boundary conditions: (1) the absence of a detectable heat flow anomaly and (2) the orientation of S_H nearly perpendicular to the strike of the fault. The first condition results from the argument that significant frictional heating should accompany the slip on a fault with friction coefficients for rocks, $\mu>0.6$, as determined in the laboratory. The second weak boundary condition causes controversy over the interpretation of inverted stress data from earthquake focal mechanisms, in particular in the near-field of the SAF.

We do not want to join these discussions nor the myriads of discussions about untested theoretical models (Hickman et al. 2004) concerning SAF being strong (Scholz 2000), weak (Chery et al. 2004), or very weak (Chery et al. 2001; Lynch and Richards 2001) due to high pore pressure (Fitzenz and Miller 2004) or low friction coefficients (Chery et al. 2004). Our concluding remark in this context is to look for strong boundary conditions in identifying the actual strength of SAF. One of these may be the appearance of talc at the San Andreas fault (Wibberley 2007). Serpentinized ultramafic rock have been associated with creeping faults in California for a long time (Irwin and Barnes 1975). Moore and Rymer (2007) reported on talc-bearing serpentinite found in cuttings from SAFOD. Talc is known as the softest natural mineral and with a friction coefficient as low as $\mu \sim 0.1$, it is discussed now as the big weakener for SAF. Maybe 2% of talc found in the cuttings is not enough to weaken an entire fault, but the weak mineral found is a stronger boundary condition for fault models as the heat flow or stress directions discussed. Physical fault models are only as good as the physical input parameters used for their setup. Another big weakener discussed in literature is melt lubrication of faults during earthquake rupture. Evidence of low friction coefficients come from high velocity friction experiments in the laboratory and natural exhumed faults (di Toro et al. 2006). Recent reviews in understanding the mechanics, structure and evolution of fault zones were given by Wibberley et al. (2008) and Ben-Zion and Sammis (2009). SAFOD is the right project to better understand the actual strength of faults and the mechanics of earthquakes in the crust.

Note-Box The state of stress at SAFOD is characterized by a gradual transition from strike-slip faulting stress regime in the upper part of the pilot hole to reverse faulting at greater depth. The azimuth of S_H shifts with depth relative to the strike of the San Andreas Fault so that the angle changes from ~44° at 850–1000 m depth to ~69° at 2050–2200 m depth. Apparent friction coefficients calculated from borehole stress data indicate values of 0.6 for shallower depth and 0.2 for greater depth. The present results give support to the hypothesis that the San Andreas Fault at depth behaves as a weak fault in a strong crust. The weak fault hypothesis results from low friction coefficients. Two big weakeners discussed are weak minerals (wet talc) and melt lubrication during earthquake rupture. Stress data in combination with other geophysical and geological data are likely to provide the final input to a long scientific controversy about the strength and deformability of a hypocentral zone of the San Andreas faulting system.

Chapter 10
Generic Stress Data

In this chapter we present and discuss in-situ stress data in terms of magnitude-depth profiles (Sect. 10.1) and stress-orientation maps (Sect. 10.2). We refer to the relationship of stress-state scaling (Sect. 10.3) and how to find the Best Estimated Stress Model for a study area (Sect. 10.4). In this sense, *generic* means commonly used mathematical relationships to present stress data and their general agreement to this relationship for different sites selected.

10.1 Magnitude Depth Profiles

As we know from reference state-elastic stress models of the Earth's crust discussed in Chap. 5, stress magnitudes increase linearly with depth as long as Byerlee's law is valid in the brittle upper crust. According to KTB results (Sect. 9.1), this law about rock friction properties is valid at least in the top third of the continental crust beneath Europe down to 9.1 km and approaching temperatures of 265°C close to the brittle ductile transition.

There are a number of ways of displaying profiles of stress magnitudes versus depth in the Earth's crust (Table 10.1). One category of data and plots deals with stress magnitudes or a combination of magnitudes (Table 10.1A). Common methods within this category are to plot the mean horizontal stress with depth or to plot the two horizontal and vertical stress components with depth. While McGarr (1980) used maximum shear-stress data, Stephansson et al. (1986) preferred mean-stress data, and Engelder (1993) focussed on octahedral shear stress versus depth (see Table 10.1A).

There is still no final statement on whether stress increases linearly with depth. As we know from Chap. 1, McGarr (1980) concluded that on average the maximum shear stress in rock increases linearly with depth in the upper 5 km of the Earth's crust. However, he makes a distinction between the maximum shear-stress gradient of crystalline and sedimentary rock. From Engelder's (1993) analysis, it follows that, also for the octahedral shear stress, the highest values came from crystalline rocks of the Appalachian Mountains (Moodus, Connecticut). The lowest values

A. Zang, O. Stephansson, *Stress Field of the Earth's Crust*,

DOI 10.1007/978-1-4020-8444-7_10,

Table 10.1 Presentation of stress with depth in the Earth's crust

A	Stress magnitudes or their combination, MPa	References
1	$\frac{1}{2}(S_h + S_H)$; mean horizontal stress	Voight (1966b), Herget (1974), Lindner and Halpern (1977)
2	S_h, S_H, S_v; principal stress components	Haimson (1977), Arjang (1989), Baumgärtner et al. (1993)
3	$\frac{1}{2}(S_1 - S_3)$; maximum shear stress	McGarr (1980)
4	$\frac{1}{3}(S_H + S_h + S_V)$; mean stress	Stephansson et al. (1986)
5	$\tau_{OCT} = \sqrt{\frac{2}{3}J_2}$; octahedral shear stress	Engelder (1993)
B	**Dimensionless stress ratios, stress coefficients**	**References**
1	$k = \frac{S_H+S_h}{2S_V}$; lateral stress coefficient	van Heerden (1976), Brown and Hoek (1978)
2	$k_H = \frac{S_H}{S_V}$; maximum horizontal stress coefficient	Bieniawski (1984)
3	$k_h = \frac{S_h}{S_V}$; minimum horizontal stress coefficient	Rummel et al. (1986), Herget (1987), Savage et al. (1992)
4	K; probabilistic stress coefficient	Sen and Sadagah (2002)

came from sedimentary rocks of the Michigan Basin. In general, stress magnitudes appear to increase with depth most rapidly in areas of active faulting (Moodus, San Andreas Fault). The least increase of stress magnitudes with depth is found in a stable continental interior (Canadian Shield, Michigan Basin, or KTB site in Bavaria).

A second category of plots of stress versus depth is using dimensionless stress ratios (Table 10.1B) rather than stress magnitudes. Among them, the lateral-stress coefficient (van Heerden 1976; Brown and Hoek 1978), the ratio of maximum horizontal to vertical stress (Bieniawski 1984) and the ratio of minimum horizontal to vertical stress (Rummel et al. 1986; Herget 1987; Savage et al. 1992) are discussed. Sen and Sadagah (2002) introduced a probabilistic approach for the lateral stress coefficient variations at any given depth.

Not all components of the stress tensor can be determined for several of the in-situ stress measurements. Information on the vertical stress component (Table 10.1A, S_V) and the lateral-stress coefficient (Table 10.1B, k), however, is usually available. Therefore, the pragmatic way suggested by Brown and Hoek (1978) interpreting collected worldwide stress data in terms of the vertical stress and the lateral-stress coefficient is one example of demonstrating generic trends in stress magnitudes versus depth in the Earth's crust (Fig. 10.1).

In Fig. 10.1a, the vertical stresses S_V do cluster around the line of the lithostatic stress model (Fig. 5.1a) that corresponds to a mean rock density of 2750 kg m^{-3}. The vertical stress component is correct in the sense of a linear best-fit regression line. Deviations are due to variations in local topography and local geologic heterogeneities. In some cases, S_V differs by about five times the predicted component. We

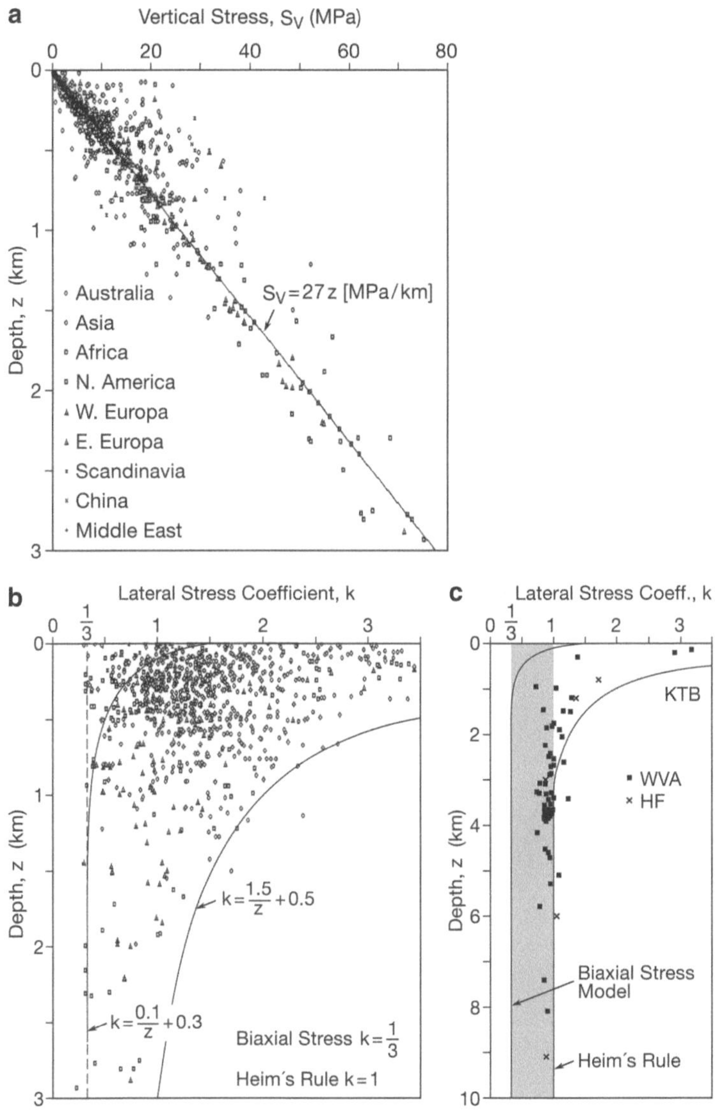

Fig. 10.1 Global compilation of **a** vertical stress and **b** lateral stress coefficients, k down to a depth of 3 km in the Earth's crust according to Brown and Hoek (1978) (modified after Brady and Brown 2004). For reference in **c**, the k values at *KTB* from borehole and core-based methods down to 9.1 km depth are shown. In **b**, the upper and lower bound of k values from Brown and Hoek (1978) are shown. In **c**, Heim's rule ($k=1$) and the biaxial stress model ($k=1/3$) are indicated by *vertical lines*

conclude that Eq. (5.1) with generic unit weight provides a good predictive estimate of the averaged vertical stress from global stress data, but certainly is not appropriate to provide a correct stress estimate at any specific location. For local stress estimates, an integration of rock density versus depth according to Eq. (4.1) has to be used instead. For this purpose, the local variation of rock density with depth has to be measured, e.g., from drill cores or cuttings while a borehole is penetrated into the Earth's crust. This implies that it is best to measure rather than estimate the vertical stress component (Sect. 10.4).

In Fig. 10.1b, the ratio of the average of the minimum S_h and maximum horizontal stress component S_H to the vertical stress component S_V, the so-called *lateral-stress coefficient* (k value, see Table 10.1B)

$$k = \frac{S_h + S_H}{2S_V} \tag{10.1}$$

is shown. At depths of 300 m or less, the k value is found to range from 1 to 3.5. At greater depth, the range of k narrows considerably, and below 4 km the observed values are generally less than one. Neither Heim's rule (Chap. 1, $k=1$) nor the biaxial stress model (Sect. 5.2, $k=\nu(1-\nu)^{-1}=1/3$) provides a good explanation for the data trends, which show a highly nonlinear decrease of the k value with depth. Therefore, Brown and Hoek (1978) fitted two envelopes that provide rough bounds to the data:

$$\frac{0.1}{z} + 0.3 \leq k \leq \frac{1.5}{z} + 0.5, \tag{10.2}$$

where z is taken in kilometres. Both envelopes are of the same algebraic form, $k=a/z+b$, as is predicted by a spherical shell model of the Earth's crust (e.g. McCutchen 1982; Sheorey 1994; Aydan 1995). In this model, the crust is taken to be a thin, elastically isotropic and homogeneous shell, resting on top of a thermo-elastic mantle and deforming due to gravity. Assuming the thickness of the crust to be very small compared to the Earth's radius (Chap. 1), the asymptotic values of 0.3 and 0.5 in Eq. (10.2) correspond to the values predicted by a spherical shell model if the Poisson ratio ν of the rock is varied between 0.23 and 0.33.

Data shown in Fig. 10.1a, b for depths down to 3 km, are of interest for most mining projects. In Fig. 10.1b, data from Brady and Brown (2004) are bounded on the lower side by $k=1/3$ (biaxial stress model). The convergence of $k$values to a value of unity at depth is consistent with the principle of time-dependent elimination of shear stress in rock (Heim's rule, $k=1$). The final conclusion from inspection of Fig. 10.1a, b, however, is that in-situ stress in rock cannot be calculated but rather must be determined through measurements.

In Fig. 10.1c, we show published stress data from the KTB boreholes at Windischeschenbach, Germany in terms of lateral stress coefficient k from surface to 9.1 km depth. For this purpose, we used hydraulic fracturing data (Fig. 10.1c, crosses, data from te Kamp et al. 1995a, b) and wave-velocity analysis data from measurement of drill cores (Fig. 10.1c, squares, data from Zang et al. 1996a). From this it follows that k values at depth greater than 4 km in fact approach unity (Heim's rule) rather than the lower bound value of $k=1/3$. With the exception of two data points from wave velocity analysis (WVA), the lateral stress coefficient at KTB is

Table 10.2 Relation between vertical and horizontal stress for simple elastic homogeneous Earth stress model, and rock mass with transversely and orthotropic anistropy

Case 1	Poisson ratio, v	Lateral-stress coefficient, $k = \frac{\nu}{1-\nu}$	$S_V = kS_H$
isotropic	$v=0$	$k=0$	$S_H=0$
	$v=1/4$	$k=1/3$	$S_V=3S_H$
	$v=1/3$	$k=1/2$	$S_V=2S_H$
	$v=1/2$	$k=1$, Heim's rule	$S_V=S_H$
Case 2*	Poisson ratios $v_{xy}=v_{yx}\equiv v=0.25$	Lateral-stress coefficient, $k = \frac{\nu_{xz}}{1-\nu}$	$S_V=kS_H$
transversely isotropic	$v_{xz}=2v$ $v_{xz}=0.5v$	$k=2/3$ $k=1/6$	$S_H=1.5S_V$ $S_H=6S_V$
Case 3#	Poisson ratios $v_{xz}\neq v_{yz}\neq v_{xy}$	Lateral-stress coefficient, $k = \frac{\nu_{xz}}{1-\nu}$	$S_h \neq S_H$
orthotropic		$k_x = \frac{\nu_{xz}+\nu_{yz}\nu_{xy}}{1-\nu_{xy}\nu_{yx}}$ $k_y = \frac{\nu_{yz}+\nu_{yx}\nu_{xz}}{1-\nu_{xy}\nu_{yx}}$	$S_V=7.5S_H$

* Sedimentary layers without fractures
\# Rock mass with three mutually perpendicular sets of fractures

within the upper bound of the Brown and Hoek (1978) envelope (Eq. (10.2)). The lower bound at KTB is a k value of 0.7 at a depth of about 1 km.

In Table 10.2, the relation between vertical and horizontal stresses for simple elastic earth models are listed according to the physical properties of the rock mass (cf. Chap. 4: isotropic, transversely isotropic, orthotropic). In the isotropic case with zero Poisson ratio, the application of vertical stress does not induce any horizontal strain and, therefore, no horizontal stress. The other end-member with the Poisson ratio of a viscous fluid, $v=0.5$ results in horizontal stresses equal the applied vertical stress and Heim's rule is valid. In between, typical Poisson ratios for rock are taken to be one-fourth (or one-third), indicating that the induced vertical stress is three times (or two times) the horizontal stress, respectively. For transversely isotropic rock, as is the case for sedimentary layered rock, the horizontal stresses may be as high as six times the vertical stress assuming the Poisson ratio perpendicular to bedding one-half Poisson ratio parallel to bedding (0.25). This anisotropic end-member scenario can result from a layered rock mass which was fractured during uplift in the Earth's crust. In the orthotropic case, three Poisson ratios of the rock have to be identified, which can be the result of three mutually perpendicular sets of discontinuities in the rock mass. For these cases, Poisson ratio can reach values larger than 0.5 (Min and Jing 2004; Jing and Stephansson 2007).

The shaded vertical bars in Fig. 10.1b, c give the range of k values from one-third to one that was predicted from linear elastic, isotropic rock models (Table 10.2). With increasing depth k values given by the envelope formula, Eq. (10.2) trend towards the lower one-third and upper one-half bound. Thus, for significant depth, simple elastic models provide some indication of the lateral stress coefficient. Highest k values are expected near the surface ($z=0$) because the vertical stress is zero at a free surface (Chap. 2, principal stress plane). With reference to Fig. 10.1a, the calculation of the vertical stress component by a simple gravity model gives a reasonable prediction of the overall trend. Horizontal stress components, however, do not follow the trends predicted by simple elasticity theory, except asymptotically for depths greater than 4 km (Fig. 10.1c). Heim's rule is valid for this depth range. The reason for k values larger than one (up to the value of six) for shallow depth is due to tectonic forces and is best discussed in terms of stress orientations, as done in Sect. 10.2.

Dimensionless horizontal stresses were also used by Rummel et al. (1986) to interpret hydraulic fracturing data from 500 tests in 100 boreholes at 30 different geographical locations. The average over all data only neglecting results with abnormal stress-depth relations (Auburn, Auriat, Bad Creek, Fjällbacka) yields

$$\begin{aligned} k_h &= \frac{S_h}{S_V} = \frac{0.15}{z} + 0.65 \\ k_H &= \frac{S_H}{S_V} = \frac{0.27}{z} + 0.98 \end{aligned}, \tag{10.3}$$

where z is the depth in kilometres. Note that it is difficult to compare stress results from Eq. (10.3) with those from Eq. (10.2) since Brown and Hoek (1978) used two envelopes of the lateral-stress coefficient, while Rummel et al. (1986) treated minimum and maximum horizontal stress coefficients separately (Table 10.1B). Averaging the two envelopes in (10.2) and calculating an average lateral stress coefficient (10.1) from (10.3), Brown and Hoek's (1978) a value of the spherical shell algebraic formula turns out to be about four times the Rummel et al. (1986) a value. In addition, Rummel et al.'s (1986) spherical shell off-set value b turns out to be about twice that of Brown and Hoek's (1978) average value. Since it is the minimum horizontal stress component which is determined most reliably in hydraulic tests (Chap. 7), the k_h value in Eq. (10.3) may be more significant. Note that the a value of k_h from Eq. (10.3) is much closer to the a value from the lower bond envelope in Eq. (10.2).

Savage et al. (1992) modified the laterally constrained isotropic and homogeneous elastic half-space model under its own weight (Chap. 5, biaxial stress model) by allowing the crust to be subjected to small horizontal (tectonic) strains, ε_h and ε_H. They found the following expressions for the horizontal stresses in the Earth's crust including the effects of gravity and horizontal strains.

$$\begin{aligned} S_h &= \frac{E}{1-\nu^2}\left(\varepsilon_h + \nu\varepsilon_H\right) + \frac{\nu}{1-\nu}S_V \\ S_H &= \frac{E}{1-\nu^2}\left(\varepsilon_H + \nu\varepsilon_h\right) + \frac{\nu}{1-\nu}S_V \end{aligned}. \tag{10.4}$$

In the limiting case of vanishing horizontal strains, these equations yield the lateral constraint conditions governed by Eq. (5.4). In addition, Savage et al. (1992)

modified (10.4) for the case of transversely isotropic and orthotropic rock material (see anisotropic Poisson ratio coefficients in Table 10.2). According to Amadei et al. (1987), the ratio of in-plane to out-of-plane moduli in transversely isotropic rocks varies from 1 to 3. Consequently, they found horizontal stresses to vary between 0.133 and 0.933 S_V for layered sedimentary rocks, which gives a maximum vertical stress value of 7.5 times the horizontal maximum stress value.

A mathematically different approach to interpret horizontal stress ratios in rock was given by Sen and Sadagah (2002). In their approach, the horizontal stress at a certain depth in the Earth's crust is assumed to have a common probability distribution function. Using Chebyshev inequality for random variables, they found an exponential relationship to predict the average stress ratio for a given depth which is contrary to the spherical shell relationships which has a parabolic form in general. Unfortunately, Sen and Sadagah (2002) when applying their probabilistic approach to Brown and Hoek's (1978) data set did not separate for different rock types, which is a pre-requisite for obtaining reliable stress estimation with depth.

Note-Box Stress magnitude-depth profiles can be separated into the linear increase of individual stress components with depth (z) and the nonlinear decrease of dimensionless stress ratios with depth ($1/z$) in agreement with simple elastic spherical shell models of the Earth's crust. At a greater depth, the range of the lateral-stress coefficient k narrows considerably and below 4 km the observed values generally approach one, Heim's rule of isotropic lithostatic stress is valid. At great depth, the lateral stress coefficient is smaller than one. Close to the surface of the Earth, horizontal stresses increase because the vertical stress decreases to zero at a free surface. Depending on the rock mass property, topography, erosion and tectonics near surface horizontal stresses can reach six times the vertical stress component.

10.2 Orientation Maps and Smoothing

We leave the generic section about stress magnitudes and focus on the display of generic stress orientations in the following. Horizontal stresses in the Earth's crust can best be visualized and rated in terms of stress orientation maps, which for the human eye should be weighted and smoothed according to the distribution and reliability of the measured in-situ stress data used. Elevated horizontal stresses are caused by density contrast (Sonder 1990), erosion (Goodman 1989), tectonics (Zoback and Zoback 2002a, b), glaciations and structural stresses. According to Chap. 4 terminology, high horizontal stresses are a superposition of the combined effects of *A1* (density gradients, irregular topography), *A2* (tectonic stress) and structural stresses (Fig. 4.1, *C*). On erosion, the removal of overburden will cause an increase in the lateral stress coefficient. One of the first elastic erosion models was presented by Goodman (1980). However, the reference state of stress during erosion is not clear. The behaviour of Earth stress in the upper crust will be complicated by the burial and uplift of rock masses at geological time scales. This subject turns out to be complex because of the prograde and retrograde modes of geological deformation, the time-dependent effects involved, and multiple erosion sequences.

The relative contribution of tectonic stresses (Fig. 4.1, *A2*, e.g. plate boundary forces), and structural stresses (Fig. 4.1, *C*) on horizontal stresses within the Earth's crust is best illustrated in stress-orientation maps like, for example, in Fig. 10.2. The consistent overall NW-SE trend of the S_H azimuth on the stress map of Western Europe determined with different methods in rocks with different lithologies and ages indicates that large-scale tectonic stress causes the *first-order stress pattern* (Chaps. 4, 11). From a structural geologist view point, the terrane tectonics of the European subcontinent (Sect. 9.1) is not very favourable to determine first-order tectonic stresses. From a geophysical-geodetic view point, the European stress map, however, is an ideal playground to demonstrate the difference and interaction mechanisms between a first-order tectonic stress pattern at plate scale, second-order tectonic stress pattern at mountain range (e.g., the Alps), and third-order tectonic stress pattern at the fault scale. As we see at the end of this section, the key to visualize stress pattern and to interpret stress orientation maps are so-called smoothers.

In Fig. 10.2a data entries from the World Stress Map (WSM 1997 release) are shown (Wehrle 1998). Each data point gives the orientation of the maximum horizontal compressive stress component S_H (Fig. 10.2a, short bar azimuth), the method used for stress determination (Fig. 10.2a, symbol in the centre of the bar; Chap. 7), and the faulting regime (Fig. 10.2a, filling of symbols for *NF*, *SS* and *TF*; Chap. 5). The length of short bars corresponds to the data quality (Fig. 10.2a, *A* quality data with standard deviation in azimuth by ±15°, *B* quality ±20°, and *C* quality ±25°). For identifying stress patterns (Chap. 4, coherent domains in which the orientation of S_H is constant), Fig. 10.2a is not very suitable because (1) there are large lateral fluctuations in data densities throughout Europe, (2) there is a wide spread in orientations at single locations where data density is high (clusters), and (3) there are error bars in any single location due to the data-quality ranking system within the World Stress Map. Because it is impossible for the human eye to take into account (1) to (3) at proper weight, the smoothing of stress orientation maps is required to identify trends.

Fig. 10.2 Stress orientation map of Western Europe. **a** Section of the World Stress Map release 1997 showing Western Europe. Short bars indicate the maximum horizontal stress azimuth. Lengths of bars indicate data quality. *Symbols in the centre of the bars* show the stress-determination method, and *filling of symbols* indicate the faulting regime (*SS* strike-slip, *NF* normal faulting, *TF* thrust faulting). **b** Gridded, smoothed stress map using input data from a and the constant search radius method by Wehrle (1998). Smoothing parameters include the search radius R=250 km, fidelity-smoothing-balance parameter $\lambda=1$, normalized tricubic weight function, and quality ranking $A=1$, $B=0.75$, C=0.5. The error bar in b is the average deviation which gives the misfit of the true S_H azimuth measured and the smoothed, gridded stress value (degrees). Note that the constant diameter of search circle *2R* differs in size for different latitudes in the Mercator projection (after Wehrle 1998)

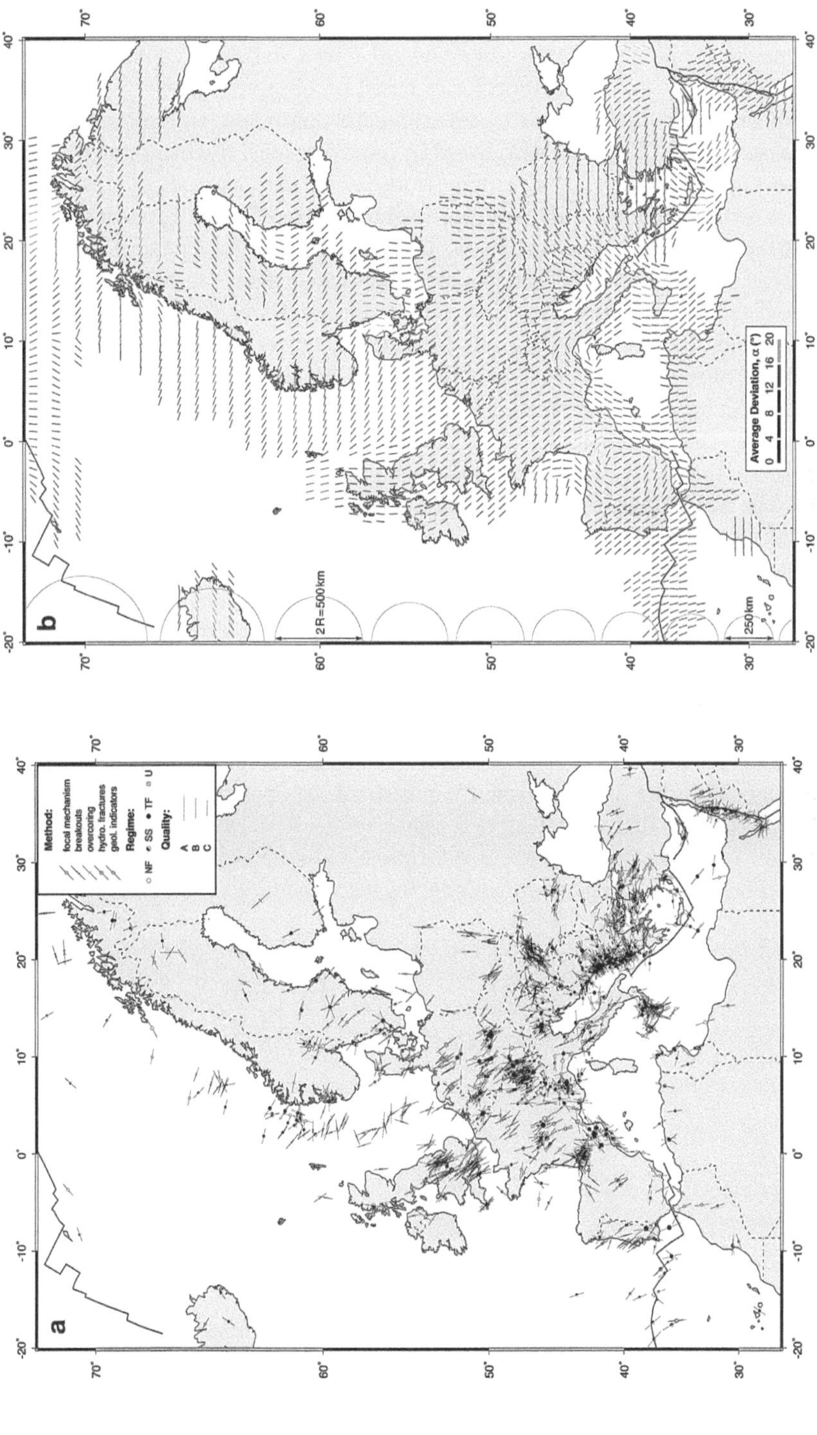
b
2R=500km
250km
Average Deviation, α (°)
0 4 8 12 16 20
a
Method:
focal mechanism
breakouts
overcoring
hydro. fractures
geol. indicators
Regime:
NF SS TF U
Quality:
A
B
C

To understand the process of smoothing and how to rate the smoothing result in Fig. 10.2b, the reader requires information about the smoothing parameters used to generate the plot. We define *stress smoother* as a statistical method for estimating stress pattern and stress trajectories based on observed stress orientation data. Smoothers aid in data analysis by revealing and enhancing pattern present in a set of measurements. This is accomplished by removing local fluctuation in data, while preserving large-scale trends. A framework for the development of smoothers for the use with directional data was proposed by Watson (1985). Hansen and Mount (1990) used Watson's approach to quantitatively predict stress pattern based on the theory of smoothers. They found an algorithm which is able to balance the fidelity to the data and the smoothness. Two runs of the statistical algorithm developed by Hansen and Mount (1990) on a small synthetic data set are illustrated in Fig. 10.3. Each short bar indicates the S_H orientation at a point. In Fig. 10.3a, the stress pattern is calculated with parameters of the algorithm set to emphasize fidelity. Stress trajectories are forced to meander around prescribed, synthetic data points. In Fig. 10.3b, the stress field is calculated with parameters set to emphasize smoothness. The trajectories indicate an almost constant fitted stress field. The degree of smoothness in the fitted stress field is controlled by the user, through the choice of the parameters F_N and λ. The scalar λ is used to place emphasis on fidelity to the data ($\lambda < 1$), or on the smoothness ($\lambda > 1$). Setting $\lambda = 1$ results in an equal weight. The fraction of data points which will be used in any one evaluation of the smoothing function, is denoted by F_N. Intuitively, it is clear that smoothness will increase with increasing F_N.

While Hansen and Mount (1990) used a constant number of the next-neighbours approach, Wehrle (1998) applied a constant search radius for next neighbours in order to weight the distance from the measured data point to the smoothed stress field. Figure 10.4 demonstrates, for a search radius R = 225 km, the tricubic weight of point $\bar{x}_i$ when evaluating stress at location $\bar{x}$. In the example, eight next neighbours (Fig. 10.4a, short bars) are influencing the smoothing process of the data point

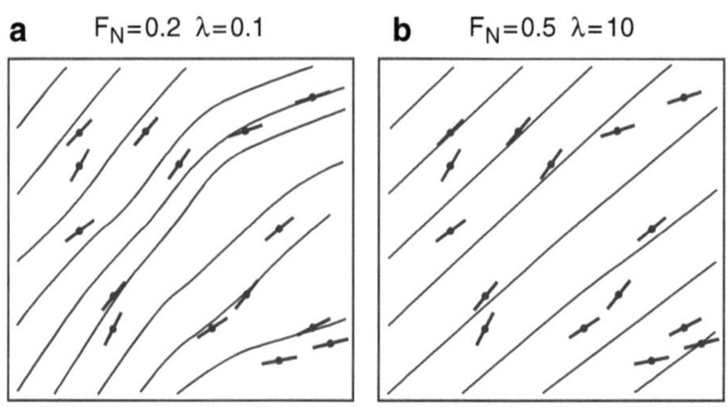

Fig. 10.3 Two runs of the statistical algorithm on a small synthetic data set illustrating the concepts of **a** fidelity to the data ($F_N = 0.2$, $\lambda = 0.1$), and **b** smoothness to the stress data ($F_N = 0.5$, $\lambda = 10$). Each *short bar* indicates the azimuth of maximum horizontal stress at a point. *Solid lines* indicate calculated stress trajectories (after Hansen and Mount 1990)

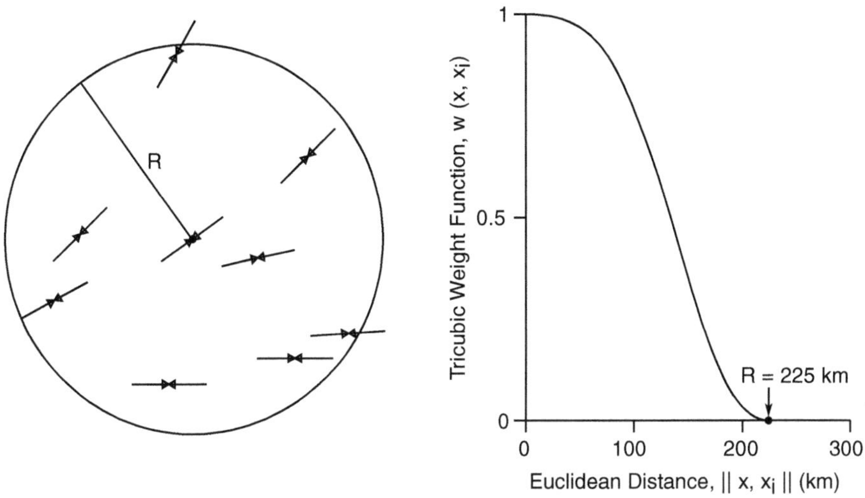

Fig. 10.4 Constant search radius approach by Wehrle (1998). **a** The eight data points in the interior of the circle with radius R are used to calculate the smoothed stress field. **b** The tricubic weight function varies from one in the centre of the circle to zero at the rim. The Euclidean distance is used as a measure on the horizontal axis

located in the centre of the circle. The influence, however, decreases with Euclidean distance from the central data point (Fig. 10.4b). In addition, Wehrle (1998) incorporated weights based on the user's knowledge of the reliability of each data point. The quality weight of the S_H orientation varies according to the World Stress Map data-ranking system ($A=1$, $B=0.75$, $C=0.5$, $D=0.25$, $E=0$). Wehrle (1998) also applied different weight functions. Normalized tricubic weight functions put emphasis on regions with a high density of data measured (cluster amplification), while summarizing next neighbours into bins allows reducing the influence of clusters in the final stress orientation map.

We are now prepared to understand the smoothed S_H-orientation map in Fig. 10.2b. The gridded map was created by Wehrle (1998) by evaluating the estimated stress field at the locations lying along a grid in the region in question, here Europe. The error is given in average deviation (degree) between the smoothed stress field and the measured S_H orientation. Zero deviation results in a solid black, short bar for the S_H azimuth, while a 20° deviation is indicated by a light grey short bar in the gridded map. For this run of the European smoothed stress map, Wehrle (1998) picked a search radius of 250 km, $\lambda=1$ for equal balance of fidelity and smoothness, the data quality *A-C* of WSM data points corresponding to $q=1$, 0.75, 0.5 and a normalized tricubic weight function. The constant search circles have a diameter of 500 km and show an increase with increasing geographic latitude due to Mercator projection. They indicate the region in which the next neighbours are considered for the smoothing operation. Compared to Fig. 10.2a, in Fig. 10.2b the fan-shaped distribution of S_H orientation in the western Alps, the E-W orientation of S_H in western Hungary and small-scale perturbations of the S_H orientation in Spain are resolved. The average deviation (error bar of the smoothed stress orientation map) show that in regions like the Pyrenees and Sicily, the smoothed stress field does not fit the observed data very well (Fig. 10.2b, light grey short bars). In these regions, the stress data either store higher-order tectonic stresses (lateral variations of stresses are smaller than the search radius) like in Sicily, or S_H orientations show large scatter like in the Pyrenees (probably indicating that stress magnitudes are close to isotropic and small changes in the stress field lead to large rotations in S_H orientation). In decreasing the search radius, second-order stress patterns with smaller wavelength can be resolved. For Western Europe (England, France, Germany, Benelux), however, the smoothed S_H-orientation map indicates the existence of a first-order stress pattern with overall NW-orientation of S_H. This Western Europe stress pattern has been identified earlier based on significant less data points (Ahorner 1970; Ahorner 1975; Illies and Greiner 1979; Grünthal and Stromeyer 1986; Klein and Barr 1986; Müller et al. 1992; Müller et al. 1997). In terms of modern plate tectonics, the European stress map is re-evaluated in Sect. 11.1.

Wehrle's (1998) smoothing algorithm can be applied to stress-data sets with strong variation in lateral data density. To vary the region affected by the smoother explicitly allows resolving stress patterns at different scales. If the deviation between a smoothed S_H orientation and an observed one is too large, however, then probably no such stress pattern exists within the diameter of the search circle. Smoothed S_H-orientation maps can be used as input for studies of plate tectonic problems. In this context, we have to calculate stress trajectory maps rather than gridded maps, as shown in Fig. 10.2b. As we know from Sect. 2.4, stress trajectory maps show a series of lines which are tangent to the estimated stress field. The number of trajectories needed to adequately describe a stress field depends on the smoothness of the field. A relatively large number of stress trajectories are required to characterise an irregular stress field (cf. Ex. 2.10 and Fig. 4.7). Wehrle's smoothing algorithm has been discussed in detail by Müller et al. (2003). Additional information about smoothers and how to select data for a study area from the World Stress Map is available on DVD, Chap. 11 WSM with CASMI = Create A Stress Map Interactively.

Exercise 10.1 Read the publication by Heidbach and Höhne (2008). Generate your own S_H orientation map by either using CASMI (Create A Stress Map Interactively) or alternatively by using the web based database interface CASMO (Create A Stress Map Online) from the *World Stress Map* project.

(a) Select a square around your birth place in the World Stress Map (WSM). Set the Map Range so that at least 20 WSM data records are shown. What is the scale of your square?
(b) Select the Data Parameter so that only A-C quality data are shown. Increase your Map Range so that at least 20 azimuths of S_H remain visible. Now, what is the scale of your square?
(c) Separate your data into S_H stress orientation from earthquake focal mechanisms and azimuths from other stress indicators. Is there a difference in stress orientation at shallow depth ($z < 6$ km) and greater depth ($z > 6$ km, earthquakes)?

(d) Create a smoothed S_H-orientation map of your study area by using guidelines from the WSM homepage.

A more recent approach for reconstruction of stress fields in a plane elastic domain from discrete data on stress orientations is given by Galybin and Mukhamediev (1999, 2004). The method was applied to determine the first-order tectonic stress pattern of Western Europe and the Australian continent (Mukhamediev et al. 2006). The direct integration of equilibrium equations leads them to a stress pattern in Western Europe which is oriented N150°E, some 5°off the average S_H orientation published earlier (Müller et al. 1992). However, second- and third-order tectonic stress patterns as indicated in Fig. 10.2b were not resolved. Like in Fig. 10.3b, the approach of Mukhamediev et al. (2006) put emphasis on the smoothness of data. On the Australian continent they identified a stress singulary point (Fig. 4.3) where stress trajectories behave asymptotically and k values approach unity. A second application of the method was shown for Antarctica and the Sunda trench (Galybin 2006), both are examples for stress fields in jointed rock regions.

The statistical computation of stress fields in the Earth's crust has many potential applications. Estimated stress orientations can be used in assigning boundary conditions to finite-element models investigating plate-boundary forces (Sect. 11.1). A comparison of fitted stress fields based on global-scale S_H-orientation compilations with absolute plate-motion vectors provides insight into the plate-driving forces (Sects. 11.2, 11.3). Stress-field orientations in conjunction with geodetic measurements have the potential to contribute to our understanding of crustal deformation processes (Argus et al. 1989; Albarello et al. 1995; Kahle et al. 1998; Campell and Nothnagel 2000). Knowledge of present-day stress fields is important in earthquake hazard and prediction studies. Of interest to the petroleum industry, orientation of hydraulic fractures and flow can be predicted over broad regions prior to exploratory drilling (e.g., stress field detector technology, Smith 2007).

Note-Box Elevated horizontal stresses are due to combined effects of crustal density contrasts, erosion and uplift, tectonic stress and rock structural stresses. Smoothed stress-orientation maps allow us to interpret the influence of tectonic and structural stresses based on the resolution of measured stress data. Smoothers aid in stress-data analysis by revealing and enhancing pattern present in a set of data. Four smoothing parameters, namely λ (the scalar quantity balancing fidelity and smoothness to the stress orientation data), F_N (the fraction of data points used in any one evaluation of the smoothing function), R (the search radius identifying the region which is affected by the smoother) and w (the weight function placing emphasis on data clusters or bins) were discussed. Deviation of smoothed and measured stress orientations is the key to rate stress maps. Using an inappropriate set of smoothing parameters, stress-orientation maps can be misleading.

10.3 Stress State-Scale Relations

By definition, the concept of stress as defined by Cauchy (Chap. 2) is a limiting concept that is applied at a point around which a small area is defined. In the limiting process, the average force per unit area converges to a vector called a stress vector and the resultant moment is assumed to be zero. In continuum mechanics, stress is expressed as a piecewise continuous function in space and time. Because of its intrinsic definition, stress in any material cannot be scale-dependent.

As we know from Chap. 7, when stress measurements are carried out on rock, a spread in the measured values is observed (Fig. 7.1). While measurement errors (Chaps. 7, 8) and statistical errors (Sects. 10.1, 10.2) were discussed earlier, we now focus on the physical reason of intrinsic errors caused by the influence of structural stresses. The spread in data by natural (intrinsic) anisotropy or heterogeneity of the rock material (Sect. 4.4) will depend on how pre-existing defects (anisotropy, heterogeneity) have affected the measured stress values. In analogy to Hudson and Harrison (2000), we define the *representative elementary volume* (REV) as a volume for any given body for which the size of the sample tested contains a sufficient number of defects for the "average" value of stress, $<\sigma>$ to be reasonably consistent with repeated testing. Initially, REV was used in characterizing hydraulic properties of rock, but has been adapted to in-situ stress (Hyett et al. 1986; Hudson and Cooling 1988; Cuisiat and Haimson 1992; Cornet 1993; Amadei and Stephansson 1997). The REV concept is illustrated in Fig. 10.5 for a rock volume containing defects and a set of discontinuities, where REV is marked as a rectangle (Fig. 10.5a). In Fig. 10.5b, the variability of stress data versus the volume of rock is shown.

With small specimen volumes, the absence and presence of defects is highly variable and stress values can be very different from point to point (Fig. 10.5b, $V \to 0$). As the specimen volume is increased, the sample of defects becomes more and

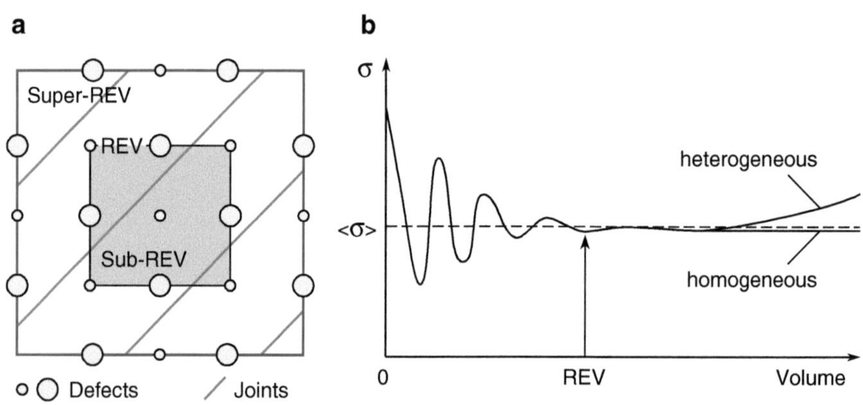

Fig. 10.5 Concept of the representative elementary volume. **a** *REV* adequately describing the rock mass by capturing different defect types and a set of discontinuities, and **b** variability in measured stress values with respect to a sample volume (modified from Hudson and Harrison 2000)

more statistically representative, until the REV is reached at the <σ> stress value on the vertical axis. This concept applies to all rock properties and conditions which are affected by defects, and is especially pertinent for stress data. The paradox however exists because stress is defined as a property at a point (Chap. 2). At zero volumes according to Fig. 10.5b, we expect large variations in measured in-situ stress values because of the unpredictable effect of defects at small volumes (Fig. 10.5a). The paradox arises because we are generally attempting to estimate the in-situ stress which is being applied to a volume greater than REV, but stress is a point property. Certainly, the far-field in-situ stress (Fig. 10.5a, super-REV stress) is the one which we would require for the tectonic stress field. In the design of underground excavation, however, a local near-field structural stress value due to defects acting on small rock volumes (Fig. 10.5a, sub-REV stress) may be critical for the stability of the structure as a whole. In conclusion, the rock volume involved in stress measurements defines the scale of the stress state estimated (cf. Table 7.2).

Nevertheless, in-situ stress data collected worldwide (WSM) show that first-order tectonic stress patterns exist over which the S_H-orientation is quite uniform despite the topography, geologic setting and rock mass properties. For example, a stress pattern oriented NW has been identified for Western Europe (Müller et al. 1992; Stephansson 1993). For well-defined geologic boundaries and adequate boundary conditions, the variation of in-situ stress is quite predictable and has nothing to do with scale effects (Amadei and Stephansson 1997).

Exercise 10.2 Representative elementary volume (REV) in rock. Consider the 2D periodic patterns of two different mineral assemblies in Fig. 10.6.

(a) In both diagrams, mark the unit cell of the pattern which stores the whole symmetry of the 2D rock model (equals REV).
(b) How many unit cells (REVs) fit in each of the models shown?

As we know from Chap. 7, stress measurements involve different rock volumes. According to the REV concept we should expect different values of the stress magnitude for the different rock volumes tested. From Fig. 10.5, it is evident that overcoring stress measurements activating a smaller rock volume compared to hydraulic fracturing from a borehole (Table 7.2) would produce more variation in stress data since the effect of rock irregularities has been reduced. There is a rock volume beyond which the stress scatter is minimal (Fig. 10.5, REV). When dealing with in-situ stress, REV has three basic characteristics (Amadei and Stephansson 1997). First, REV must be small enough so that the in-situ stress is constant and stress gradients can be neglected. Second, REV must consist of an equivalent continuum material where the rock mass is homogenized. Third, REV is useless in understanding sub-REV phenomena (Fig. 10.5a). The variability of the state of stress with sample size has strong implications for stress-measurement strategies, data reduction and presentation (Sect. 10.4). It immediately suggests the idea of measuring stress on the super-REV scale. Figure 10.5 suggests that numerical stress analyses of fractured rock is consistent with continuum methods for rock volumes $V>$REV and discrete methods for $V<$REV. How do we reconcile the state of stress at a point with what we measure in-situ from different rock volumes?

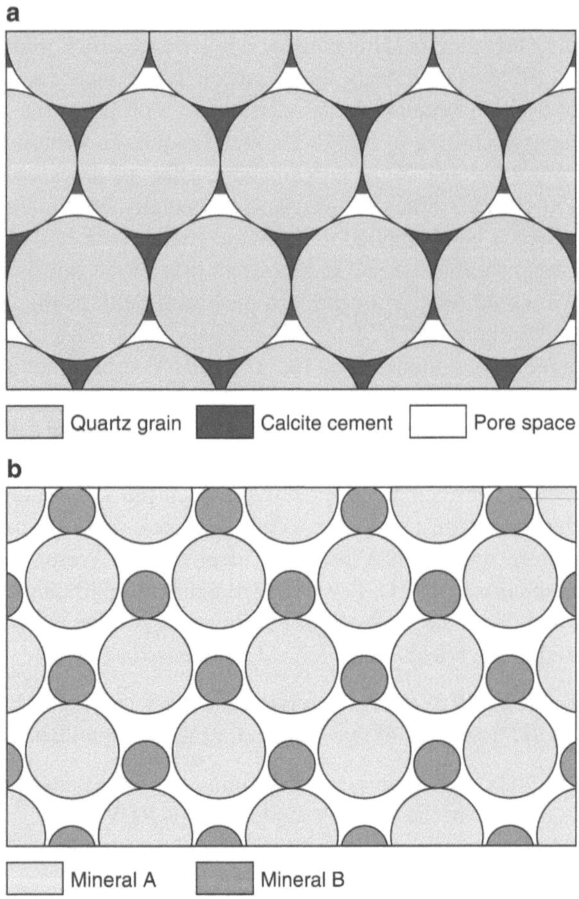

Fig. 10.6 Two two-dimensional periodic rock models are shown for illustrating the unit cell and the concept of the representative elementary volume (*REV*) in rock

Based on the fact that the zone of influence of a defect depends on the size of the defect (Sect. 4.4), we can speculate about the type of relation that exists between the stress state and the scale of rock mass under consideration. Taking into account rock geological history, this results not only in alteration to the existing far-field stress, but also leads to the presence of residual stress which is superimposed at different scales (Sect. 4.3). Residual stresses vary with the volume of rock considered, up to a certain volume where they are in equilibrium (Sect. 4.3, equilibrium volume). According to Hyett et al. (1986), the amount of residual stress in rock decreased as the volume of rock tested increased. As we know from Chap. 4, this trend is attributed to the fact that as rock volumes increase, discontinuities are more likely to be found. Rock volumes bounded by discontinuities and separated from the host rock are unable to transmit any residual stresses.

Hudson and Harrison (2000) attempted to illustrate this combined effect on stress magnitude in Fig. 10.7. The horizontal axis in Fig. 10.7 denotes the domain size (zone of influence), while the vertical axis stands for the stress magnitude normalized to the applied stress. Indicated are stresses on the scale of millimetre (resid-

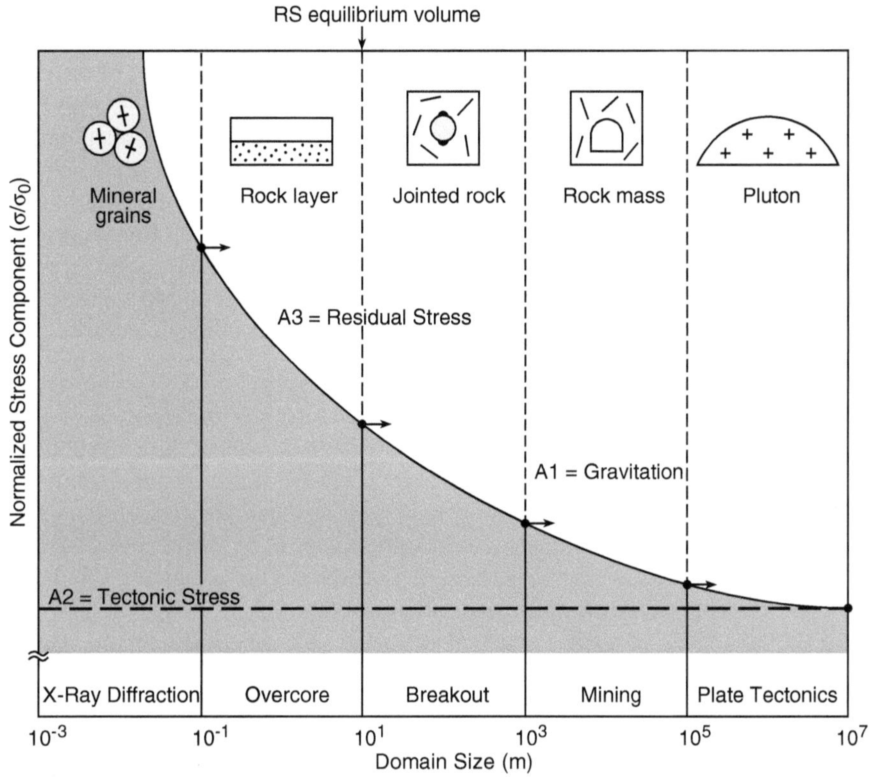

Fig. 10.7 Stress state scale relations (modified after Hudson and Harrison 2000)

ual stress in mineral grains), centimetre (overcore strain gauge), metre (borehole), 100 m (mine excavation), kilometre (granite pluton) up to 1000 km (tectonic plate). Note that the curve is an envelope and demonstrates that the spread in the results of stress determination programmes must be expected but this spread will reduce as the size of the sampled volume increases.

The far-field value of the normalized stress component is defined as the tectonic stress component A2 (Fig. 10.7, asymptotic horizontal line). Contributions to the A2 stress value for the smaller rock volumes tested come from the combined effects of topography and erosion induced stresses A1 and at an even smaller scale from residual stresses A3. Note that for rock volumes larger than the equilibrium volume, the effect of A3 on tectonic stresses is negligible. A relationship between equilibrium volume of residual stresses (V > EQV; $\sigma_{RS} \rightarrow 0$) and representative elementary volume (V > REV; $\langle\sigma\rangle = \text{const.}$) has not been quantified.

To convert strain to stress–rock properties are needed. However, these rock properties are scale-dependent and can cause a scale effect in rock-stress measurements (Amadei and Stephansson 1997). One such effect is the dependence of rock tensile strength and the borehole diameter in the analysis of breakdown pressures for hydraulic fracturing (Chap. 7). Enever et al. (1990) found a decrease in hydraulic fracture tensile strength (difference in crack initiation pressure and crack reopening pressure) with an increase in the borehole diameter.

Another effect in context to borehole breakouts is the scaling effect of the ratio of tangential stress to the compressive strength necessary for a borehole wall failure (Guenot 1987; Haimson 1990; Cuisiat and Haimson 1992). As pointed out by Martin et al. (1994), the scale effect is important for borehole diameters smaller than 75 mm. Comparing laboratory and borehole breakout data in the field Martin et al. (1994) found different scaling laws. In this case study, Lac du Bonnet granite from the underground research laboratory URL in Canada was investigated. In summary, this stress-scaling effect suggests that the classical breakout theory (Sect. 7.3) for large borehole diameters underestimates the magnitudes of in-situ stresses.

Note-Box With respect to stress scaling, we need to know the representative elementary volume of rock, the stress determination method, the volume of rock involved in stress measurements and the type of tectonic stress regime. The asymptotic value of the normalized stress component versus rock volume defines the tectonic stress component. Contributions of residual stresses drop off as the equilibrium volume is reached. The effect of the topography diminishes when plate tectonic dimensions are reached.

10.4 Best-Estimate Stress Model

An exact prediction of the in-situ state of rock stress and its spatial variation at a site or a region is very difficult and, for practical purposes, impossible since the current state of stress is the end-product of an often long series of past geological events (Chap. 4). The rock mass has experienced several phases or cycles of physiochemical, thermal and tectonic processes which have all contributed to the current state of stress. Each of the processes can act individually but usually they are coupled and act simultaneously in space and time.

Estimating in-situ stresses requires detail knowledge of the site morphology and geology. Often the estimation of stress is conducted as a part of a site investigation or an investigation of an area. The aim of a site characterization is to produce a three-dimensional model of the site or the area containing information about topography, soil cover, rock mass lithology, structural geology and rock stresses. Numerical models can be of assistance in estimating the 3D stress model for a site. Such models are of great help in analyzing the cause and effect of faults and fracture zones intersecting the model. Although it is impossible to know all the details of the geological evolution of a site, it is worth the effort to try to ascertain the stress state from the bulk knowledge of the site morphology, topography and geology and, if possible, to verify the information with additional data from boreholes and drill cores.

Prior to any in-situ stress measurements at a site or an area, establishment of the Best Estimate Stress Model (BESM) is recommended (Fig. 10.8, Column 1). A BESM is generated from collecting existing stress data in archives, analyzing morphology, topography, geology in the field and stress information from borehole and drill cores. The established model is the result of the integrated study and the outcome is used in selecting the appropriate stress measurement technique and assist in planning the measurements in the field. After BESM is established, stress measurement methods (SMM) have to be conducted on site (Fig. 10.8, Column 2). In the Integrated Stress Determination (ISD) available stress data from BESM and measured, new stress data from SMM are merged and analyzed by means of, e.g., a least-squares criterion, Monte Carlo simulation or generic algorithms (Fig. 10.8, Column 3). At this stage, numerical rock-stress modelling is required to improve the establishment of the most likely state of stress at the site. Results of the information from previous steps (BESM, SMM, ISD) all together with the rock stress-scaling relationship lead to the Final Rock Stress Model (FRSM) as presented in the fourth column of Fig. 10.8. The arrow in the figure illustrates the sequence of deriving the final rock stress model (Stephansson 2003).

10.4.1 Data Extraction and Perturbation

At present, no rigorous methods are available to exactly estimate virgin stresses but Amadei and Stephansson (1997) presented existing limits, effects, methods and knowledge that are available to estimate the state of stress without using any of the existing in-situ rock-stress measurement techniques in boreholes or core-based methods. When stress data are needed, an estimation of the state of stress, both in magnitude and orientation, should be performed prior to any in-situ or laboratory measurement. The result of the measurements can later support or overthrow the prediction.

The section about data extraction contains information related to classes of stress and the S_H-orientation of stresses from the World Stress Map Project (Sect. 11.2). Before estimating the state of stress at a site or an area, we need to know what type of rock stresses is likely to exist. Here the rock-stress terminology presented and discussed in Fig. 4.1 will certainly help us to determine the classes of stress. For applied rock mechanics and rock engineering purposes, gravitational (mass of overburden) and tectonic stresses and structural stresses are by far the most important ones.

When estimating the state of stress at any depth in the rock mass we make the first assumption that the state of stress can be described by three components: a vertical component due to the weight of the overburden at that depth and two horizontal components which are larger or smaller than the vertical stress. For the variation of vertical stress with depth, there has been a long series of in-situ stress measurements conducted and several data compilations done (Herget 1974; Brown and Hoek 1978; Amadei and Stephansson; 1997) proving that, in most cases, the magnitude of the vertical stress can be explained by the overburden weight only

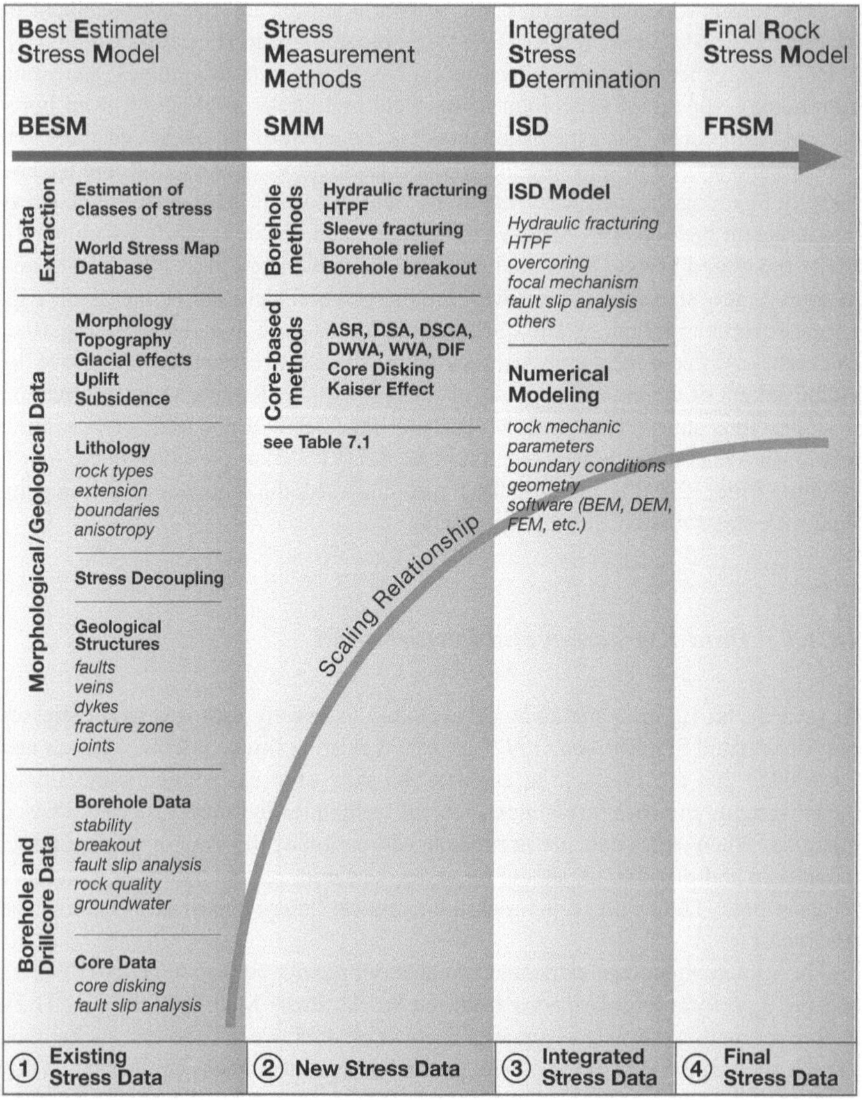

Fig. 10.8 Combination of available stress data from the Best Estimate Stress Model (*BESM*), new stress data from stress measurement methods on site (*SMM*), integrated stress determination (*ISD*) using previous data plus numerical modelling in order to generate the Final Rock Stress Model (*FRSM*) at a site or an area

(Sects. 4.1, 10.1). Deviations from this rule exist and in particular in areas of young tectonics and volcanism, areas of strong topographic variations and adjacent to major discontinuities in the rock mass.

A second major step in determining BESM is to extract the corresponding data from the World Stress Map (WSM). Note, however, that the WSM data set is focussed on S_H-orientations and type of faulting regime and, at the moment, does not release any stress magnitudes (see DVD). From Exercise 10.1 we can create our own stress map of the region in question to get a first impression of the S_H orientation and also the type of faulting terrain the area belongs to. Once the WSM data have been extracted, S_H orientation at a site can be obtained by averaging stresses from a region with the same geological conditions (Sects. 10.2, 11.2).

The effect of topography on estimating in-situ stress is of particular interest when conducting rock engineering and mining projects in mountainous area, near valley slopes and at the top of high mountains (see Fig. 4.2). The slopes and valley walls can create unbalanced stress concentrations at underground excavations located at the toe of the slopes and valleys and cause rock burst, spalling and other types of rock failure (Chap. 3). It is a difficult task to determine analytically the in-situ stress field in a rock mass or a region with an irregular surface using the theory of linear elasticity. Bipolar coordinate transformation, exact conformal mapping and the perturbation method are three different analytical methods that have been applied to study the effect of topography on the stress state for a site or region (Chap. 4). All the derived analytical expressions predict tensile stress in a valley bottom and this is supported by the observations from the field in terms of a zone of discontinuous and loose rock masses and tendencies of up-warping phenomena in the bottom of the valleys.

An understanding of the geological history of a site or a region is important as it can be used to determine the evolution of the stress regime in which the site or the area of interest is situated. Estimating in-situ stresses requires a detailed characterization of the site geology like lithology and lithological boundaries, its tectonic history, critical structures, erosion, uplift, influence of glaciation, neotectonics and others. In-situ stresses can vary significantly from one lithological unit to the next, depending on the relative stiffness between the individual rock masses. In addition, abrupt stress changes are likely to appear at the contacts between different lithologic units. In general, one expects to find larger stress magnitudes in the more competent strata as stresses tend to concentrate in hard rocks surrounded by less competent rocks subjected to the same far-field stress system (Sect. 4.4).

Some measured variations of stress with depth have indicated *stress decoupling* (Haimson 1980; Stephansson 1993; Martin and Chandler 1993; Roth and Fleckenstein 2001; Ask and Stephansson 2003) where stresses at shallow depth might be entirely different from stresses at great depth. Stress decoupling is valid for both stress magnitude and orientation. A typical example where stress decoupling develops is an area when crystalline basement rocks are overlaid by sedimentary rocks in a sedimentary basin.

Roth and Fleckenstein (2001) have published an interesting study related to stress decoupling in the eastern North German Basin (eNGB). From data collected in the World Stress Map project, it has been known that Central West Europe is dominated by a NW-SE to NNW-SSE S_H orientation by ridge push from the North Atlantic and the northward drift of Africa (Sects. 10.2 and 11.1). From new analysis of four-arm-dipmeter and televiewer loggings at intervals of 1500–6700 m in deep boreholes and comparison with hydraulic fracturing stress measurements

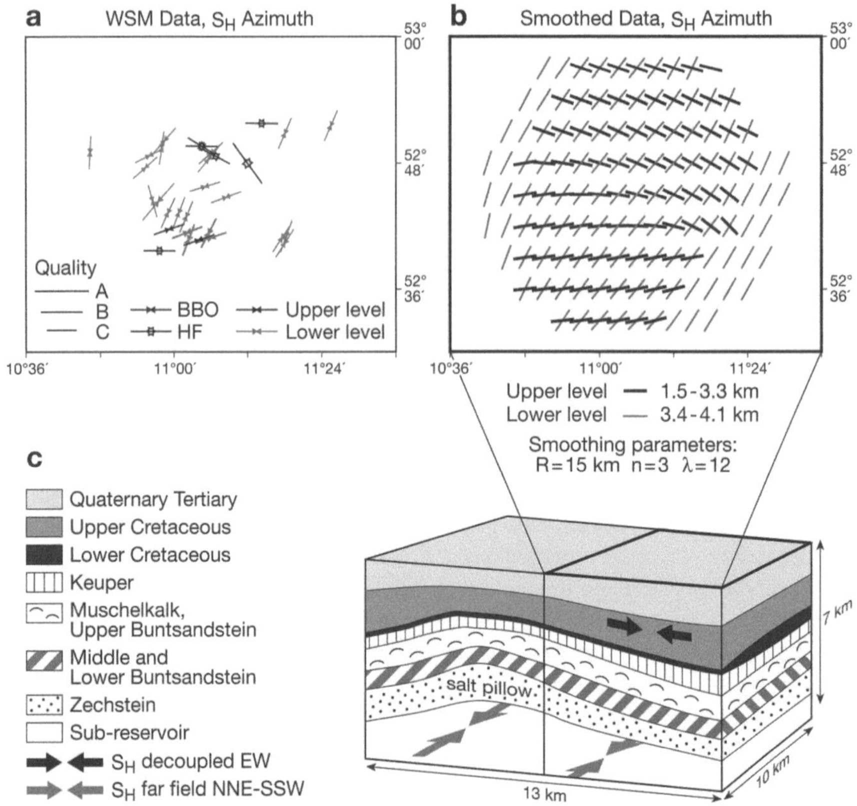

Fig. 10.9 Stress decoupling in the eastern part of the North German Basin. **a** Stress data entries from WSM, **b** smoothed S_H orientations, **c** Schematic block diagram of the North German Basin. In all figures, S_H orientations are presented in two different depth levels. The upper depth level (post-Zechstein at intervals of 1.5–3.3 km, *black symbols* in (**a**), *black short bars* in (**b**), *black arrows* in (**c**)) indicate the decoupled E-W orientation of S_H in the overlaying sediments. The lower depth level (intervals of 3.4–4.1 km, *light grey symbols* in (**a**) *light grey short bars* in (**b**), *light grey arrows* in (**c**)) indicate the NNE-SSW orientation of S_H that corresponds to the Western Europe far-field stress (modified after Heidbach et al. 2007)

from the eNGB region, the substrata below the more than 1000 m thick Zechstein salt formation is dominated by a NNE-SSW striking S_H orientation (Fig. 10.9). In Fig. 10.9a, S_H orientations from WSM in the eastern part of the North German Basin are extracted. In Fig. 10.9b, smoothed S_H orientations calculated from WSM entries are shown for the same area. In both figures, S_H orientations are separated according to depth levels (upper level 1.5–3.3 km depth, lower level 3.4–4.1 km depth). In the block diagram (Fig. 10.9c), the difference in S_H orientation above and below the detachment zone of the Zechstein evaporite sequences in the eNGB is indicated schematically by arrows. The far-field stresses pointing NNE-SSW in the sub-salt unit (Fig. 10.9c, light grey arrows) differ by 90° from the S_H orientation determined in the post-Zechstein upper level formations (Fig. 10.9c, black arrows). This vertically decoupled stress field in the eNGB (Fig. 10.9c, right half of block), is not visible in the central and western part of the North German Basin (Fig. 10.9c, left half of block). The stress decoupling mechanism is restricted to eNGB.

Roth and Fleckenstein (2001) have suggested three different reasons for this decoupling: (1) displacements along large ancient suture zones (Trans-European Fault Zone and Elbe Fault System), (2) stresses due to postglacial lithosphere flexure (compressive stresses outside the edge of the Weichselian and earlier Fennoscandian ice sheets caused the reorientation of the stress field in the sub-saline formations), and (3) a strong lithospheric barrier below the northern margin of the eNGB, derived from rheology-depths profiling, which proves that stresses are attracted and reoriented to the observed N-S orientation. As there is no indication for stress differences from the plate boundaries in the area, the stress decoupling in the eNGB is likely to be due to contrast in rigidity between the eNGB and Fennoscandia. In conclusion, the stress decoupling excursion eNGB is an excellent example for third-order stress pattern discussed in Sect. 11.2 (cf. Fig. 11.6).

Geological structures such as faults, folds, dykes, veins, sills, fault striation or slickensides have long been used by structural geologists to indicate the paleostress, i.e. the state of stress prevailing at the time of genesis of the structure. Since the stresses that created the structure may have been modified due to later tectonic events, erosion, uplift, glaciations, etc., the structure and fabric might not be correlated at all with the current stress field. In order to determine the contemporary stress state we have to seek the most recent geological structures and use them as stress orientation indicators. As an example, different volcanic vent alignments and inversion of fault-slip data are used for stress orientation in the World Stress Map database (Zoback 1992). Fault-slip analysis as develop by Angelier (1990) and his co-workers for stress analysis of recent geological formations or inversion of data from slickensides on fracture surfaces in oriented drill core samples (Hayashi and Masuoka1995), are powerful tools in estimating the stress state at a site or a region and is recommended for the establishment of BESM.

The existence of geological structures and heterogeneities will effect orientation and magnitude of in-situ stresses and make near-field stresses different from the far-field stresses (Chap. 4). The amount of perturbation is very much governed by the strength and deformability of the discontinuity (cf. Fig. 4.10). If it happens that the discontinuity is open, the stresses cannot transect (Fig. 4.10a). If it has the same properties as the surrounding rocks the stresses are unaffected (Fig. 4.10b). If the material in the discontinuity is more rigid than the surrounding rock mass the maximum principal stress is diverted perpendicular to the discontinuity (Fig. 4.10c). If it is less rigid, the maximum stress will tend to divert parallel with the discontinuity. The classical example of the last situation is the stress field in the surrounding of the central part of the San Andreas Fault system often referred to as a weak fault in a strong crust (Hickman and Zoback 2004). As we know from Sect. 9.3, in the presence of faults, we are faced with the problem of a lack of strength and stiffness data about these large geological structures and sometimes with the difficulty of delineating their orientation in space.

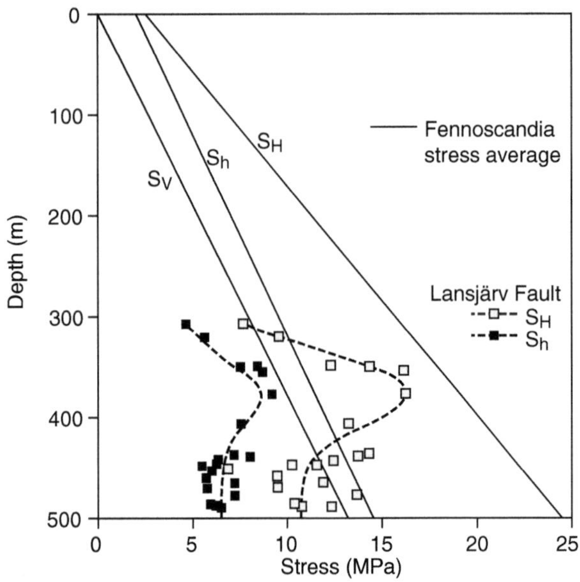

Fig. 10.10 Hydraulic stress measurements adjacent to the Landsjärv neotectonic fault, Northern Sweden. Average hydraulic fracturing stress data from Fennoscandia (*solid lines*) are shown to illustrate the stress anomaly at the fault (modified from Amadei and Stephansson 1997)

Stress relieve from neotectonic faulting in the northern parts of the Fennoscandian Shield has been reported by Amadei and Stephansson (1997). Measured stresses using the hydraulic fracturing method in a borehole adjacent to the neotectonic Landsjärv fault in Northern Sweden show a marked stress anomaly compared to the average state of stress in Fennoscandia. Magnitude of both minimum and maximum horizontal stress is reduced to half the expected value close to the fault at about 500 m depth (Fig. 10.10).

10.4.2 Stress-Measurement Methods

Rock stress measurements should be performed after the establishment of the best-estimate rock-stress model. The data and information collected for BESM can also be used in selecting the most actual measurements in the borehole or on the drill core if available (Ljunggren et al. 2003). Amadei and Stephansson (1997, their Chap. 3) presented the list of factors needed to consider when planning a detail program for the stress measurements where one or several of the methods presented in Chaps. 7 and 8 are applied.

Borehole breakout (Sect. 7.3) is now an established method to estimate the orientation of minimum horizontal principal stress S_h in the plane perpendicular to the borehole axis. The shape and depth of the breakouts depend on the type of rock, its strength and the magnitude of stresses. Hard rocks and high stresses tend to generate deep breakouts with relative small breakout angle. Soft, low porosity sedimentary rocks and moderate stresses generate more shallow breakouts with larger breakout angles. The method requires the logging of the borehole with a dip meter; televiewer or some sort of formation scanner and therefore borehole breakouts belong to the category of borehole methods (Fig. 10.8, SMM). If data of borehole breakouts exist from a site, the information is of great value for delineating the stress orientation of the stress model (BESM) and the information can be used as input to the integrated stress determination (ISD) to generate the final rock-stress model.

Once a drill core is available from a site, the search for an analysis of core disking should be included in the stress estimation program. Core disking is often an indication of high horizontal stress magnitude. The geometry of the disks and the orientation of the disk saddle are indicators of stress orientation. The core breaks up into disks that are usually curved with the center of curvature oriented towards the bottom of the borehole. Laboratory testing and later numerical modelling have shown that once the radial stress in the core trunk during drilling exceeds the compressive strength of the rock core, disking starts to develop. Haimson and Lee (1995) proposed that thinner disks are indicative of higher horizontal stresses and that the trough axis of saddle-shaped core disks are often aligned with the virgin S_H orientation. Less regular core disking might also develop due to existing discontinuities or fabrics in the rock mass. Application of high thrust during the drilling operation can generate too high sub-horizontal tensile stress at the root of the drill core so that extensile microcracks are formed and coalescence to generate core disking (Kutter 1993; Hakala 1999). Matsuki et al. (2004) estimated the directions of 3D in-situ stresses from the height at the periphery of the end surface of the core disks investigated. They also obtained stress magnitudes from analyzing tensile stresses in the central part of the core and determination of rock tensile strength.

10.4.3 Integrated Stress and Modelling

The method of integrating the results of various stress-measurement techniques in order to obtain a more reliable assessment of the virgin state of stress at a site or an area, was introduced in the mid 1980s and is still under development. Cornet (1993) presented the hydraulic testing on pre-existing fractures (HTPF) stress-determination method together with the Integrated Stress Determination (ISD). Data from hydraulic fracturing and HTPF were integrated in order to obtain a better indication of the regional stress field. The integration method is based on a least-square criterion (Tarantola and Valette 1982) where all measurements are assumed to obey a Gaussian statistical distribution. Cornet (1993) integrated hydraulic fracturing and HTPF for two sites in southern France and Ask et al. (2001) and Ask and Stephansson (2003) did the same for the Äspö Hard Rock Laboratory, Sweden and for the geothermal project on Björkö, Sweden. Further, Ask (2001) and Ask et al. (2001, 2003) made an integrated stress analysis of hydraulic and overcoring rock stress data from the Äspö region, Sweden (Ask 2006). Different combinations of in-situ stress data and other stress indicators such as focal mechanisms from earthquakes, fault-slip data and borehole breakouts alone or together with information from the best-estimated rock-stress model can be applied in ISD to constrain the in-situ stress field (Fig. 10.8, ISD). The inversion technique used to establish ISD for various combinations of stress data can also provide information regarding the confidence level of the final solution of stresses.

Numerical analyses with a variety of numerical techniques (e.g., BEM, DEM, FEM) have been used in an attempt to predict or explain the in-situ stress field and in illustrating the effect of topography, stress distribution in a blocky rock mass subjected to a 2D stress field (Su and Stephansson 1999), estimation of regional stress, separation of gravitational and tectonic components and comparison with stress measurement results and modelling of the influence of large-scale structures on the in-situ stress (Jing and Stephansson 2007). Inside and in the vicinity of major fractures and fracture zones, both the magnitude and orientation of stresses will vary from point to point. Stress prediction in these areas is more uncertain and the variations of stresses will be large. Stress modelling helps in obtaining an overall understanding of the state of stress at a site or for a region and assist in explaining the stress variation in magnitude and orientation found from measurements in boreholes and on cores, and supports in predicting the stresses in points or regions between measurements. The modelling results contribute to the estimation of the variability and uncertainty in presenting the final rock stress model.

10.4.4 Final Rock Stress Model

Several possibilities exist to derive the Final Rock Stress Model (FRSM). The most complete model contains results from each of the consecutive steps presented below the solid arrow in Fig. 10.8. The least accurate result consists of only the Best Estimate Rock Stress Model, BESM. If stress measurements are added to BESM, the precision in defining the stress state at a site or an area increases. Additional analysis by means of ISD and rock-stress modelling will reduce uncertainty and give additional confidence in establishing the FRSM.

Hakami et al. (2002) have presented a flow chart illustrating an alternative process of building an in-situ stress prediction, Fig. 10.11. Their stress-model approach starts with overcoring and hydraulic fracturing stress measurements. If data from the measurements do not scatter and no major fracture zones intersect the area or site, a linear trend of mean-stress value with depth is predicted. Hakami et al. (2002) introduced two categories of uncertainty parameters, the 'u-parameter' (uncertainty) and the 'v-parameter' (variability). The u-parameter is meant to cover the uncertainties in the geological model, the tectonic regimes, lack of measurements, systematic measurement errors and bias. The v-parameter corresponds to the expected spatial variability of in-situ stress around the average magnitude and orientation (Sect. 10.3, REV). Confidence in the stress magnitude will be dependent on measurement results and complexity of the site. The mean orientation of, e.g., the maximum horizontal stress may be predicted with a fairly high degree of certainty because both regional S_H orientation and the S_H orientations from stress measurements at the site can be used.

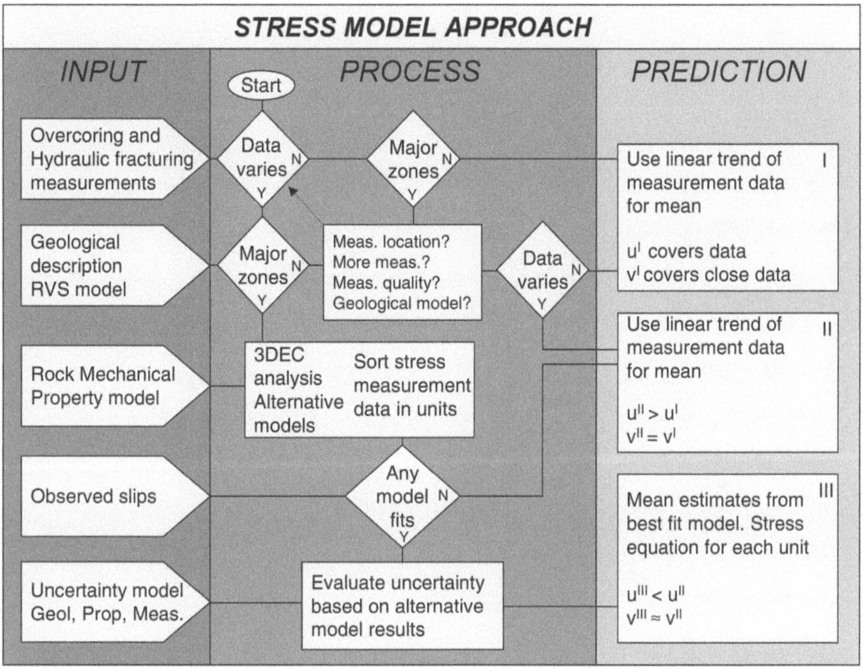

Fig. 10.11 Flow-chart illustrating the steps in deriving at a rock stress model of an engineering site starting with overcoring and hydraulic fracturing measurements (after Hakami et al. 2002)

If stress measurement data varies, major fracture zones from the geological rock visualization model (Fig. 10.11, *RVS*) are imposed and discrete-element modelling by the 3DEC code (Itasca Inc.) is performed for a series of alternative fracture models. If the results from modelling are reliable and show no or slight influence from discontinuities, the prediction is completed. This corresponds to prediction stage II in Fig. 10.11 (right column). If stresses from the numerical modelling vary, rock stress modelling with input data from rock mechanical testing is introduced in combination with geological observations to cover the uncertainty in geological and rock mechanical parameters. The final result in prediction step III is the mean estimates from the best-fit model. Hakami et al.'s (2002) approach is particularly developed to suite the development of a rock mechanics site-descriptive model for a potential deep geological repository of radioactive waste.

Note-Box In order to reach the final rock stress model at the location in question, we have to proceed in steps. (1) Collect all available stress data of your location and surroundings. (2) Include topography, lithology and faults as well as borehole and core-based stress data. (3) Measure stresses at your site and determine vertical and horizontal stresses versus depth. (4) Combine available and measured on-site stress data with earthquake and fault related stresses to obtain integrated stresses. (5) Generate a 3D model with rock parameters measured, appropriate boundary conditions and solve the resulting momentum equations with appropriate numerical techniques. (6) Synthesize and rate your final near-field rock stress model in context to the far-field stress pattern.

Chapter 11
Global Stress

Even if I have got the idea from the similar contours of the coast lines, of course, the proof must be based on the observed results of geology. Here we are forced to assume a land connection, for example, between South America and Africa which at a certain time was broken. This event we can consider in two versions: (1) through subsidence of a connecting continent "Archhelenis" or (2) through the drift from a big fault. Till now, one has, based on the unaltered position of each land mass, only considered (1) and (2) was ignored. However, (1) conflicts with the modern science of Isostacy and even our physical principles. A continent cannot sink because it is lighter than what it is floating on. [...] Why should we hesitate to throw the old view to the winds?
–Alfred Wegener, letter written to his mentor Wladimir Köppen, dated 6 November 1911

In this last chapter of the book we present and interpret stress data in Europe (Sect. 11.1). In the following, stress orientations from the World Stress Map data set are displayed as a hybrid stress map and are interpreted based on global mantle flow (Sect. 11.2) and in terms of plate-tectonic sources (Sect. 11.3).

11.1 European Stress

The name *Europe* is deduced from the word "*ereb*" (land of sunset) the Phoenician used to call the west coast of the Aegean Sea. For the east coast they used the word "*asu*" (land of the rising sun) which later became *Asia*. In terms of tectonic plates, the sources of the European stress field are due to ridge push of the mid-ocean rift in the North Atlantic and by collision forces between the Eurasian (Europe plus Asia) Plate and the African Plate. Sometimes the European part of the Eurasian plate is called the European Plate. Due to the decoupling of the western and eastern European Plate, it is common practice in numerical models to define a fixed-model boundary at the eastern margin of the western European Plate in order to match measured and modelled data of the European stress field (Gölke and Coblentz 1996).

A. Zang, O. Stephansson, *Stress Field of the Earth's Crust*,
DOI 10.1007/978-1-4020-8444-7_11,

Variations of stress magnitudes with depth beneath Europe are shown in Fig. 11.1. Data are taken from Aydan and Kawamoto (1997) where measured in-situ stress magnitudes are covered for a number of European countries, namely Austria, France, Germany, Italy, Norway, Sweden, Switzerland and the United Kingdom. On average, the European gradients of vertical, minimum horizontal and maximum horizontal stress component with depth (Fig. 11.1a–c) are 25, 37 and 45 MPa km^{-1}, respectively. The data base ISMEAS (in-situ stress measurements) used by Aydan and Kawamoto (1997) contains about 700 entries. As we know from Sect. 10.1, these stress gradients can serve as a rule of thumb but have to be validated by stress measurements at specific locations (Sect. 10.4). The dimensionless stress coefficients (k_h, k_H, k) are calculated from crustal stress magnitudes (Sect. 10.2) and are shown in Fig. 11.1d–f. Near-surface k values of

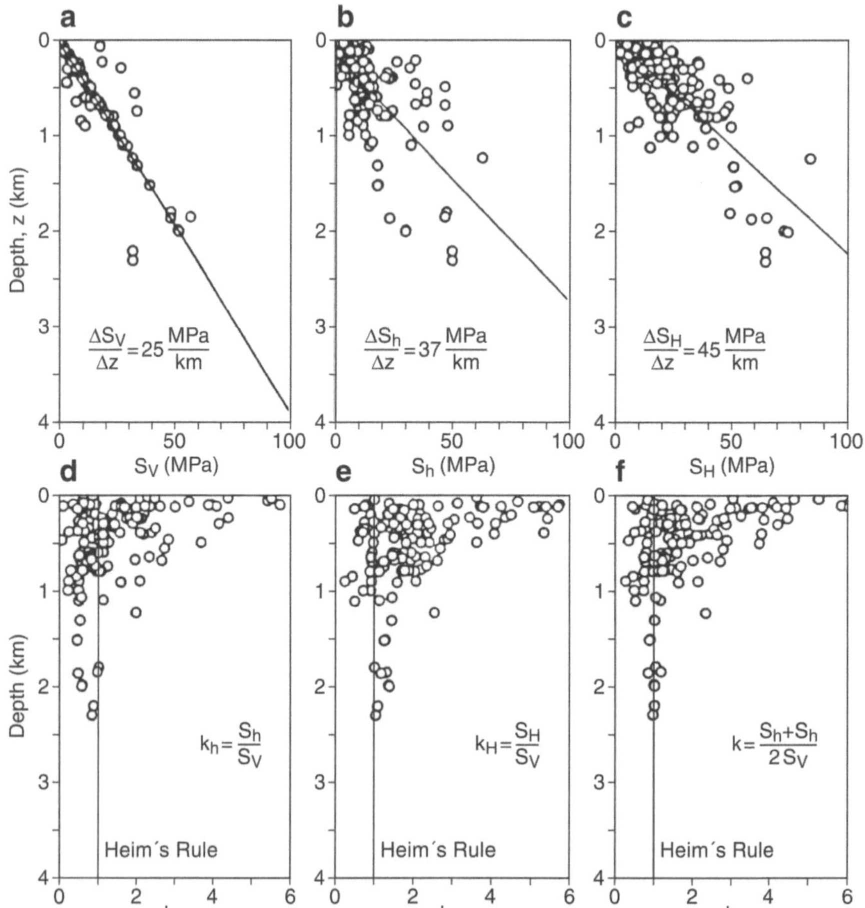

Fig. 11.1 Variation of stress magnitudes **a** vertical S_V, **b** minimum horizontal S_h, **c** maximum horizontal S_H and dimensionless stress coefficients **d** k_h, **e** k_H, **f** lateral stress coefficient k with depth beneath Europe (after Aydan and Kawamoto 1997)

six are observed (Sect. 10.1) which decrease with increasing depth, e.g. reach k values below two at 1 km depth and k values below one at depth greater than 2 km in the European crust.

The orientation of maximum horizontal compressive stress in Europe is taken from the World Stress Map (WSM) 2005 database release published by Heidbach et al. (2007). For Western Europe, the WSM 2005 release provides 3188 data records with 1721 entries having data quality A to C (Fig. 11.2). In-situ stress measurements,

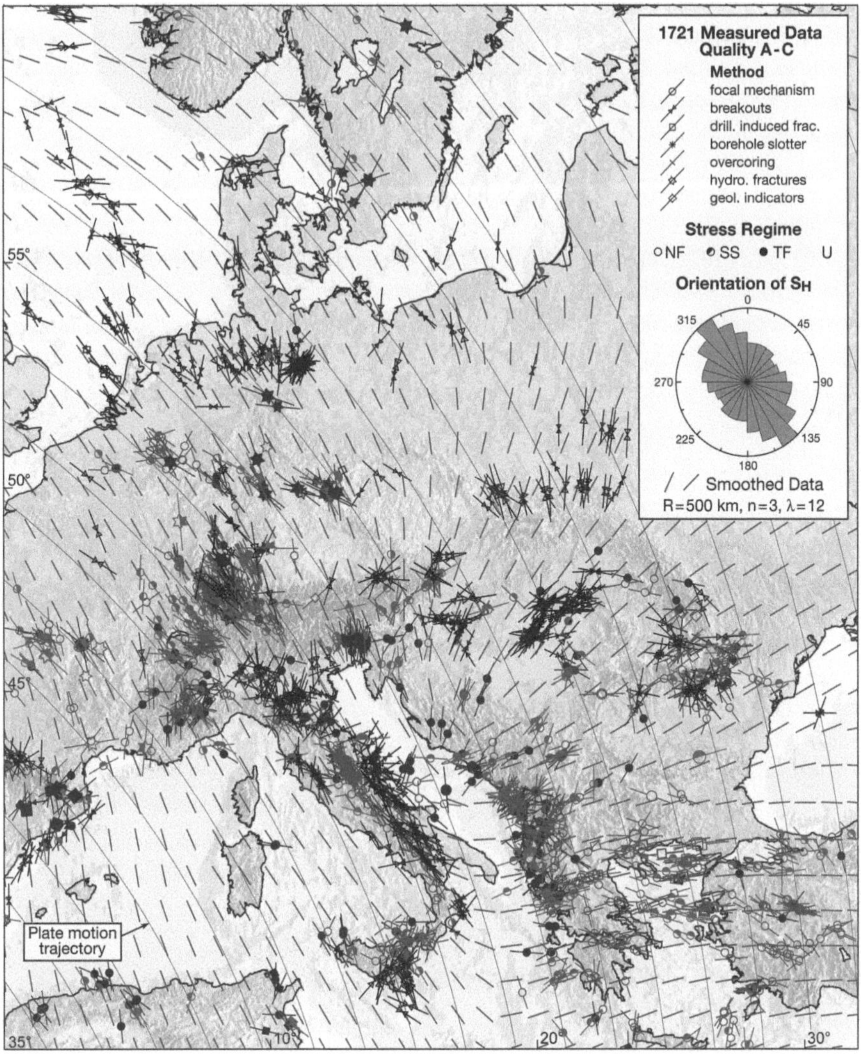

Fig. 11.2 Smoothed maximum horizontal stress direction map of Western Europe (*short bars*). *Thin grey lines* show relative plate motion trajectories of the African plate with respect to the Eurasian plate (DeMets et al. 1994). Topography is based on data from the National Geophysical Center including bathymetry from Smith and Sandwell (1997) (modified from Heidbach et al. 2007)

geological indicators and earthquake fault-plane solutions indicate that the S_H-orientation in Western Europe is NW-SE. This orientation is mainly controlled by plate boundary forces, in particular by ridge push of the Mid-Atlantic Ridge and by collision forces in the Africa-Europe convergence zone.

From Fig. 11.2 it can be stated that the prevailing, smoothed S_H-direction (Fig. 11.2, short bars) is parallel to the relative motion of the African plate with respect to the Eurasian Plate (Fig. 11.2, light grey lines). Deviations from this overall first-order trend occur in the Aegean-Anatolian region, where the slab rollback at the Hellenic arc induces a E-W orientation of S_H in the back-arc region (Heidbach and Drewes 2003), and in the Pannonian Basin where the NE-SW orientation of S_H is due to the collision in the Dinarides (Bada et al. 1998, 2007). The smoothed orientation of S_H across Italy (Fig. 11.2, short bars) is similar to that in Western Europe. Local deviations near the coast and towards the Alps and Dinarides can result from lateral density contrast, topography, rotation of the Adriatic block relative to Eurasia, and the collision resistance along the NW-SE strike of the Dinarides Mountains (Heidbach et al. 2007).

In 1992, Müller et al. subdivided the European intraplate stress field into three *stress provinces* with more or less homogeneous orientation of S_H. The first stress province is the northern part of Europe where stress indicators at Fennoscandia show considerable scatter with a mean S_H azimuth N120°E ± 45°. There seems to be no correlation between the stress orientation and post-glacial uplift (Gregersen 1992). However, according to Stephansson (1993), isostatic response from ice loading of the thick crust, effect of topography and shear strength variability of faults may be responsible for the large scatter in stress orientation. The physical properties of geologically old shield areas (e.g., Fennoscandia) are characterized by a thick lithosphere and low heat flow. This reduces the mean stress level of the lithosphere and permits local effects due to lateral heterogeneities in the crust such as density or strength to have an important influence on the stress field.

The second stress province lies in the western and central part of Europe where the consistent NW-SE orientation prevails (N145°E ± 26°) which has been identified earlier (Ahorner 1975; Illies and Greiner 1979; Grünthal and Stromeyer 1986; Klein and Barr 1986). The NW-SE trend of S_H azimuth progressively rotates clockwise towards eastern Europe (Grünthal and Stromeyer 1995), because of the decoupling of the Western and Eastern European Plate. The third stress province is located in the southern part of Europe, the Aegean-Anatolian region which is characterized by a S_H orientation N85°E ± 27° and N-S extension. Normal faulting dominates the homogeneous stress orientation in the NW part of this region. In the eastern part of Aegean-Anatolia, strike-slip faulting regimes dominate, having more diffuse stress orientations.

The second stress province was interpreted at higher resolution in terms of decoupling of upper crustal fragments from the lithospheric mantle (Müller et al. 1997). In their study the constant search radius algorithm (Sect. 10.2) was used to identify smaller-scale stress provinces (second-order effects). Müller et al. (1997) replaced the constant strain rate model of strength envelopes in the crust (Chap. 5) by a constant shear stress model. This is because the constant strain rate model can only develop when the lithosphere as a whole deforms through plastic deformation (Kusznir 1991). In the constant shear-stress model, the effective viscosity drops from the top of the lower crust to the Moho by six orders of magnitude. The decoupling zone turns out to be a ~2 km thin layer within which substantial strain rates develop. The crust of Western Europe is described as being in frictional faulting equilibrium.

As pointed out by Müller et al. (1992), the uniform Western Europe stress field coincides with a medium thick lithosphere (50–90 km) and high heat flow (80 mW m^{-2}). Only major geological structures like the Alps and the Dinarides Mountains produce second-order stress patterns. In this context, we refer to a recent reference model for the thickness of the European crust (Tesauro et al. 2008). The model is based on seismic refraction, reflection and receiver function studies and covers the area 35°N–71°N, 25°W–35°E. The depth map of the European Moho is shown in Fig. 11.3. The Moho depth beneath Finland is deeper (z_M> 60 km) than suggested by previous studies (Bassin et al. 2000, Kozolovskaya et al. 2004). The Moho depth decreases to z_M~40 km beneath Sweden and to

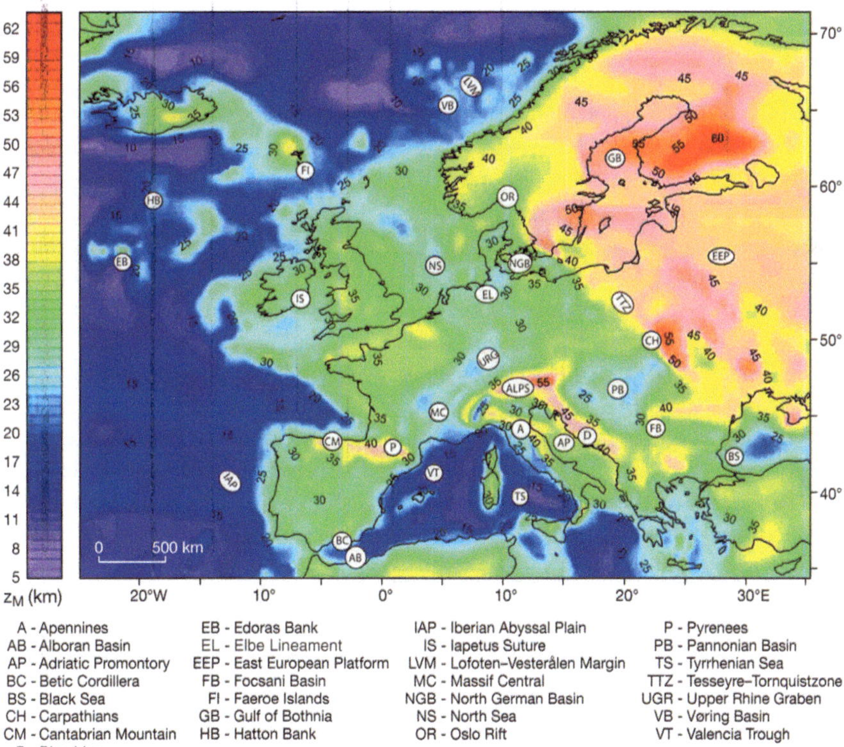

Fig. 11.3 Contour plot of Moho depth beneath Europe based on a compilation of seismic refraction, reflection and receiver-function studies (modified after Tesauro et al. 2008)

$z_M \sim 30$ km beneath the Oslo rift and the Norwegian coast line. West of the Norwegian coast the Moho is less than 20 km thick. In the study area of the smoothed stress direction map (Fig. 11.2), the Moho contours show almost constant values around 30 km (Fig. 11.3). Exceptions are the Alps with Italy and the Pyrenees.

A comparison of different finite-element models to explain the contemporary stress field of Europe can be found in Jarosinski et al. (2006). In the plane-stress model of Grünthal and Stromeyer (1992), the stress-orientation map for Central Europe is calculated with resolution 2° × 2°. Orientation of the first-order stress pattern, again is explained by North Atlantic sea-floor spreading and the northward motion of the African Plate relative to the Arabian Plate. In the elastic Finite-Element Model (FEM) of Gölke and Coblentz (1996), both principal-stress magnitudes and orientations in Europe are calculated with resolution 1° × 1° (3117 triangular finite elements) taking into account topography, Atlantic Ridge push (3×10^{12} Nm^{-1}) and collision forces between Africa and Eurasia (0.5–2×10^{12} Nm^{-1}). In the two-dimensional elastic finite element model of Jarosinski et al. (2006), the plate-scale model of Europe is refined comprising 24 tectonic blocks, 16 faults and 12 geological boundaries with a maximum resolution of 0.5° × 0.5° (3963 triangular finite elements) taking into account Adriatic indenter forces (9×10^{12} Nm^{-1}). If the reader is interested in second-order stress patterns in the southern part of the European Plate, the work by Jarosinski et al. (2006) must be referred to. This work highlights the shapes and friction properties of the Dinaric suture zone which controls the stress transfer from the Adriatic indenter into Central Europe. In the following, we restrict ourselves to the simplified finite-element model of Gölke and Coblentz (1996), because the main features of stress orientation and surface stress magnitude in Europe are captured in their approach including the effect of topography and excluding the effect of major faults.

The tectonic stresses at the surface of the European crust as predicted by the 2D finite-element model of Gölke and Coblentz (1996) for the simplest combination of tectonic forces and boundary conditions are shown in Fig. 11.4. In this model, a line ridge force of 2.5×10^{12} Nm^{-1} equal to a horizontal deviatoric stress of 25 MPa averaged over a 100 km-thick lithosphere was applied along the Mid-Atlantic Ridge. Line forces (Fig. 11.4, SE pointing arrows) act in the direction of relative plate motion (Fig. 11.2, thin grey lines) but are concentrated at the ridge itself. The principal stresses in Fig. 11.4 (solid bars = compressive, open arrows = tension) vary between 30 MPa in the western part of Europe to about 12 MPa in the south-eastern part of Europe. The stress field azimuth is characterized by NW-SE compression throughout most of Europe, as indicated also by the WSM data set (Figs. 10.2, 11.2). Boundary effects related to the pinned southern and eastern plate margins are encountered in the south-western and northern parts, resulting in tensile stresses (Fig. 11.4, open arrows). The choice of boundary conditions (line ridge push versus distributed ridge push force along the western margin, pinned boundary versus collisional boundary forces along the southern margin) affects the calculated orientation of maximum principal stress only slightly. The choice of zero displacement on the eastern margin is unrealistic but is justified by the fact

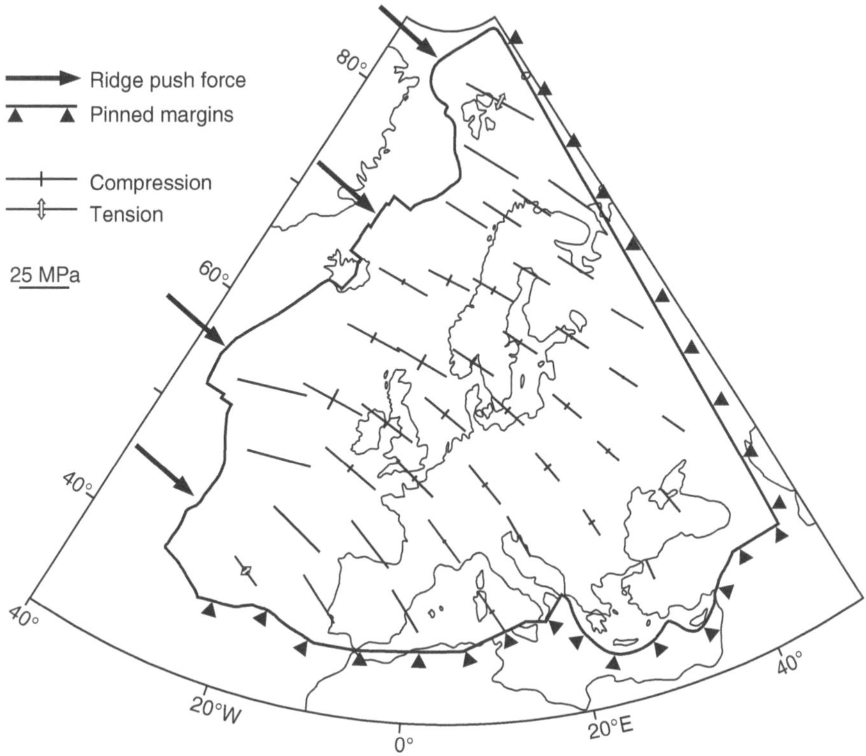

Fig. 11.4 Predicted tectonic surface stresses in Europe from the 2D elastic finite-element model of Gölke and Coblentz (1996) using line-ridge forces which are equivalent to a horizontal deviatoric stress of 25 MPa over a 100 km-thick lithosphere. Displacements are prescribed to zero along the pinned eastern and southern boundary of the study area. Principal stresses (*solid bars* = compression, *open arrows* = tension) are shown at selected Gaussian integration points

that the modelled and measured (smoothed) stress directions in the European Plate are reasonably consistent.

Note that all FEM results require boundary conditions to be prescribed along the entire margin of the region of interest. Even if residuals between observed and calculated stress directions (or magnitudes) tend to zero, this does not guarantee the uniqueness and validation of the stress field inferred. Therefore, the direct integration of elasticity equations or the solution obtained from two Cauchy problems (Mukhamediev et al. 2006) can be an alternative approach. In the direct integration approach, "only" a reliable pattern of stress trajectories is needed. In the first Cauchy solution they prescribed boundary stresses along the Mid-Atlantic ridge, in the second Cauchy solution they described boundary stresses along the Africa-Eurasia convergence zone. In both cases, no stresses need to be prescribed along the eastern boundary of the European Plate.

Note-Box Sources of the contemporary tectonic stress field of Europe are ridge push of the Mid-Atlantic Ridge in the NE and collision forces between the African and the Eurasian Plate in the SW. The measured, smoothed stress direction map of Western Europe aligns with relative plate motion trajectories. For most part of Western Europe, the stress field shows a nearly uniform NW-SE compression with surface magnitudes from finite-element modelling in the range of 10 to 20 MPa averaged over a 100 km-thick lithosphere. Magnitude and orientation of the modelled stress field are largely invarant to the boundary conditions used to represent tectonic forces along the western and southern margins. Average measured European stress gradients are 37 MPa km^{-1} for the minimum horizontal stress and 45 MPa km^{-1} for the maximum horizontal stress.

11.2 World Stress Map

In this section we present a new *hybrid stress map* consisting of single data points of S_H azimuth mostly located at ocean bottoms where data density is poor and smoothed sections of the World Stress Map (WSM) database mostly located on continents close to continental margins where data density is high. Stress patterns with constant S_H azimuth of first, second and third order and their spatial resolution are defined. On the global scale, stress patterns of first order can cover entire continents and are discussed in terms of global mantle-flow models.

The digital WSM database release 2008 (Heidbach et al. 2008), including instructions and comments in preparing specific stress maps, is attached to the DVD in this book, WSM (http://www.world-stress-map.org). A recent published version of the WSM can be found in Heidbach et al. (2007) where the present-day maximum horizontal in-situ stress direction within the Earth's crust is displayed based on the WSM database release 2005. In Fig. 11.5, we present a *hybrid stress direction map* based on the data published by Heidbach et al. (2007). For this purpose, single stress-data points (Fig. 11.5, short bars with symbols in the centre) and smoothed stress-field data (Fig. 11.5, short bars) are compiled in one and the same figure. Single measured stress-data points are picked in areas where data density is poor, i.e. ocean bottoms, continental areas like Africa, Arabia, eastern South America and Eastern Europe. Smoothed stress field data are calculated for continental areas where data density allows the application of the quality- and distance-weight smoothing algorithm of Wehrle (1998). For their smoothed global stress-field approach, Müller et al. (2003) used a search radius of $R=750$ km with a minimum of 10 data points within distances $r<R$ (Sect. 10.2) for calculating the mean-stress orientation at the gridded data points (Fig. 11.5, short bars).

Each measured single stress data point in Fig. 11.5 (short bar with symbol) contains information on (1) the azimuth of S_H (orientation of bar), (2) the quality of S_H azimuth (length of bar), (3) the type of stress measurement technique (symbols in

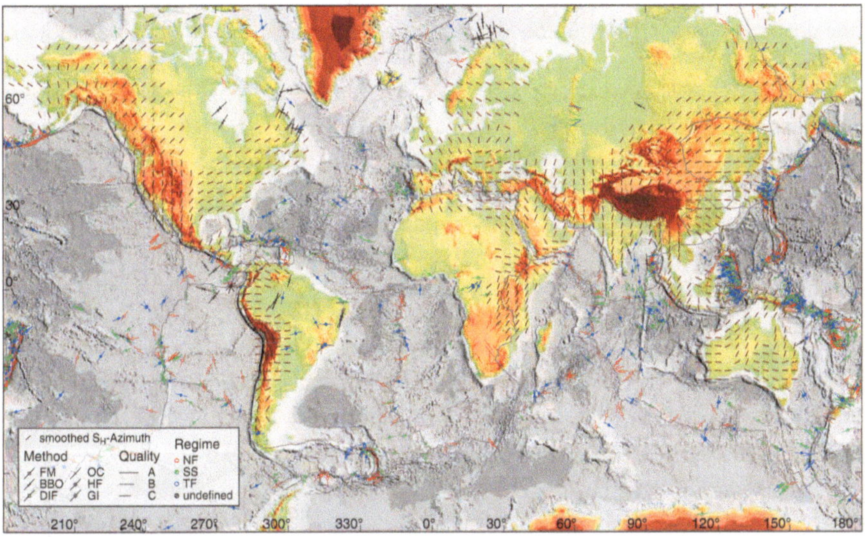

Fig. 11.5 Global hybrid stress orientation map showing the azimuth of maximum horizontal compressive stress as single data points (*short bars with symbols*) in regions where data density is poor and smoothed stress data (*short bars*) from Müller et al. (2003) in regions where data density is high. Plate boundaries are taken from Bird (2003). Topography is based on data from the National Geophysical Data Center including bathymetry data of Smith and Sandwell (1997) (modified after two figures in Heidbach et al. 2007)

the centre of the bar), (4) the location (visible) and the depth (not visible) of the stress data point, (5) the tectonic stress regime according to the Anderson faulting (open symbol=*NF*, half-solid=*SS*, solid symbol=*TF*), and (6) the reference of the published stress-data point. Within the upper 6 km of the Earth's crust, the stress field is mapped by a wide range of methods (focal mechanisms of earthquakes, borehole breakouts (BBO), drilling-induced fractures (DIF), hydraulic fracturing (HF), overcoring (OC) and geological indicators (GI) as fault slip inversion and volcanic vent alignments) with BBO as a major contributor. Below 6 km depth, earthquake focal mechanisms are the only stress indicators available, except for ultra-deep wells like Kola and KTB. Among the A-C quality of data points (12046) used in Fig. 11.5, the focal mechanisms (9278) contribute to ~77% of the WSM data set. A quality *A* data point requires a standard deviation of S_H-azimuth of less than 15°, quality *B* of less than 20° and quality *C* of less than 25°, respectively.

The smoothed part of the hybrid stress direction map in Fig. 11.5 (short bars without symbols) is dominated by two non-connected regions. The smoothed "western territory" covers stress directions on continents close to continental margins from Tierra del Fuego in the south to Alaska in the north. In this territory, the smoothed S_H azimuth is mostly oriented perpendicular to the plate boundaries (e.g. the subducting oceanic plate offshore South America). The second interconnected, smoothed "eastern territory" covers Europe through the Mediterranean down to the east Kenya Rift, Arabia and Asia including the Indian subcontinent, as well as Indonesia and the Australian continent. The eastern territory belt ends in the northeast with stresses in the Siberian Plate. A brief summary of explanations of global stress patterns in continents is given by Amadei and Stephansson (1997, their Chap. 11). A number of world-wide first-order stress patterns have recently been discussed by Zoback (2007, his Chap. 9). Global crustal-stress patterns based on the 2008 World Stress Map database release are discussed in Heidbach et al. (2009).

For the reader, we have two remarks at the end of this world stress section. The first is concerned with the appropriate identification of stress patterns of different spatial scales evident from the stress maps discussed. The second remark is focussed on the physical reason of the present-day stress field of the Earth's crust in terms of a thermodynamically self-balancing planet.

In Fig. 11.6, we visualize different scales of crustal stress patterns based on a table published by Heidbach et al. (2007). The first- and second-order stress patterns have been defined and discussed earlier (Zoback 1992). According to these publications, intraplate stresses are the result of the same forces that drive plate motion, in particular ridge push, slab pull, trench suction, collision forces and traction at the base of the lithosphere (Fig. 11.6, *first-order stress pattern*). Stress patterns at second-order are due to lateral density contrasts caused by continental rifting, isostatic compensation and topography, deglaciation effects as well as lithospheric flexure (Fig. 11.6, *second-order stress pattern*). Stress patterns at third-order (Heidbach et al. 2007) are explained by faults, seismic induced stress changes due to large earthquakes and volcanic eruptions, as well as local density contrasts, e.g., from salt diapers or detachment horizons (Fig. 11.6, *third-order stress pattern*).

Physical causes for first-order stress patterns discussed so far are plate tectonic mechanisms (Zoback 1992), but more general models include the effect of global mantle flow (Hager and O'Connell 1981; Ricard et al. 1984; Bai et al. 1992). In the following, we refer to a publication by Steinberger et al. (2001) in which large-scale lithospheric stresses are calculated from global mantle convection. Their simplified model is based on the major distribution of world-wide hotspots. It turns out that the distribution of hotspots fits the overall pattern of measured, smoothed stress direction maps quiet well.

In global mantle-flow models, the solid Earth is divided rheologically into the lithosphere (Chap. 1, top 100 km) and the viscous mantle beneath. The lithosphere is broken into 14 plates (see Fig. 11.5, black solid lines representing plate boundaries)

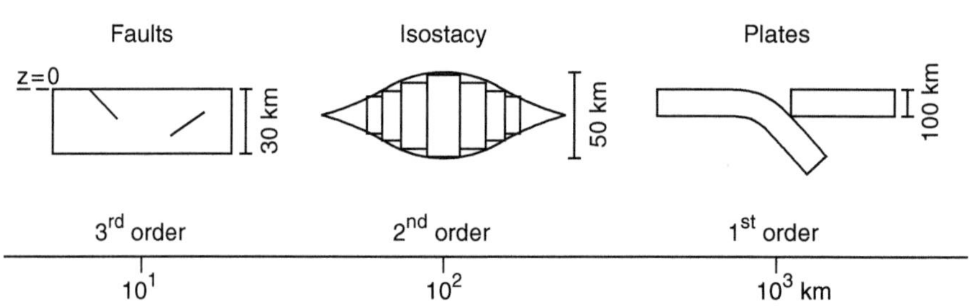

Fig. 11.6 Physical sources and spatial scales of crustal tectonic stresses

that are assumed to be rigid. Convective flow is caused by density variations in the viscous mantle. Mantle flow causes anomalies of normal σ_{rr} and tangential stresses $\sigma_{r\varphi}, \sigma_{r\vartheta}$ at the base of the lithosphere.

Steinberger et al. (2001) define a *stress anomaly* in their global flow model as

$$\begin{pmatrix} \sigma_{rr} & \sigma_{r\vartheta} \\ \sigma_{r\vartheta} & \sigma_{\vartheta\vartheta} \\ \sigma_{r\varphi} & \sigma_{\vartheta\varphi} \end{pmatrix} := T_{ij} - pI_{ij}, \tag{11.1}$$

where T_{ij} is the total in-situ stress tensor, p is the pressure of the reference Earth model which only depends on elevation relative to the geoid and I_{ij} is the unity tensor (Chap. 2). Tensor elements are given in spherical coordinates with basic vectors $(\bar{e}_r, \bar{e}_\vartheta, \bar{e}_\varphi)$. Since only density variations enter into the calculations, directly stress anomalies are observed. From the *horizontal stress anomaly tensor*

$$\begin{pmatrix} \sigma_{\vartheta\vartheta} & \sigma_{\vartheta\varphi} \\ \sigma_{\vartheta\varphi} & \sigma_{\varphi\varphi} \end{pmatrix}, \tag{11.2}$$

the *scalar stress anomaly* can be calculated

$$\sigma_1 + \sigma_2, \text{ with } \begin{pmatrix} \sigma_1 & 0 \\ 0 & \sigma_2 \end{pmatrix} \tag{11.3}$$

which is tensile if $\sigma_1+\sigma_2>0$ and compressive if $\sigma_1+\sigma_2<0$. In Fig. 11.7, Steinberger et al. (2001) compare their calculated stress anomaly (Eq. (11.3)) in the lithosphere (Fig. 11.7a) to a published interpolation of the WSM by Bird and Li (1996), Fig. 11.7b. Short bars in Fig. 11.7 indicate the orientation of S_H. The length of the bars is proportional to the difference between the two horizontal stresses. At the equator, 2° length corresponds to 10 MPa stress difference; towards the poles correspondingly more, as the scale of the map, but not the scale for the stresses increases. Contours indicate the scalar stress anomaly from Eq. (11.3). Dots are locations of hot spots, with the area of dots proportional to estimates of anomalous mass flux (Steinberger 2000).

Stress anomalies in Fig. 11.7a are dominated by two major regions under tension caused by mantle plumes, namely the Pacific and that surrounding Africa, while other regions of the world are under compression. Calculated stress magnitudes (~100 MPa) are substantially larger than the shear stress expected at the base of the lithosphere (~10 MPa), see Sect. 11.3. Highest-tensile stresses are predicted in a region stretching N-S in East Africa in agreement with current continental rifting and in a region of the South Pacific in agreement with intraplate volcanism. The difference in predicted to observed stress magnitudes is due to strong vertical strength variations of the lithosphere. The computed stress orientations from the mantle flow model (Fig. 11.7a), however, as a first-order approximation, agree well with the interpolated WSM data set (Fig. 11.7b). Agreement is best in regions where lithospheric stress is well constrained and the resolution of seismic tomography

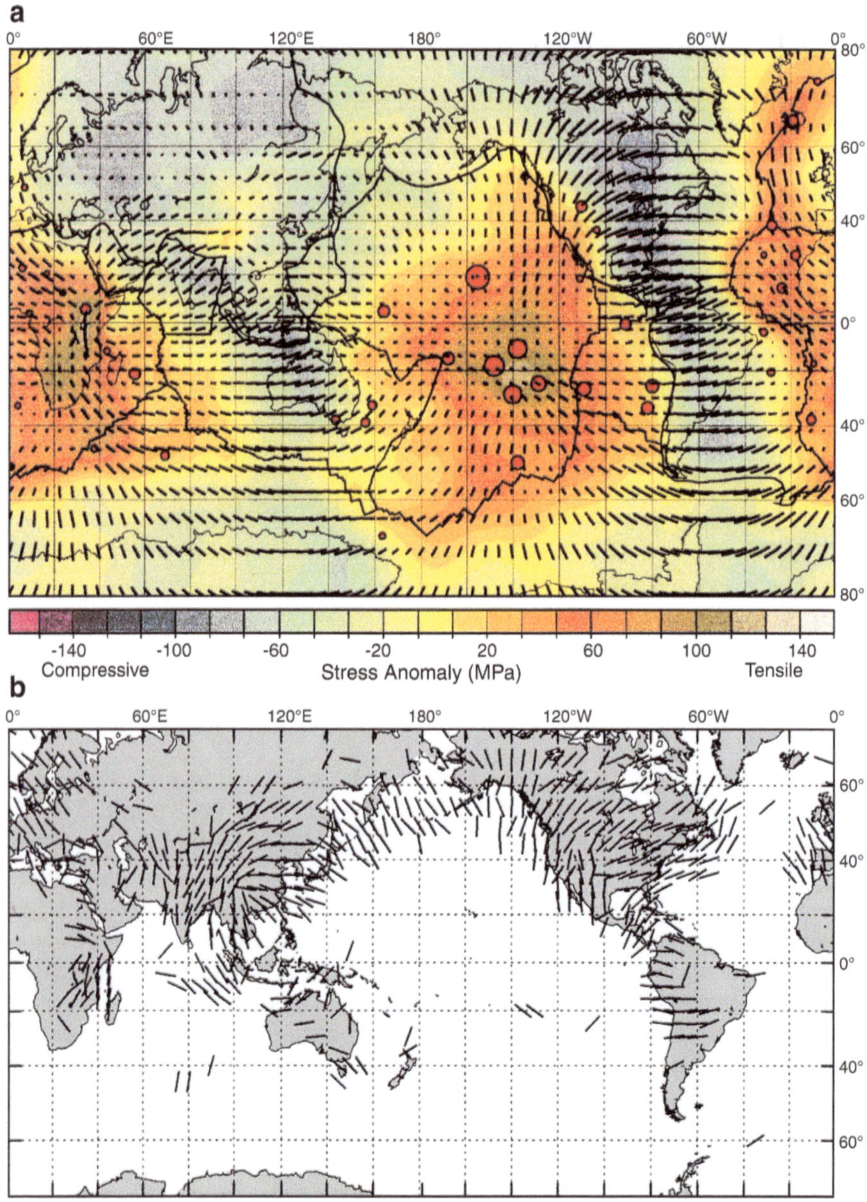

Fig. 11.7 Comparison of calculated scalar stress anomalies in an elastic lithosphere by Steinberger et al. (2001) based on a global mantle-flow model with hot spots (*dots*) with **b** a published interpolation of the World Stress Map according to Bird and Li (1996). Bar length is proportional to the difference between the two horizontal stresses

is high (e.g., North America, Europe, East Asia). The predicted stress orientations form a radiating pattern around the two large-scale regions where tensile scalar-stress anomalies dominate (South Africa, South-Central Pacific), and a roughly ring-shaped pattern around the region of Eastern Europe where a compressive scalar stress anomaly is predicted (Steinberger et al. 2001).

In some parts of Asia, the predictions (Fig. 11.7a) give a poor match to the observations (Figs. 11.5, 11.7b), e.g. the Tibetan region and the Lake Baikal region. Disagreement is due partially because the subducted lithosphere in East Asia pulls mainly at the Pacific Plate, thus associated stress anomalies on the Asian continent are less compressive than in the mantle flow model. Also, part of the isostatic compensation can occur at deeper levels than assumed (Steinberger et al. 2001).

At present, the understanding of lithospheric rheology is still poor (Karato 2008) and models of mantle convection with self-consistently generated plate tectonics are only beginning to emerge (Richards et al. 2000). More detailed models take into account lateral density variations of mantle viscosity, vertical layering of lithospheric strength and an appropriate treatment of plate boundaries (Lithgow-Bertelloni and Guynn 2004; Conrad and Lithgow-Bertelloni 2006). New generation mantle flow models will be able to explain significant parts of global lithospheric stress patterns. An absolute reference model for global plate motion, however, is certainly needed (Torsvik et al. 2008; Plag and Pearlman 2009).

Note-Box In the global hybrid stress map, single stress data are combined with smoothed stress directions from the World Stress Map data base. In the "western territory" of the hybrid map a belt with smoothed S_H azimuth oriented mostly perpendicular to the plate boundaries prevails. In the "eastern territory", very different orientations of the smoothed S_H azimuth occur, which can be explained by the different scale of crustal stresses ranging from faults (third-order stress patterns) via stress provinces like in Western Europe (second-order stress patterns) to tectonic plates (first-order stress patterns). Large-scale lithospheric stresses calculated from global mantle convection including the effect of major global hotspot distribution fit the overall directions of smoothed stress maps quiet well, while stress magnitudes estimated are off by a factor of 10.

11.3 Plate Tectonic Interpretation

The scientific criss-cross path from Alfred Wegener's first idea of continental drift to modern plate tectonics might have proceeded in a different way had the final personal remarks in his last sketchbook not been lost in the ice cap of Greenland in 1930. Alfred Wegener died probably November 16, 1930—his 17th wedding anniversary and a few days after his 50th birthday—on his way back from "Eismitte" to the Weststation during the German Greenland expedition. His body was found with a frozen smile on the face 189.5 km from Weststation (Wutzke 1997). After shovelling Wegener's temporary tomb in the ice (Fig. 11.8, cross), his companion, the 22-year-old Greenlander Rasmus Villumsen, took his sketch book back to the main station for safekeeping. Despite an intense search, the body of the Greenlander, and the important document of the continental drift, was never found.

After the death of Alfred Wegener, it took another 7 years before the South African Geologist Du Toit (1937) presented fossil evidence that the super-continent Pangaea was split into two continents, *Laurasia* to the north (which became North America and Eurasia) and *Gondwanaland* to the south (which became the remaining continents) about 200 million years ago. Thereafter it became quiet in the scientific community about the drift of continents until the 1950s, see Table 11.1.

In the mid-1950s, several fundamental discoveries of importance for the development of seafloor spreading and continental drift were made, see Table 11.1. The American seismologist Benioff (1954) and the Japanese seismologist Kiyoo Wadati studied the depth distribution of earthquakes and found that for certain regions the earthquakes were grouped along a descending plane called the Benioff zone, later named the Wadati-Benioff zone. The earthquakes in these zones are now thought to represent the upper surface of a descending plate in areas of plate collision. At this time, paleomagnetic studies of dated rocks became important and Runcorn (1955) and his colleagues in the UK could show that rocks of the same age and origin were

Fig. 11.8 Alfred Wegener participated in three expeditions through Greenland and Iceland: (1) the Mylius Erichsen or Danmark Expedition (1906–1908), (2) the Koch-Wegener Expedition (1912–1913), and (3) the German Greenland Expedition. During the last expedition, Wegener died on his way back from Eismitte to Weststation. His temporary tomb, 1930, in Greenland is marked (modified after Schwarzbach 1980)

Table 11.1 Summary of major steps in the history of plate tectonics

Alfred Wegener, 1911 and 1930	Alfred Wegener 1911 presents his theory of *wandering continents*, and dies in 1930 on the Greenland ice cap.
Alexander Du Toit, 1937	The South African geologist Du Toit presents fossil evidence for the splitting of the supercontinent *Pangea* into the southern continent Gondwana and the northern Laurasia in support of Wegener's ideas.
Hugo Benioff, 1954	Shallow focus earthquakes in the overriding plate are principally produced by motions on thrust faults, indicating compression (converging plates). A plane of earthquake foci descends beneath the overriding plate. The farther from the trench, the deeper the earthquakes. Kiyoo Wadati in Japan and Hugo Benioff study these type of earthquakes and in 1954 the latter proposes that the plane of descending earthquakes is the result of the seafloor subducting beneath the continent. These earthquakes of the Benioff zone (or *Wadati-Benioff zone*) are now believed to delineate the upper surface of the descending plate (or slab).
Keith S. Runcorn, 1955	Results of *paleomagnetic investigations* of dated rocks from the European and North American continents show clear evidence that once in time the two continents were joined and thereafter started to drift apart and form the present Atlantic Ocean.
Maurice W. Ewing and B.C. Heezen, 1956	W.M. Ewing and his colleagues at Lamont Doherty Geological Observatory of Columbia University present *seismic profiles and topographical maps over ocean ridges* and adjacent areas.
Robert S. Dietz, 1961; Harry H. Hess, 1962	The American geophysicist R.S. Dietz and the geologist H.H. Hess formulate the theory of *seafloor spreading* where hot magma is rising at the mid-ocean ridges and forces the oceanic crust apart and forms new oceanic crust.
Frederick J. Vine and Drummond H. Matthews, 1963	The two British geophysics map the *magnetic anomalies* across the Mid-Atlantic Ridge south of Iceland and find a symmetric pattern on each side of the centre of the ridge which supports seafloor spreading.
Tuzo J. Wilson, 1965	The Canadian geophysicist T.J. Wilson proposes that there must be a third type of plate boundary to connect the oceanic ridges and trenches, which he notes can end abruptly and "transform" into major faults that slip horizontally. Unlike ridges and trenches, *transform faults* offset the crust horizontally, without creating or destroying crust. A well-known example of such a transform-fault boundary is the San Andreas Fault.
Xavier Le Pichon, 1968	Based on the known rate of spreading at the ocean ridges, the French geophysicist X. Le Pichon computes a consistent *global kinematic model* that accounts for the spreading of the mid-ocean ridges and the shortening along deep-sea trenches and young mountain belts. The model is demonstrated for the six largest rigid plates of the Earth. It became the *birth of "plate tectonics"*.
B.L. Isacks, J. Oliver and L.R. Sykes, 1968	The American seismologists use the *plate tectonic model* by Le Pichon to demonstrate that most of the seismic phenomena occurring on the Earth can be explained by the six-plate model. The *plate tectonic model* of today contains 14 larger plates.

magnetized by the same magnetic field but later, spreading resulted in a new orientation with respect to the present field.

The 1950s was also the time when geoscientists started to explore the sea floors of the Earth and discovered the large system of ocean ridges and many other structures (Ewing and Heezen 1956). The mapping of the symmetric pattern of magnetic anomalies on each side of the mid-ocean ridges by Vine and Matthews (1963) and the explanation behind the existence of the large transform faults in the Earth's crust by Wilson (1965) established the concept of seafloor spreading.

Plate tectonic theory, a key ingredient in understanding the stress field of the Earth's crust by mantle flow processes (Sect. 11.1), arose out of the hypothesis of *continental drift* proposed by Alfred Wegener in 1911 (Chap. 11, heading text) and expanded in his 1915 book "Die Entstehung der Kontinente und Ozeane" (Wegener 1924, English translation). The 31-year-old German geophysicist and meteorologist contended that around 200 million years ago the single protocontinent called *Pangea* (Greek meaning "all lands") began to split apart through the drift of subunit continents. Wegener was widely ridiculed for his suggestion that the continents had moved, mainly because he did not propose a reasonable driving mechanism for the motion of the continents. The forerunner to the theory of plate tectonics, the continental drift, was put into cold storage shortly after the death of Alfred Wegener (see Table 11.1). Today, much of the physical evidence concerning plate tectonics is acquired from outer space via satellite technology (Chap. 1, GPS).

Not to repeat historical facts about plate tectonics (Oreskes 2001), we limit ourselves to three viewgraphs summarizing the principal features of plate tectonics starting with the block diagram of Isacks et al. (1968), via the diagram of Forsyth and Uyeda (1975) putting emphasis on the plate driving forces and, finally, the diagram of Zoback et al. (1989) putting emphasis on the forces responsible for tectonic stresses. The block diagram of Isacks et al. (1968) is enriched with specific historical milestones from Table 11.1 (Fig. 11.9a). While the mantle flow (Holmes 1928) and the seismological mapping of a subducting slab (Benioff 1954) are affecting the deeper asthenosphere, all other historical evidence for continental drift is related to the seafloor and its spreading (Ewing and Heezen 1956; Dietz 1961; Vine and Matthews 1963; Wilson 1965). The rigid six-plate approximation of Le Pichon (1968) was the basic model of plate tectonics working on a global scale, the basis of the seismological study published by Isacks et al. (1968) 6 months later, and the realization of Alfred Wegener's "frozen idea" of continental drift.

In Fig. 11.9a, arrows on the lithosphere represent relative motions of the plates (plate velocity ~5–10 cm year^{-1}). Arrows in the asthenosphere represent complementary flow in the mantle (velocity ~20 cm year^{-1}). Within this basic theory, tectonic plates are considered to be rigid and to act as efficient stress guides. Deformation takes place only at the plate margins. Plates are mechanically decoupled from each other, although plate margins are in intimate contact. The block diagram illustrates the three different types of plate boundaries existing in nature (divergent margins (Dietz 1961; Hess 1962; Vine and Matthews 1963)), convergent margins (subduction zone (Benioff 1954)), and conservative margins (transform faults (Wilson 1965)).

The first significant update of the plate tectonics view from Isacks et al. (1968) was given by Forsyth and Uyeda (1975) putting emphasis on the forces acting on

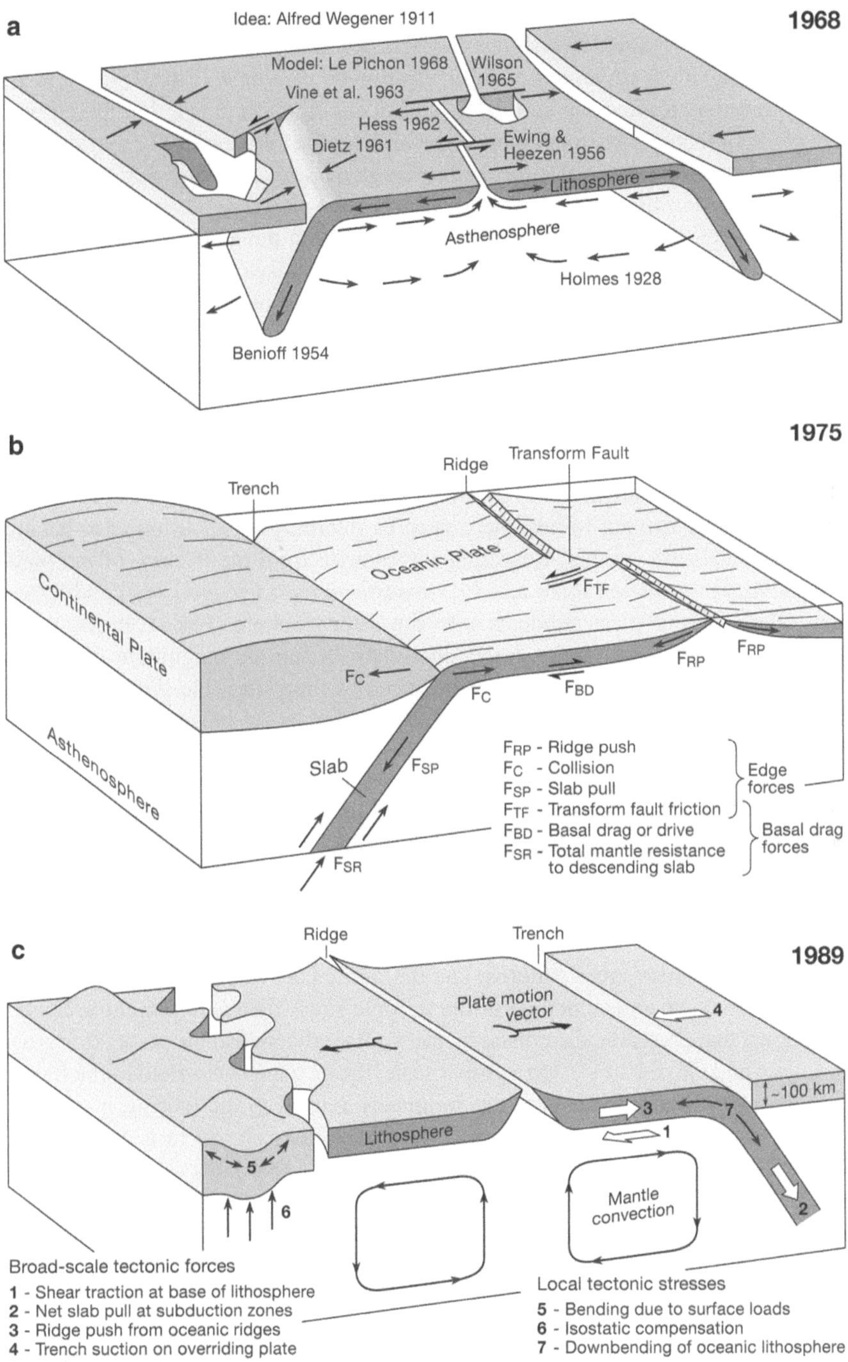

Fig. 11.9 Block diagrams summarizing principal features of plate tectonics in chronological order **a** Isacks et al. (1968) enriched by historical milestones from Table 11.1, **b** Forsyth and Uyeda (1975) highlighting forces acting on plates, and **c** Zoback et al. (1989) summarizing forces responsible for tectonic stresses

the plates. They subdivide plate driving forces into *edge forces* and *basal forces*, some of which will intensify, whilst other will retard motion. The sum of forces acting at or near ridges (divergent margin) is termed *ridge-push* (Fig. 11.9b, F_{RP}). F_{RP} arises from the elevated topography at mid-ocean ridges and is a distributed pressure gradient that acts normal to the strike of the ridge (Wilson 1993). Physically, it is more accurate to refer to this mechanism as *gravitational sliding*. Another driving force is the pull of the subducting plate, termed *slab-pull* (Fig. 11.9b, F_{SP}). The subducting, cooler, denser tectonic plate sinks by pulling the attached horizontal part of the plate after it. F_{SP} originates from negative buoyance of the down-going dense oceanic lithosphere at subduction zones and is proportional to the excess mass of the old slab compared with the mass of the warmer displaced mantle (Spence 1987). The amount of net force actually transferred to the horizontal plate is still controversial. While Conrad and Lithgow-Bertelloni (2004) suggest as much as 90% of slab pull force is transferred to the horizontal plate, Schellart (2004) suggests as little as 10%. A basal force (Fig. 11.9b, F_{BD}=*basal drag* or drive) will arise by viscous drag whenever there is relative motion between a plate and the mantle beneath.

In addition to driving forces, there must be retarding forces in order to balance finite plate-velocity vectors. Retarding forces arise from the friction of the plates (Fig. 11.9b, F_C=*collision force* and F_{TF}=*transform fault friction force*) and by viscous drag where a plate subducts into the asthenosphere (Fig. 11.9b, F_{SR}=total *mantle resistance force* to the descending slab). Since the magnitude of the viscous retarding forces increases with the speed of the plates, the plates will move at a steady speed in which retarding and driving forces are balanced. This balance makes it difficult to determine the magnitude of forces acting on any plate.

At this point, the reader may be wondering what connection there is between the stress field of the Earth's crust, plate driving forces and mantle convection. First, the relative motion between mantle flow and plate motion affects the lithospheric stress guide. Second, gravity sliding and slab pull affecting lithospheric stress are processes related to mantle convection. As mentioned by Lyzenga et al. (1991), the crustal stress field, in concert with deeper mantle convection, drives both the movement of the plates (first-order stress patterns) and the brittle behaviour of faults (third-order stress patterns). An understanding of the tectonic stress field is important because it is the agent that preserves the crustal memory of geodynamical processes (Lithgow-Bertelloni and Guynn 2004). On a global scale, the tectonic stress field of the Earth's crust and its variation with time are intimately related to the driving mechanism of plate tectonics. Mantle-flow models have previously shown to account for the Earth's geoid anomaly (Ricard et al. 1993), Cenozoic true polar wander (Richards et al. 1997) and plate motions (Lithgow-Bertelloni and Richards 1998).

The World Stress Map project (Sect. 11.2) was able to outline first-order pattern of tectonic stress in the lithosphere. Based on this, Zoback et al. (1989) identified two groups of forces, shown in Fig. 11.9c, as being the major sources of tectonic stresses. The first group, so-called *broad-scale tectonic forces* act on lithospheric plate boundaries such as shear tractions at the base of the lithosphere (Fig. 11.9c, 1), slab pull at subduction zones (2), ridge push from oceanic ridges (3) and trench suction (4). The second group, so-called *local tectonic forces* are related to bending of the lithosphere due to surface loads (5), isostatic compensation (6) and downbending of ocean lithosphere (7). Plate motion velocities at ridges, so-called spreading rates, range from 15 mm $year^{-1}$ (SW Indian Ocean Ridge) to 156 mm $year^{-1}$ (East Pacific Rise), while the flow velocity of the mantle is in the order of 200 mm $year^{-1} = 6 \times 10^{-9}\,m\,s^{-1}$.

The current view of plate tectonics is that tectonic plates are able to move because of the relative density of oceanic lithosphere and the relative weakness of the asthenosphere. Dissipation of heat from the mantle is the physical origin of energy driving the tectonic plates. When a plate breaks and forms at mid-ocean ridges, the oceanic lithosphere is initially less dense than the underlying asthenosphere, but it becomes denser with age as it cools and thickens. The greater density of the lithosphere relative to the underlying asthenosphere allows it to sink into the deep mantle at subduction zones, providing most of the driving forces for the plate motions. The weakness of the asthenosphere allows the plates to move towards the subduction zone. This, however, cannot be the only force on plates, since there are plates on Earth which are nowhere being subducted.

Today, two types of forces are thought to drive plate motions: *friction* (basal drag, slab suction) and *gravitation* (ridge push, slab-pull). In the diagram of Conrad and Lithgow-Bertelloni (2004) in Fig. 11.10, two mechanisms by which slab material can drive surface-plate motions are shown. If the slab is detached from the subducting plate (Fig. 11.10a, slab suction mechanism), the downward motion of the slab induces mantle flow that exerts tractions on the base of nearby plates, driving both subducting and overriding plates towards the subduction zone. If the slab remains attached to the subducting plate as it subducts (Fig. 11.10b, slab pull mechanism), it exerts a direct pull force on the subducting plate, drawing this plate rapidly towards the subduction zone. In this case, the mantle flow induced by the motion of the subducting plate exerts shear tractions on the base of the overriding plate, driving it slowly away from the subduction zone. By compiling slab pull from upper mantle slabs with slab suction from lower mantle slabs, Conrad and Lithgow-Bertelloni (2004) explain the observation that *subducting plates* currently move about four times faster than *non-subducting plates*.

The shear tractions that mantle flow exerts on the base of the Earth's lithosphere contribute to plate-driving forces and lithospheric stresses. In an effort to elucidate the origin of the present state of stress in the lithosphere (crust plus upper mantle material), Lithgow-Bertelloni and Guynn (2004) used a global finite-element model of the lithosphere to calculate stresses induced by mantle flow, crustal heterogeneity and topography, and compared these to observations of intraplate stresses as given by the World Stress Map database. While Steinberger et al. (2001) use a thin shell approximation (Sect. 11.2) Lithgow-Bertelloni and Guynn (2004) solve the full momentum equation for an elastic lithosphere incorporating topography effects as boundary conditions. Their treatment of crustal and topographic contributions is more complete using recent compilations of global crustal thickness and examining both isostatic and non-isostatic models. Plate velocities are determined by the balance of driving and resisting torques, i.e., the *net tractions* (the sum of driving and resisting tractions at each point) over the surface of the plate. It is the net trac-

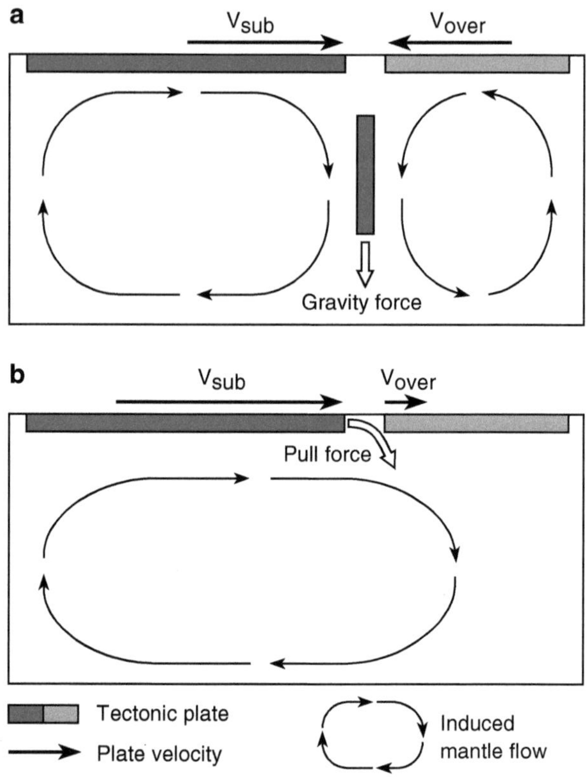

Fig. 11.10 Diagram showing two mechanisms by which subducted slab material can drive plate motions: **a** slab suction and **b** slab pull plate driving mechanism (after Conrad and Lithgow-Bertelloni 2004)

tion that generates the stresses in the lithosphere. The net tractions due to mantle flow in the presence of plates are computed to spherical harmonic degree of order 20, corresponding to a wavelength of about 2000 km. The net tractions include both the effects of density heterogeneity and plate motions. Observed plate motions are not imposed. Self-gravitation is not included. Stresses are resolved with the commercial finite element code ABAQUS on a $2° \times 2°$ surface mesh of the Earth (9408 finite elements) with a constant lithospheric thickness of 100 km.

Lithgow-Bertelloni and Guynn (2004) investigated two different density heterogeneity fields, one derived from a seismic tomographic model (Grand et al. 1997) and the other derived from the history of subduction (Lithgow-Bertelloni and Richards 1998). The rheological structure of the model uses a lithosphere 10 times as viscous, and a lower mantle 50 times as viscous as the upper mantle. These viscosity contrasts yield the best fit to observed geoid anomalies and Cenozoic plate motions for an absolute upper mantle viscosity of 4.2×10^{20} Pa s. The corresponding net shear and radial tractions at the base of the lithosphere for the subducted slab history model are shown in Fig. 11.11. As stated by Lithgow-Bertelloni and Guynn (2004), prominent features of the *radial tractions* are broad upwellings in the Atlantic basin, in the middle of the Pacific plate and under the western and southern parts of the African plate. Focused downwellings (radial tractions are posi-

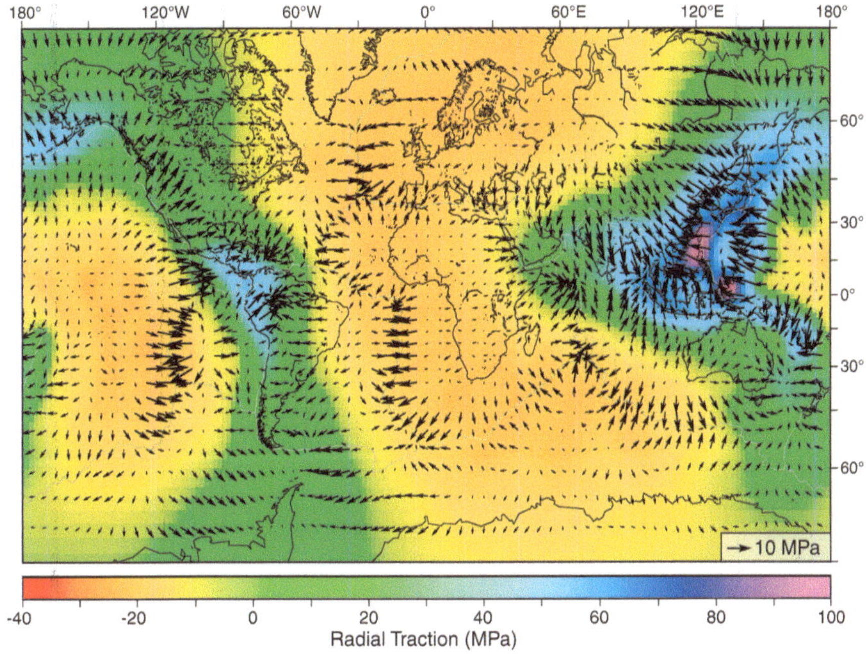

Fig. 11.11 Net tractions at the base of the lithosphere for the history of the subduction model (Lithgow-Bertelloni and Richards 1998). *Arrows* indicate the direction and magnitude of horizontal tractions; *color contours* give the value of the radial tractions (downward positive) (after Lithgow-Bertelloni and Gynn 2004)

tive downward) are observed near current and extinct trenches like the large area around southeastern Asia. Magnitudes of radial tractions range from 0 to 50 MPa, with downwellings much stronger than upwellings. This is because the upwellings in this model are passive and the result of return flow from slabs (Fig. 11.10). The horizontal net tractions are greatest in the vicinity of current or past subduction and at ridges. The former are due to the driving forces associated with sinking slabs, while the latter are due to resisting forces, since the driving forces are small at ridges in the subduction history model.

Lithgow-Bertelloni and Guynn (2004) found that the combined modelling of mantle flow and lithospheric heterogeneity (topography, crustal thickness and density structure) yields the best match to the observed WSM stress field although there are many regions where agreement between observed and predicted stress is poor. Their combined model in Fig. 11.12 indicates that the total stress field is dominated by the contribution from mantle flow (on average a factor 3 greater than stresses from lithospheric heterogeneity). In many regions, the two contributions to the stress field add constructively since continental geomorphology and mantle flow are correlated.

The comparison of the combined mantle-flow-lithospheric-heterogeneity model (Fig. 11.12) and the hybrid stress map (Fig. 11.5) is not straight forward for a number

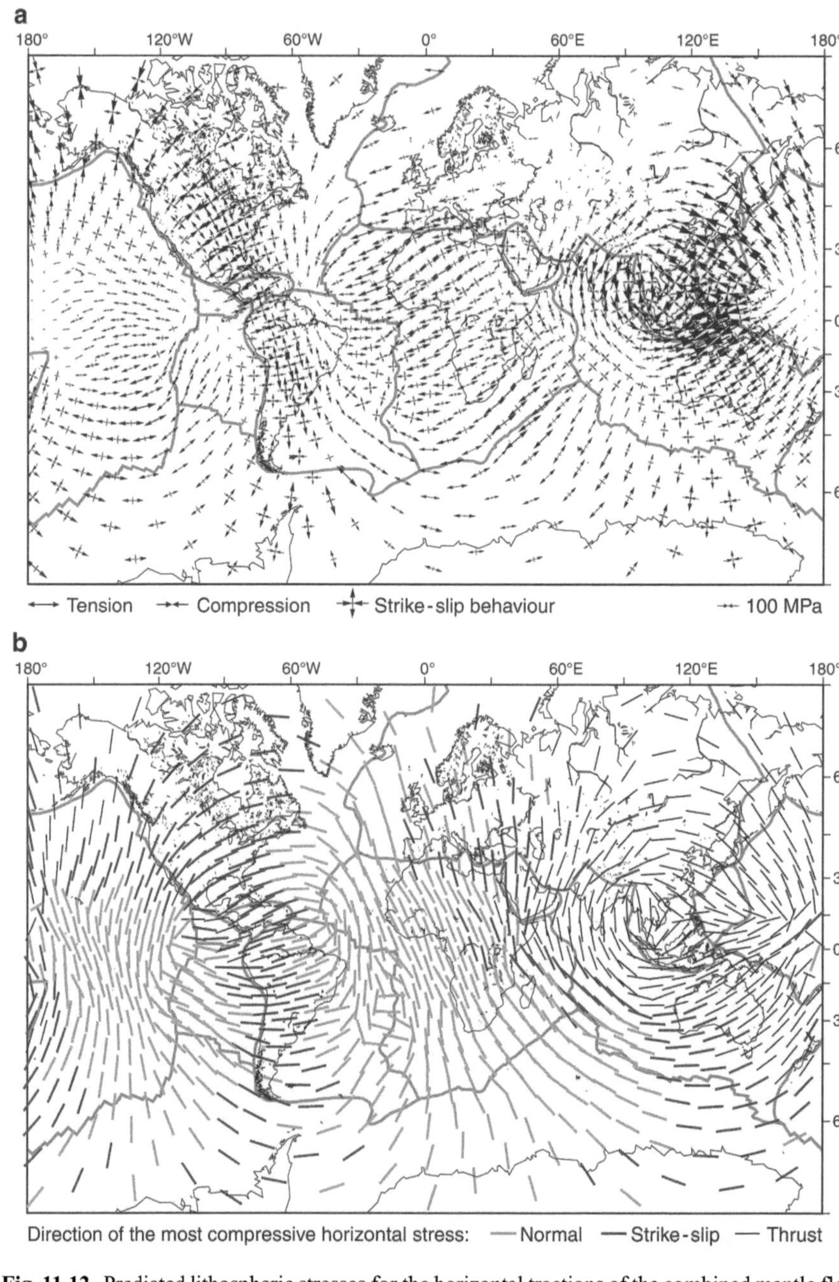

Fig. 11.12 Predicted lithospheric stresses for the horizontal tractions of the combined mantle flow and lithospheric density model of Lithgow-Bertelloni and Guynn (2004). **a** *Arrows* indicate principal stress directions (*outward*=tensile, *inward*=compression, *length*=magnitude). The *longest compressive arrows* or *shortest tensional arrows* indicate the direction of maximum compressive principal stress, S_H. **b** Orientations of S_H with tectonic stress regimes indicated by *grey scales*

of reasons. First, the depth of conventional crustal stress measurements is limited to the top third of the continental crust (Sect. 9.1), while earthquake focal mechanisms may cover the second third of the crust down to 20 km. No stress data, however, are available for the remaining third of the crust and the upper mantle lithosphere. Therefore, WSM stress data may not be representative for the lithosphere as a whole. Second, stress orientations from WSM have a large degree of uncertainty from ±15° for quality *A* to ±25° for quality *C* data, and caution must be exercised in comparing predicted stress orientations to measured stress orientations based on a local small data set. It has been argued that the data of the WSM may only represent stress in the very top of the crust (Liu et al. 1997) or that stresses may vary with depth due to rheological differences (Lynch and Richards 2001). Nevertheless, the WSM data base shows regions in which stress orientations are constant over length scales that are much longer than the thickness of the lithosphere (Zoback 1992). At least within these first-order stress patterns, measurements of "shallow stresses (~10 km)" may be representative of stresses prevailing in the lithosphere as a whole (~100 km).

Lithgow-Bertelloni and Guynn (2004) argue that the coverage of the Earth's surface from WSM data is minimal, 21%. The large gaps in the stress data make it hard to discern long wavelength patterns, such as those generated by mantle tractions, and leave open the possibility that isolated stress measurements may be strongly influenced by local density or topography. The relative small number of oceanic stress data is particularly troublesome, as this represents the majority of the Earth's surface and the fastest moving plates have little, if any, continental material. This is especially unfortunate since the homogeneity of oceanic crust and lithosphere would result in less local effects on the stress orientations and would help in resolving the different relative contributions to the stress field. There is also a concentration of stress data from earthquakes near plate boundaries, which may reflect kinematics more than the forces on the plate. Even in continental areas, such as Western Europe, most stress measurements give consistent stress orientations, but all types of faulting are found in the same area.

More recent models include variations in the actual lithospheric thickness and lateral as well as vertical variations in rheological properties (Conrad and Lithgow-Bertelloni 2006). The difficulty is not in the implementation but the determination of appropriate material parameters (Karato 2008). Including vertical variations in the strength of the lithosphere can help to ascertain whether a single stress data point measured represents the state of stress of a tectonic plate or the upper crust only. Such models would allow us to further examine the degree of coupling between the crust, the non-crustal lithosphere and the upper mantle. Given the 21% coverage of WSM stress data over most of the tectonic plates and the uncertainties in the data, it is not possible at this stage to resolve for the different contributions to the stress field or to distinguish between models of plate driving forces with the available stress data any better than using other global geophysical observables (Lithgow-Bertelloni and Guynn 2004).

Finally, earthquake prediction relies on the stress field of the Earth's crust and how it changes with time. We know that there are methods to estimate stresses from satellite data, e.g., from surface displacements detected by GPS. Hori and Kameda (2001) proposed an inversion method that determines stresses in a body without full knowledge of the material constitutive relations. The method assumes elasto-plastic material and relies on high-order derivatives of high accuracy displacement measurements. Iinuma et al. (2005) applied this method to estimate stress changes within the crust from 3 years of GPS observations at the surface in Japan. Strain-rate tensors were calculated by Kostrov's (1974) method, boundary tractions were computed from plate-boundary earthquake parameters (stress drop, recurrence cycle, slip), and modelling of Japanese island were done with FEM. Maximum shear stress distribution has an average value 50 kPa year^{-1} (i.e., 5 MPa change in shear stress per 100 years). By mapping the difference between long-term residual geodetic strain and intraplate seismic strain, it may be possible to forecast the strain to be released in future intraplate earthquakes (Sone and Zoback 2008). While GPS data are restricted to the surface, WSM data come from borehole measurements and earthquakes. However, GPS data should be correlated with creep-strain data in boreholes, if possible at depth level below 1 km. To convert borehole creep strains to stress, drill-core parameters are required.

Note-Box Based on Alfred Wegener's idea of continental drift in 1911 and later, the theory of plate tectonics developed in the 1950s and 1960s by identifying processes like seafloor spreading (birth of plates) and the Wadati-Benioff zone (death of plates). While two decades ago the prediction of the stress field related to whole mantle flow was poor in matching Word Stress Map data (Zoback 1992), today finite-element models calculating lithospheric stress induced by mantle flow, crustal heterogeneity and topography fit better with observations from intraplate stresses given by the WSM data set. On a global scale, the crustal stress field and its variations are intimately related to the driving mechanisms of plate tectonics. The physical origin of the energy driving the plates and, therefore, the present-day stress field, is the dissipation of heat from the Earth's mantle while our planet Earth and its crust are approaching a new state of equilibrium during cooling from big bang subsequent accretion.

Epilogue

What did you learn from reading the book *Stress Field of the Earth's Crust?*

1. You can follow Alfred Wegener's first ideas of *continental drift* to modern *plate tectonics* to understand the physical reason of tectonic stress in the Earth's crust. May be you are lucky enough to find his lost sketch book in the melting ice sheet of Greenland.
2. You are able to use the correct concept of stress to follow discussions in both material science and Earth sciences. You know how to average a *stress tensor,* how to find *principal stresses,* and how to *visualize stress* in two and three dimensions.
3. You are proficient in applying *rock failure criteria* based on phenomenological (relating fracture orientation to stress) and mechanistic theories (quantifying the crack tip stress field). You are not afraid of *linear fracture mechanics* formalism, or of process zone models in *nonlinear fracture mechanics.*
4. You internalized a modern *rock stress terminology* and you know that all components of stress are perturbed by openings and geological structures at all scales. You are aware of simple *inclusion models* for *heterogeneous rock* and you know how to incorporate *joint closure curves* for analysis of *anisotropic rocks.*
5. You know about simple models for drawing stress profiles through the Earth's crust taking into account pore water pressure and constraints derived from laboratory tests. More important, you are able to attach error bars on *crustal strength profiles* depending on water content, rock material or strain rates used for extrapolation to Earth conditions.
6. You know that it is impossible to directly measure stress, but you also know how to convert from measured strain to stress. To obtain strain data, you know that the rock mass has to be perturbed and its response has to be quantified. To calculate the stress, you need strain data and rock material property.
7. You know how to *classify rock stress measurement* techniques based on the underlying physical principle and the rock volume involved. *Overcoring* methods rely on an elastic response of the rock mass. The most important methods based on rock failure in boreholes are the instantaneous stress release technique *hydraulic fracturing,* and the time-dependent release method *borehole breakouts.*

8. Compared to many people ignoring stress data from drill cores, you are informed about *anelastic strain recovery* (ASR) and the *Kaiser effect* (KE). While ASR infers stress data from time-dependent relaxation behaviour, KE as well as other *crack-based methods* (DSA = differential strain analysis, DRA = deformation rate analysis, WVA = wave velocity analysis) rely on repressurizing a part of the drill core in a pressure vessel.
9. You know that an *integrated stress measurement strategy* taking into account both borehole and core-based stress data is our favourite approach in obtaining reliable in-situ stresses. From *KTB*, you learned that Byerlee's friction law is valid in the top third of the continental crust. From *Olkiluoto*, you remember that the least-principal stress is oriented perpendicular to the foliation of gneisses. From *SAFOD*, you take away that the San Andreas is a weak fault in a strong crust.
10. You know about how to present *generic stress data* as stress *magnitude versus depth profiles*, *dimensionless stress ratios* or how to prepare a *smoothed stress orientation map*. You know how to *scale your stress data* according to the rock *representative elementary volume* and how to find the *best estimated stress model* in your study area.
11. Finally, you are able to create your own *stress map from World Stress Map* data. You know how to place your stress map in the context of *geodynamics*, and the current view of *plate tectonics* to understand the *Stress Field of the Earth's Crust* as a result of heat dissipation from the Earth's mantle while our planet is approaching a new state of equilibrium.

References

Stress References

Abass H, Neda J (1988) Rock mechanics in wellbore construction. In: Economides MJ, Watters LT, Dunn-Normann S (eds) Petroleum well construction. Wiley, Chichester, pp 143–173

Abou-Sayed AS, Brechtel CE, Clifton RJ (1978) In situ stress determination by hydrofracturing: a fracture mechanics approach. J Geophys Res 83:2851–2862

Acock A, Orourke T, Shirmboh D, Alexander J, Andersen G, Kaneko T, Venkitaraman A, Lopez-de-Cardenas J, Nishi M, Numasawa M, Yoshioka K, Roy A, Wilson A, Twynam A (2004) Practical approaches to sand management. Oilfield Rev 16(1):10–27

Addis MA (1997) The stress-depletion response of reservoirs. Society petroleum engineers paper 38720. In: Proceedings of the SPE annual technical conference and exhibit, San Antonio, pp 55–65

Ahorner L (1970) Seismo-tectonic relations between the graben zones of the Upper and Lower Rhine valley. In: Illies JH, Mueller S (eds) Graben problems. Schweitzerbart, Stuttgart, pp 155–166

Ahorner L (1975) Present-day stress field and seismotectonic block movements along major fault zones in central Europe. Tectonophysics 29:233–249

Alava MJ, Nukala PKVV, Zapperi S (2006) Statistical models of fracture. Adv Phys 55(3–4):349–476

Albarello D, Mantovani E, Babucci D, Tamburelli C (1995) Africa-Eurasia kinematics: main contraints and uncertainities. Tectonophysics 243:25–36

Albers GA, Gephart JW (2001) Direct inversion of earthquake first motions for both the stress tensor and focal mechanisms and application to southern California. J Geophys Res 106:26523–26540

Albright JN, Pearson CF (1982) Acoustic emissions as tool for hydraulic fracture location: experience at the Fenton Hill hot dry rock site. Soc Pet Eng J 22:523–530

Amadei B (1985) Applicability of the theory of hollow inclusion for overcoring stress measurements in rock. Rock Mech Rock Eng 18:107–130

Amadei B (1996) Importance of rock anisotropy when estimating and measuring in situ stress in rock. Int J Rock Mech Min Sci 33(3):293–325

Amadei B, Goodman RE (1982) The influence of rock anisotropy on stress measurements by overcoring techniques. Rock Mech 15:167–180

Amadei B, Savage WZ, Swolfs HS (1987) Gravitational stresses in anisotropic rock masses. Int J Rock Mech Min Sci and Geomech Abstr 24:5–14

Amadei B, Savage WZ, Swolfs HS (1988) Gravity-induced stresses in stratified rock masses. Rock Mech Rock Eng 21(1):1–20

Amadei B, Stephansson O (1997) Rock stress and its measurements. Chapman & Hall, London

Anders HM, Wiltschko DV (1994) Microfracturing, paleostress and the growth of faults. J Struct Geol 16:795–815

Anderson EM (1951) The dynamics of faulting and dyke formation with applications to Britain. Oliver and Boyd, Edinburgh

Andersson J, Ahokas H, Hudson JA, Koskinen L, Luukonen A, Löfman J, Keto V, Pitkänen P, Mattila J, Ikonen, ATK, Ylä-Mella MM (2007) Olkiluoto site description 2006. Technical report Posiva 2007-03, Posiva Oy, Olkiluoto

Andreev GE (1995) Brittle failure of rock material. AA Balkema, Rotterdam

Angelier J (1990) Inversion of field data in fault tectonics to obtain the regional stress—3. A new rapid direct inversion method by analytical means. Geophysical Journal International 103: 363–376

Angelier J (2002) Inversion of earthquake focal mechanism to obtain the seismotectonic stress IV—a new method free of choice among nodal planes. Geophys J Int 150:588–609

Angelier J, Tarantola A, Valette B, Manoussis S (1982) Inversion of field data in fault tectonics to obtain the regional stress—1. Single phase fault populations: a new method of computing the stress tensor. Geophys J R Astron Soc 69:607–621

Argus DF, Gordon RG, DeMets C, Stein S (1989) Closure of the Africa-Eurasia-North America plate motion circuit and tectonics of the Gloria fault. J Geophys Res 94:5585–5602

Arjang B (1989) Pre-mining stresses at some hard rock mines in the Canadian shield. In: Proceedings 30th US symposium of rock mechanics, Morgantown, Balkema, Rotterdam, pp 545–551

Arnadottir T, Jiang W, Feigl KL, Geirsson H, Sturkell E (2006) Kinematic models of plate boundary deformation in southwest Iceland derived from GPS observations. J Geophys Res 111, B07402. doi:10.1029/2005JB003907

Arnold R, Townend J (2007) A Bayesian approach to estimate tectonic stress from seismological data. Geophys J Int 170(3):1336–1356. doi:10.1111/j.1365-246X.2007.03485.x

Arnold R, Townend J, Vignaux T (2005) Mapping tectonic stress using earthquakes. In: KH Knuth, AE Abbas, RD Morris, Castle JP (eds) Bayesian interference and maximum entropy methods in science and engineering. American Institute of Physics, pp 475–482

Ashby MF, Hallam SD (1986) The failure of brittle solids containing small cracks under compressive stress states. Acta Metall 34(3):497–510

Ashby MF, Sammis CG (1990) The damage mechanics of brittle solids in compression. Pure Appl Geophys 133(3):489–521

Ask D (2001) Inversion and interpretation of hydraulic and overcoring stress measurements in the Äspö region. Licentiate thesis, Royal Institute of Technology, Stockholm

Ask D (2006) New developments in the integrated stress determination method and their application to rock stress data at the Äspö HRL, Sweden. Int J Rock Mech Min Sci 43(1):107–126

Ask D, Cornet F-H, Fontbonne F, Nilsson T, Jönssen L, Ask MVS (2009) A quadruple packer tool for conducting hydraulic stress measurements in mines and other high stress settings. Technical Note, Int J Rock Mech Min Sci 46(6):1097–1102. doi:10.1016/j.ijrmms.2009.04.004

Ask D, Stephansson O (2003) Hydraulic stress measurements in borehole BJO01, Björkö impact structure, Lake Mälaren, Sweden. In: Sugawara K, Obara Y, Sato A (eds) Proceedings international symposium on rock stress, Kumamoto, Japan, Balkema, Lisse, pp 115–121

Ask D, Stephansson O, Cornet FH (2001) Integrated stress analysis of hydraulic and overcoring rock stress data in the Äspö region. Analysis of hydraulic fracturing stress measurements and HTPF in boreholes KAS02, KAS03, and KLX02. SKB international progress report IPR-01-26, Swedish Nuclear Fuel and Waste Management Co., Stockholm

Ask D, Stephansson O, Cornet FH (2003) Integration of CSIR- and CSIRO-type of overcoring rock stress data at the Zedex Test Site, Äspö HRL, Sweden. In: Handley M, Stacey D (eds) Proceedings 10th international congress on rock mechanics of ISRM, Johannesburg, Balkema, Rotterdam, pp 63–68

Atkinson BK (1984) Sub-critical crack growth in geologic materials. J Geophys Res 89:4077–4113

Atkinson BK (1987) Introduction to fracture mechanics and its geophysical applications. In: Atkinson BK (ed) Fracture mechanics of rocks. Academic Press, London, pp 1–26

Atkinson BK, Meredith PG (1987) The theory of subcritical crack growth with applications to minerals and rocks. In: Atkinson BK (ed) Fracture mechanics of rock. Academic Press, London, pp 111–166

Austin NJ, Evans B (2007) Paleowattmeters: a scaling relation for dynamically recrystallized grain size. Geology 35(4):343–346

Aydan Ö (1995) The stress state of the Earth and the Earth's crust due to gravitational pull. In: Daemen JJK, Schulz RA (eds) Proceedings 35th US symposium on rock mechanics, Lake Tahoe, AA Balkema, Rotterdam, pp 237–243

Aydan Ö, Kawamoto T (1997) The general characteristics of the stress state in the various parts of the Earth's crust. In: Sugawara K, Obara Y (eds) Proceedings international symposium on rock stress, Balkema, Rotterdam, pp 369–373

Azzam R, Bock H (1987) A new modified borehole jack for stiff rock. Rock Mech Rock Eng 20:191–211

Babcock EA (1978) Measurement of subsurface fractures from dipmeter logs. AAPG Bull 62(7):1111–1126

Backers T (2005) Fracture toughness determination and micromechanics of rock under Mode I and Mode II loading. Dissertation, University of Potsdam, scientific technical report 05/05, GFZ German Research Centre for Geosciences, Potsdam

Backers T, Stephansson O, Rybacki E (2002) Rock fracture toughness testing in Mode II—punch trough shear test. Int J Rock Mech Min Sci 39:755–769

Bada G, Coetingh S, Gerner P, Horvath F (1998) Source of recent tectonic stress in the Pannonian region: interferences from finite element modelling. Geophys J Int 134:87–101

Bada G, Horvath F, Dövenyi P, Szafian P, Windhoffer G, Cloetingh S (2007) Present-day stress field and tectonic inversion in the Pannonian basin. Glob Planet Change 58:165–180

Bai Q, Mackwell SJ, Kohlstedt DL (1991) High-temperature creep of olivine single crystals 1. Mechanical results for buffered samples. J Geophys Res 96:2441–2463

Bai W, Vigny C, Ricard Y, Froidevaux C (1992) On the origin of deviatoric stresses in the lithosphere. J Geophys Res 97(B8):11729–11737

Baisch S, Harjes H-P (2003) A model for fluid injection induced seismicity at KTB, Germany. Geophys J Int 152:160–170

Baisch S, Bohnhoff M, Ceranna L, Tu Y, Harjes H-P (2002) Probing the crust to 9 km depth: fluid injection experiments and induced seismicity at the KTB superdeep drilling hole, Germany. Bull Seismol Soc Am 92:2369–2380

Baisch S, Weidler R, Vörös R, Wyborn D, de Graaf L (2006). Induced seismicity during the stimulation of a geothermal HFR reservoir in the Cooper Basin, Australia. Bull Seismol Soc Am 96:2242–2256

Bandis SC, Lumsden AC, Barton N (1983) Fundamentals of rock joint deformation. Int J Rock Mech Min Sci Geomech Abstr 20:249–268

Barenblatt GI (1962) The mathematical theory of equilibrium cracks in brittle fracture. Adv Appl Mech 7:55–129

Barr SP, Hunt DP (1999) Anelastic strain recovery and the Kaiser effect retention span in the Carnmenellis granite, UK. Rock Mech Rock Eng 32(3):169–193

Barth A, Reinecker J, Heidbach O (2008) Stress derivation from earthquake focal mechanisms. guidelines: focal mechanisms, world stress map project, 12 pp (see DVD enclosed)

Barton N (2007) Rock quality, seismic velocity, attenuation and anisotropy. Taylor & Francis, London

Barton N, Lien R, Lunde J (1974) Engineering classification of rock masses for the design of tunnel support. Rock Mech 6:183–236

Bass JD, Schmitt D, Ahrens TJ (1986) Holographic in situ stress measurements. Geophys J R Astron Soc 85:13–41

Bassin C, Laske G, Masters G (2000) The current limits of resolution for surface wave tomography in North America. EOS Trans Am Geophys Union 81(48), Fall Meeting Supplement, Abstract F897

Baumann H (1993) In situ stress in the German KTB pilot hole deduced from differential strain analysis. Sci Drill 3:263–270

Baumgärtner J, Zoback MD (1989) Interpretation of hydraulic fracturing pressure. Time records using interactive analysis methods. Int J Rock Mech Min Sci Geomech Abstr 26:461–469

Baumgärtner J, Healy JH, Rummel F, Zoback MD (1993) Deep hydraulic fracturing stress measurements in the KTB (Germany) and Cajon Pass (USA) scientific drilling projects—a summary. In: Proceedings 7th congress international society of rock mechanics, vol 3, Aachen, Balkema, Rotterdam, pp 1685–1690

Bazant ZP, Planas J (1998) Fracture and size effect in concrete and other quasibrittle materials. CRC Press, Boca Raton

Becker TW, Hardebeck JL, Anderson G (2005) Constraints on fault slip rates of the southern California plate boundary from GPS velocity and stress inversions. Geophys J Int 160:634–650

Bell JS, Gough DI (1979) Northeast-southwest compressive stress in Alberta: evidence from oil wells. Earth Planet Sci Lett 45:475–482

Benioff H (1954) Orogenesis and deep crustal structure—additional evidence from seismology. Bull Seismol Soc Am 66:385–400

Ben-Zion Y, Sammis C (eds) (2009) Mechanics, structure and evolution of fault zones. Pure Appl Geophys, Topical Issue, Birckhäuser, Basel, ISBN 978-3-0346-0137-5

Berckhemer H (1968) Vom Erdkern bis zur Magnetosphäre. In: Murawski H (ed) Umschau Verlag, Fankfurt/Main, pp 131–146

Bergmann-Schaefer (1978) Lehrbuch der Experimentalphysik, Bd 3 Optik. Walter de Gruyter, Berlin

Berest P, Blum P-A, Durup G (1992) Effects of the moon on underground caverns. In: Proceedings 33rd US symposium on rock mechanics, Balkema, Rotterdam, pp 421–428

Bielenstein HU, Barron K (1971) In situ stresses. A summary of presentations and discussions given in Theme I at the conference of structural geology to rock mechanics problems. Department of Energy, Mines and resources, Mines Branch, Ottawa, Internal report MR71

Bieniawski ZT (1984) Rock mechanics design in mining and tunnelling. AA Balkema, Rotterdam

Bieniawski ZT (1989) Engineering rock mass classification. Wiley, Chichester

Biot MA (1941) General theory of three dimensional consolidation. J Appl Phys 12:155–164

Birch F (1960) The velocity of compressional waves in rock to 10 kilobars, Part 1. J Geophys Res 65:1083–1102

Birch F (1961) The velocity of compressional waves in rock to 10 kilobars, Part 2. J Geophys Res 66:2199–2224

Bird P (2003) An updated digital model for plate boundaries. Geochem, Geophys, Geosyst (3G) 4(3):1027. doi:10.1029/2001GC000252

Bird P, Li Y (1996) Interpolation of principal stress directions by nonparametric statistics: global maps with confidence limits. J Geophys Res 101:5435–5443

Bjarnason B, Ljunggren C, Stephansson O (1989) New developments in hydrofracturing stress measurements at Lulea university of technology. Int J Rock Mech Min Sci Geomech Abstr 26:579–586

Blanpied ML, Lockner DA, Byerlee JD (1995) Frictional slip of granite at hydrothermal conditions. J Geophys Res 100:13045–13064

Blanton TL (1983) The relation between recovery deformation and in-situ stress magnitudes. Symposium on low permeability gas reservoirs. Society of Petroleum Engineers (SPE)/ US Department of Energy (DOE) 11624, Denver, Colorado, pp 213–218

Blanton TL, Teufel LW (1986) A critical evaluation of recovery deformation mesurements as a means of in-situ stress determination. Unconventional gas technology symposium, Louisville, Ky, Society of Petroleum Engineers 15217, pp 121–131

Bloch M, Freitas SMS, Custodio FR (2006) A fully coupled poroviscoelastic model for in-situ stress determination in the oil industry. In: Van Cotthem A, Charlier R, Thimus J-F, Tshibangu J-P (eds) EUROCK—multiphysics coupling and long term behaviour in rock mechanics. Taylor & Francis Group, London, pp 219–226

Bock H (1993) Measuring in-situ rock stresses by borehole slotting. In: Hudson JA (ed) Comprehensive rock engineering, vol 3. Pergamon Press, Oxford, pp 433–443

Bohnhoff M, Baisch S, Harjes HP (2004) Fault mechanisms of induced seismicity at the superdeep German Continental Deep Drilling Program (KTB) borehole and their relation to ault structure and stress. J Geophys Res 109, B02309. doi:10.1029/ 2003JB002528

Boness NL, Zoback MD (2004) Stress-induced seismic velocity anisotropy and physical properties in the SAFOD pilot hole in Parkfield. Geophys Res Lett 31, 4, L12S11. doi:10.1029/2003GL019020

Boness NL, Zoback MD (2006) Mapping stress and structurally controlled crustal shear velocity anisotropy in California. Geology 34(19):825–828

Borsetto M, Martinetti S, Ribacchi R (1984) Interpretation of in situ stress measurements in anisotropic rocks with the doorstopper method. Rock Mech Rock Eng 17:167–182

Bott MHP (1959) The mechanics of oblique slip faulting. Geol Mag 96:109–117

Bowes C, Procter R (1997) Drillers stuck pipe handbook. Guidelines & Drillers handbook credits. Schlumberger, Ballater

Brace WF (1964) Brittle fracture of rocks. In: Judd WR (ed) State of stress in the Earth's crust. Elseveier, New York, pp 110–178

Brace WF, Bombolakis EG (1963) A note on brittle crack growth in compression. J Geophys Res 68(12):3709–3713

Brace WF, Kohlstedt DL (1980) Limits on lithospheric stress imposed by laboratory experiments. J Geophys Res 85:6248–6252

Brace WF, Paulding BW, Scholz C (1966) Dilatancy in the fracture of crystalline rocks. J Geophys Res 71:3939–3953

Brady BHG, Brown ET (2004) Rock mechanics: for underground mining. Kluwer Academic Publishers, Dordrecht

Bram K, Draxler J, Hirschmann G, Zoth G, Hiron S, Kühr M (1995) The KTB borehole—Germany's superdeep telescope into the Earth's crust. Oilfield Rev 7(1):4–22 (January)

Bredehoeft JD, Wolff RG, Keys WS, Shuter E (1976) Hydraulic fracturing to determine the regional in situ stress field, Piceance Basin, Colorado. Geol Soc Am Bull 87:250–258

Brereton NR, Chroston PN, Evans CJ, Hudson JA, Whitmarsh RB (1990) Stress in the oceanic crust: anelastic strain recovery and elastic properties of basaltic rocks. British geological survey, Technical report WK/90/10, p 28

Brereton NR, Chroston PN, Evans CJ (1995) Pore pressure as an explanation of complex anelastic strain recovery. Rock Mech Rock Eng 28(1):59–66

Brock WG, Engelder JT (1977) Deformation associated with the movement of the Muddy Mountain overthrust in the Buffington window, southeastern Nevada. Bull Geol Soc Am 88:1667–1677

Brown ET, Hoek E (1978) Trend in relationships between measured in situ stresses and depth. Int J Rock Mech Min Sci Geomech Abstr 15:211–215

Brown GC, Musset, AE (1981) The inaccessible Earth. George Allen & Unwin, London

Brown SR, Scholz CH (1985) Closure of random elastic surfaces in contact. J Geophys Res 90:5531–5545

Brown SR, Scholz CH (1986) Closure of rock joints. J Geophys Res 91:4939–4948

Brown DW, Brouke MAM, Clausen B, Holden TM, Tome CN, Varma R (2003) A neutron diffraction and modelling study of uniaxial deformation of polycrystalline beryllium. Metall Mater Trans A 34:1439–1449

Brudy M, Zoback MD (1999) Drilling-induced tensile well-fractures: implications for determination of in-situ stress orientation and magnitude. Int J Rock Mech Min Sci 36:191–215

Brudy M, Fuchs K, Zoback MD (1993) Stress orientation profile to 6 km depth in the KTB main borehole. In: Bram K, Draxler JK (eds) KTB-report 93-1, NLfB, Hannover, pp 281–300

Brudy M, Zoback MD, Rummel F, Fuchs K (1995) Application of the integrated stress measurement strategy to 9 km depth in the KTB boreholes. In: Workshop on rock stresses in the North Sea. NTH and SINTEF Publication, Trondheim, pp 154–164

Brudy M, Zoback MD, Fuchs K, Rummel F, Baumgärtner J (1997) Estimation of the complete stress tensor to 8 km depth in the KTB scientific drill holes: implications for crustal strength. J Geophys Res 102:18453–18475
Brune JN, Henyéy TL, Roy RF (1969) Heat flow, stress, and rate of slip along the San Andreas fault, California. J Geophys Res 74:3821–3827
Bruner W (1984) Crack growth during unroofing of crustal rocks: effects on thermoelastic behaviour and near-surface stresses. J Geophys Res 89:4167–4184
Budiansky B (1965) On the elastic moduli of some heterogeneous materials. J Mech Phys Solids 13:223–227
Byerlee JD (1978) Friction of rock. Pure Appl Geophys 116:615–626
Byerlee JD (1993) Model of episodic flow of high-pressure water in fault zones before earthquakes. Geology 21:303–306
Byerly P (1928) The nature of first motion in the Chilean earthquake of November 11, 1922. Am J Sci 16:232–236
Cai M, Qiao L, Yu J (1995) Study and tests of techniques for increasing overcoring stress measurement accuracy. Int J Rock Mech Min Sci Geomech Abstr 32:375–384
Campell J, Nothnagel A (2000) European VLBI for crustal dynamics. J Geodyn 30:321–326
Carder DS (1945) Seismic investigations in the Boulder dam area, 1940–1944 and the influence of reservoir loading on local earthquake activity. Bull Seismol Soc Am 35(4):175–192
Carlson RL, Gangi AF (1985) Effect of cracks on the pressure dependence of P-wave velocities in crystalline rock. J Geophys Res 90:8675–8684
Carlson SR, Wu M, Wang HF (1990) Micromechanical modelling of thermal cracking in granite. In: Duba A et al (eds) The Brittle-Ductile transition in rocks. Geophysical monograph 56. American Geophysical Union, Washington DC, pp 37–48
Cathles ML (1975) The viscosity of the Earth's mantle. Princeton University Press, Princeton
Celerier B (2008) Seeking Anderson's faulting in seismicity: a centennial celebration. Rev Geophys 46, RG4001. doi:10.1029/2007RG000240
Chandler NA (1993) Bored raise overcoring for in situ stress determination at the Underground Research Laboratory. Int J Rock Mech Min Sci Geomech Abstr 30:989–992
Charlez PA (1997) Rock mechanics, vol 2. Petroleum applications. IFP Publications Editions Technip, Paris
Cheng CH, Toksöz MN (1984) Generation, propagation and analysis of tube waves in a borehole. In: Toksöz MN, Stewart RR (eds) Vertical seismic profiling: advanced concepts. Geophysical Press, London, pp 276–287
Chernyshev SN, Dearman WR (1991) Rock fractures. Butterworth-Heinemann, Newton
Chester FM (1994) Effects of temperature on friction: constitutive equations and experiments with quartz gouge. J Geophys Res 99:7247–7262
Chery J, Zoback MD, Hassani R (2001) An integrated mechanical model of the San Andreas fault in central and northern California. J Geophys Res 106:22051–22066
Chery J, Zoback MD, Hickman S (2004) A mechanical model of the San Andreas fault and SAFOD Pilot Hole stress measurements. Geophys Res Lett 31, L15S13. doi:10.1029/2004GL019521
Clark JB (1949) A hydraulic process for increasing the productivity of wells. Pet Trans Am Inst Min Eng T.P. 2510:1–8
Colmenares LB, Zoback MD (2002) A statistical evaluation of intact rock failure criteria constrained by polyaxial test data for five different rocks. Int J Rock Mech Min Sci 39:695–729
Conrad CP, Lithgow-Bertelloni C (2004) The temporal evolution of plate driving forces: importance of "slab suction" versus "slab pull" during the Cenozoic. J Geophys Res 109, B10407. doi:10.1029/2004JB002991
Conrad CP, Lithgow-Bertelloni C (2006) Influence of continental roots and asthenosphere on plate-mantle coupling. Geophys Res Lett 33, L05312. doi:10.1029/2005GL025621
Cook NGW (1992) Jaeger memorial dedication lecture—natural joints in rock: mechanical, hydraulic and seismic behaviour and properties under normal stress. Int J Rock Mech Min Sci Geomech Abstr 29:198–223

Cook BK, Lee MY, DiGiovanni AA, Bronowski DR, Perkins ED, Williams JR (2004) Discrete element modelling applied to laboratory simulation of near-wellbore mechanics. Int J Geomech 4(1):19–27

Cornet FH (1983) Interpretation of hydraulic injection tests for in situ stress determination. In: Zoback MD, Haimson BC (eds) Hydraulic fracturing stress measurements. National Academic Press, Washington DC, pp 149–158

Cornet FH (1986) Stress determination from hydraulic tests on preexisting fractures—the HTPF method. In: Stephansson O (ed) Rock stress and rock stress measurements. Centek Publishers, Lulea, pp 301–312

Cornet FH (1993) Stresses in rocks and rock masses. In: Hudson JA (ed) Comprehensive rock engineering, vol 3. Rock testing and site characterization. Pergamon Press, Oxford, pp 297–327

Cornet FH (2007) Introduction to the special section on induced seismicity. Int J Rock Mech Min Sci 44:1077–1078

Cornet FH, Valette B (1984) In situ stress determination from hydraulic injection test data. J Geophys Res 89:11527–11537

Costin LS (1983) A microcrack model for the deformation and failure of brittle rock. J Geophys Res 88:9485–9492

Costin LS (1987) Deformation and failure. In: Atkinson BK (ed) Fracture mechanics of rock. Academic Press, London, pp 167–215

Cox JW (1970) The high resolution dipmeter reveals dip-related borehole and formation characteristics. Society of Professional Well Log Analysts, 11th annual logging symposium, CA Transactions, Los Angeles, pp D1–D25

Cox SJD, Scholz CH (1988a) On the formation and growth of faults: an experimental study. J Struct Geol 10 (4):413–430

Cox SJD, Scholz CH (1988b) Rupture initiation in shear fracture of rocks: an experimental study. J Geophys Res 93:3307–3320

Crampin S (1977) A review of the effects of anisotropic layering on the propagation of seismic waves. Geophys J R Astron Soc 49:9–27

Crampin S (1978) Seismic wave propagation through a cracked solid: polarisation as a possible dilatancy diagnostic. Geophys J R Astron Soc 53:467–496

Crampin S, Kirkwood SC (1981) Velocity variations in systems of anisotropic symmetry. J Geophys 49:35–42

Crampin S, McGonigle R, Bamford D (1980) Estimating crack parameters from observation of P-wave velocity anisotropy. Geophysics 45(3):345–360

Crampin S, Booth DC, Krasnova MA, Chesnokov EM, Maximov AB, Tarasov NT (1986) Shear wave polarisations in the Peter the first range indicating crack-induced anisotropy in a thrust-fault regime. Geophys J R Astron Soc 84:401–412

Crouch SL, Fairhurst C (1967) A four-component borehole deformation gage for the determination of in situ stresses in rock masses. Int J Rock Mech Min Sci Geomech Abstr 4:209–217

Cuisiat FD, Haimson BC (1992) Scale effects in rock mass stress measurements. Int J Rock Mech Min Sci Geomech Abstr 29:99–117

Cundall PA (1971) A computer model for simulating progressive, large-scale movements in block rock systems. In: Proceedings symposium of the international society of rock mechanics, Nancy, France paper II.8

Cundall PA (2001) A discontinuous future for numerical modelling in geomechanics? Geotech Eng 149(1):41–47

Darling TW, TenCate JA, Brown DW, Clausen B, Vogel, SC (2004) Neutron diffraction study of the contribution of grain contacts to nonlinear stress-strain behavior. Geophys Res Lett 31, L16604. doi:10.1029/2004GL020463

Daymond MR (2006) Internal stresses in deformed crystalline aggregates. In: Wenk H-R (ed) Neutron scattering in earth sciences, reviews in mineralogy and geochemistry, vol 63 (Chap 16). Mineralogical Society of America, pp 427–434

DeMets C, Gordon RG, Argus DF, Stein S (1990) Current plate motions. Geophys J Int 101:425–478

DeMets C, Gordon RG, Argus DF, Stein S (1994) Effect of recent revisions to the geomagnetic reversal time scale on estimates of current plate motions. Geophys Res Lett 21(20):2191–2194

Detournay E, Carbonnel R (1997) Fracture mechanics analysis of the breakdown process in minifrac or leakoff tests. SPE Prod Facil 12(3):195–199

Detournay E, Cheng AH-D (1988) Poroelastic response of a borehole in a non-hydrostatic stress field. Int J Rock Mech Min Sci Geomech Abstr 25:171–182

Detournay E, Cheng AH-D (1992) Influence of pressurization rate on the magnitude of the breakdown pressure. In: Tillerson JR, Warwersik WR (eds) Rock mechanics. Balkema, Rotterdam, pp 325–333

Detournay E, Cheng AH-D (1993) Fundamentals of poroelasticity (Chap 5). In: Hudson JA (ed) Comprehensive rock engineering, vol 2. Pergamon Press, Oxford, pp 113–169

Detournay E, Cheng AH-D, Roegiers JC, McLennan JD (1989) Poroelasticity considerations in in situ stress determination by hydraulic fracturing. Int J Rock Mech Min Sci Geomech Abstr 26:507–514

Dey T, Brown D (1986) Stress measurements in deep granitic rock mass using hydraulic fracturing and differential strain curve analysis. In: Stephansson O (ed) Rock stress and rock stress measurements. Centek Publishers, Lulea, pp 351–357

di Toro G, Hirose T, Nielson S, Pennacchioni G, Shimamoto T (2006) Natural and experimental evidence of melt lubrication of faults during earthquakes. Science 311:647–649

Dietrich JH (1979) Modeling of rock friction-1. Experimental results and constitutive equations. J Geophys Res 84:2161–2168

Dietz RS (1961) Continent and ocean basin evolution by spreading of the sea floor. Nature 190:854–857

Digby PJ (1981) The effective elastic moduli of porous granular rock. J Appl Mech—ASME 48:803–808

Dight PM (2006) Determination of in-situ stress from oriented core. In: Lu M, Li CC, Kjorholt H, Dahle H (eds) In-situ rock stress. Taylor & Francis Group, London, pp 167–175

Dight PM, Dyskin AV (2008) On the determination of rock anisotropy for stress measurements. In: Potvin Y, Carter J, Dyskin A, Jeffrey R (eds) 1st southern hemisphere international rock mechanics symposium programme (SHIRMS), Australian Centre for Geomechanics, Perth, pp 575–585. ISBN 978-0-9804185-5-2

Doe TW (1981) Determination of the state of stress at the Stripa Mine, Sweden. In: Proceedings of the United States Geological Survey workshop on hydraulic fracturing stress measurements, Natural Science Foundation, Monterey, California, pp 305–331

Dugdale DS (1960) Yielding of steel sheets containing slits. J Mech Phys Solids 8:100–104

Dumstorff P, Yoon J, Zang A, Dresen G, Meschke G (2006) Numerical simulation of crack paths in laboratory compressed Aue granite. Eur Geosci Union. Geophys Res Abstr 8:07361

Durelli AJ, Phillips EA, Tsao CH (1958). Introduction to the theoretical and experimental analysis of stress and strain, McGraw Hill, New York

Du Toit A (1937) Our wandering continents. Oliver and Boyd, Edinburgh and London

Duvall WI (1974) Stress relief by center hole. Appendix in US Bureau of Mines report of investigation RI 7894

Dvorkin J, Mavko G, Nur A (1991) The effect of cementation on the elastic properties of granular material. Mech Mater 12:207–217

Dvorkin J, Nur A, Yin H (1994) Effective properties of cemented granular materials. Mech Mater 18:351–366

Dyke CG (1988) In situ rock stress indicators for rock at great depth. Ph.D. Thesis, University of London, Imperial College of Science and Technology, London, p 361

Dyksterhuis S, Müller RD (2004) Modelling the contemporary and palaeo stress field of Australia using finite-element modelling with automatic optimisation. Explor Geophys 35(4):236–241

Dyskin AV (1997) Crack growth criteria incorporating non-singular stresses: size effect in apparent fracture toughness. Int J Fract 83:191–206

Eberhardt E (2001) Numerical modelling of three-dimension stress rotation ahead of an advancing tunnel face. Int J Rock Mech Min Sci 38:499–518

Economides MJ, Nolte KG (2000) Reservoir stimulation. Wiley VCH, Weinheim
El Rabaa AW (1989) Determination of the stress field and fracture direction in the Danian chalk. In: Maury VM, Fourmaintraux D (eds) Rock at great depth. Balkema, Rotterdam, pp 1017–1024
El Rabaa AWM, Meadows DL (1986) Laboratory and field applications of the strain relaxation method. SPE paper 15072, 56th SPE California regional meeting, Oakland, CA, pp 259–272
Emery CL (1964) Strain energy in rocks. In: Judd WR (ed) State of stress in the Earth's crust. Elsevier, Amsterdam, pp 235–279
Emmerman R, Lauterjung J (1997) The German continental deep drilling program KTB: overview and major results. J Geophys Res 102(B8):18179–18201
Emmerman R, Wohlenberg J (eds) (1989) The German continental deep drilling program (KTB). Springer-Verlag, Berlin, p 553
Enever JR, Wooltorton BA (1983) Experience with hydraulic fracturing as a means of estimating in situ stress in Australia coal basin sediments. In: Zoback MD, Haimson BC (eds) Hydraulic fracturing stress measurements. National Academic Press, Washington DC, pp 28–43
Enever JR, Walton RI, Wold MB (1990) Scale effects influencing hydraulic fracture and overcoring stress measurements. In: Proceedings international workshop on scale effects in rock masses, Loen, Norway, AA Balkema, Rotterdam, pp 317–326
Engelder T (1984) The time-dependent strain relaxation of Algerie granite. Int J Rock Mech Min Sci Geomech Abstr 21(2):63–73
Engelder T (1993) Stress regimes in the lithosphere. Princeton University Press, Princeton
Engeser B, Huenges E, Kessels W, Kück J, Wohlgemuth L (1993) The 6000 m hydrofrac test in the KTB main borehole: design and preliminary results. In: Bram K, Draxler JK (eds) KTB-report 93-1, NLfB. Hannover, pp 301–355
Erdogan F, Sih GC (1963) On the crack extension in plates under plane loading and transverse shear. J Basic Eng 86:519–527
Eshelby JD (1957) The determination of the elastic field of an ellipsoidal inclusion, and related problems. Proc R Soc Lond A 241:376–396
Evans DM (1966) The Denver area earthquakes and the rocky mountain arsenal disposal well. Mt Geol 3:23–36
Evans AG, Heuer AH, Potter DL (1977) The fracture toughness of ceramics. In: International conference on fracture 4, vol 1, pp 529–556
Evans B, Kohlstedt DL (1995) Rheology of rocks. In: Ahrens TJ (ed) Rock physics and phase relations: a handbook of physical constants. Reference shelf, vol 3. American Geophysical Union, Washington DC, pp 148–165
Ewing WM, Heezen BC et al (1956) Some problems of Antarctic submarine geology. In: Carey A, Gould LM, Hulburt EO, Odishaw H, Smith WE (eds) American Geophysical Union monograph 1. pp 75–81
Ewy RT, Cook NGW (1990a) Deformation and fracture around cylindrical openings in rock -I. Observations and analysis of deformation. Int J Rock Mech Min Sci Geomech Abstr 27:387–407
Ewy RT, Cook NGW (1990b) Deformation and fracture around cylindrical openings in rock-II. Initiation, growth and interaction of fractures. Int J Rock Mech Min Sci Geomech Abstr 27:409–427
Feissel-Vernier M, de Viron O, Le Bail K (2007) Stability of VLBI, SLR, DORIS, and GPS positioning. Earth Planet Space 59:475–497
Feves M, Simmons G, Siegfried RW (1977) Microcracks in crustal igneous rocks: physical properties. In: Heacock JG (ed) The Earth's crust: it's nature and properties. Geophysical monograph series, vol 20. AGU Washington DC, pp 95–116
Filimonov YL, Lavrov AV, Shkuratnik VL (2001) Prospects of memory effects for stress measurement applications in rock salt. In: Proceedings of EUROCK 2001 symposium, AA Balkema, Rotterdam, pp 59–63

Filimonov YL, Lavrov AV, Shafarenko Y, Shkuratnik VL (2002) Observation of post-failure Kaiser effect in a plastic rock. Pure Appl Geophys 159:1321–1331

Filimonov YL, Lavrov AV, Shkuratnik VL (2005) Effect of confining stress on acoustic emission in ductile rock. Strain 41(1):33–35

Fischer T, Hainzl S, Eisner L, Shapiro SA, Le Calvez J (2008) Microseismic signatures of hydraulic fracture growth in sediment formations: observations and modelling. J Geophys Res 113, B02307. doi:10.1029/2007JB005070

Fitzenz DD, Miller SA (2004) New insights on stress rotations from a forward regional model of the San Andreas fault system near its Big Bend in southern California. J Geophys Res 109, B08404. doi:10.1029/2003JB002890

Fjaer E, Ruisten H (2002) Impact of the intermediate principal stress on the strength of heterogeneous rock. J Geophys Res 107. doi:10.1029/2001JB000277

Fjaer E, Holt RM, Horsrud P, Raaen AM, Risnes R (2008) Petroleum related rock mechanics. Development in petroleum sciences, vol 53. Elsevier Sciences, Amsterdam

Flesch LM, Holt WE, Haines AJ, Shen-Tu B (2000) Dynamics of the Pacific-North American plate boundary in the western United States. Science 287:834–836

Forsyth DW, Uyeda S (1975) On the relative importance of the driving forces of plate motion. Geophys J R Astron Soc 43:163–200

Fowler CMR (1990) The solid earth—an introduction to global geophysics. Cambridge University Press, Cambridge

Fredrich JT, Wong T-f (1986) Micromechanics of thermally induced cracking in three crustal rocks. J Geophys Res 91:12743–12764

Freund LE (1998) Dynamic fracture mechanics. Cambridge University Press, New York

Friedman M (1967) Measurement of the state of residual strain in quartzose rocks by means of X-ray diffractometry. Norelco Reporter 14:7–9

Friedman M (1972) Residual elastic strain in rocks. Tectonophysics 15:297–330

Friedman M, Handin J, Alani G (1972) Fracture surface energy of rocks. Int J Rock Mech Min Sci 9:757–766

Frischbutter A, Neov D, Scheffzük Ch, Vrana M, Walther K (2000) Lattice strain measurements on sandstones under load using neutron diffraction. J Struct Geol 22:1587–1600

Frischbutter A, Janssen Ch, Scheffzük Ch, Walther K, Ullemeyer K, Behrmann JH, Nikitin AN, Ivankina TI, Kern H, Leiss B (2006) Strain and texture measurements on geologic samples using neutron diffraction at IBR-2, Joint Institute for Nuclear Research, Dubna (Russia). Physics of Particles and Nuclei 37(Suppl 1):S45–S68

Frocht MM (1941) Photoelasticity. Wiley, New York

Fu YF, Tang CA, Lin P (2003) Numerical test on AE Kaiser effect in rock fracture process. Key Eng Mater 243, 244:339–344

Fuchs K, Müller B (2001) World stress map of the Earth: a key to tectonic processes and technological applications. Review article, Naturwissenschaften 88:357–371

Fuchs K, Zoback MD, Rummel F, England P (1991) A stress measurement profile to mid-crustal depth in KTB Scientific Drilling Project, Southeastern Germany: scientific questions and technological challenges (and discussion). Philos Trans Phys Sci Eng 337(1645):25–28

Fung YC (1965) Foundations of solid mechanics. Prentice Hall, New York

Galera JM (2006) Natural stress field evaluation using borehole ovalisation analysis and its comparison with hydrofrac measurements. In: Lu M, Li CC, Kjorholt H, Dahle H (eds) In-situ rock stress. Taylor & Francis Group, London, pp 241–247

Gallagher JJ (1971) Photomechanical model studies relating to fracture and residual elastic strain in granular aggregates. Ph.D. Thesis, Texas A & M University, College Station TX, 127 pp

Galybin AN (2006) Stress fields in jointed elastic regions: modelling based on discrete stress orientations. In: Lu M, Li CC, Kjorholt H, Dahle H (eds) In-situ rock stress. Taylor & Francis Group, London, pp 193–199

Galybin AN, Mukhamediev SA (1999) Plane elastic boundary value problem posed on orientation of principal stresses. J Mech Phys Solids 47:2381–2409

Galybin AN, Mukhamediev SA (2004) Determination of elastic stresses from discrete data on stress orientations. Int J Solids Struct 41(18, 19):5125–5142

Gao H (1991) Stress concentrations at slightly undulating surfaces. J Mech Phys Solids 39:443–458

Gölke M, Coblentz D (1996) Origins of the European regional stress field. Tectonophysics 266:11–24

Goetze C, Evans B (1979) Stress and temperature in the bending lithosphere as constrained by experimental rock mechanics. Geophys J R Astron Soc 59:463–478

Goodier J (1933) Concentration of stress around spherical and cylindrical inclusions and flaws. J Appl Mech 1(1):39–44

Goodier JN (1968) Mathematical theory of equilibrium cracks. In: Liebowitz H (ed) Fracture—an advanced treatise, vol 2. Academic Press, San Diego, pp 1–66

Goodman RE (1963) Subaudible noise during compression of rocks. Geol Soc Am Bull 74:487–490

Goodman RE (1980) Introduction to rock mechanics. Wiley, New York

Goodman RE (1989) Introduction to rock mechanics, 2nd edn. Wiley, New York

Gorodkov S, Li L, Holt RM (2006) Stress path during coring: a discrete particle modelling approach. In: Lu M, Li CC, Kjorholt H, Dahle H (eds) In-situ rock stress. Taylor & Francis Group, London, pp 541–549

Gottschalk RR, Kronenberg AK, Russell JE, Handin J (1990) Mechanical anisotropy of gneiss: Failure criterion and textural sources of directional behaviour. J Geophys Res 95:21613–21634

Gough DI, Gough WI (1987) Stress near the surface of the Earth. Ann Rev Earth Planet Sci 15:545–566

Gräsle W, Kessels W, Kümpel H-J, Li X (2006) Hydraulic observations from a 1 year fluid production test in the 4000 m deep KTB pilot borehole. Geofluids 6:8–23

Grand S, van der Hilst RD, Widiyantoro S (1997) Global seismic tomography: a snapshot of convection in the Earth. GSA Today 7:1–7

Gregersen S (1992) Crustal stress regime in Fennoscandia from focal mechanisms. J Geophys Res 97:11821–11827

Griffith AA (1920) The phenomena of rupture and flow in solids. Philos Trans R Soc Lond Ser A 221:163–198

Griffith AA (1924) The theory of rupture. In: Biezeno CB, Burgers JM (eds) Proceedings 1st international congress applied mechanics, Delft Tech. Boekhandel en Drukkerij J Walter Jr, pp 54–63

Gross D, Seelig Th (2006) Fracture mechanics. Springer-Verlag, Berlin

Grünthal G, Stromeyer D (1986) Stress pattern in Central Europe and adjacent areas. Gerlands Beiträge zur Geophysik 95(5):443–452

Grünthal G, Stromeyer D (1992) The recent crustal stress field in Central Europe: trajectories and finite element modeling. J Geophys Res 97(B8):11805–11820

Grünthal G, Stromeyer D (1995) Rezentes Spannungsfeld und Seismizität des baltischen Raumes und angrenzender Gebiete—ein Ausdruck aktueller geodynamischer Prozesse. Brandenburgische Geowissenschaftliche Beiträge Kleinmachnow 2:69–76

Gueguen Y, Palciauskas V (1994) Introduction to the physics of rocks. Princeton University Press, Princeton

Guenot A (1987) Stress and rupture conditions around oil wellbores. In: Proceedings 6th international congress of rock mechanics, Montreal, vol 1. AA Balkema, Rotterdam, pp 109–118

Guenot A (1989) Borehole breakouts and stress fields. Int J Rock Mech Min Sci Geomech Abstr 26:185–195

Gunzburger Y, Cornet FH, Wileveau Y (2006) From a natural stress profile to rheological properties at geological scale. In: Van Cotthem A, Charlier R, Thimus J-F, Tshibangu J-P (eds) EUROCK 2006—Multiphysics coupling and long term behaviour in rock mechanics. Taylor & Francis Group, London, pp 287–292

Guo F, Morgenstern NR, Scott JD (1993) Interpretation of hydraulic fracturing pressure: a comparison of eight methods used to identify shut-in pressure. Int J Rock Mech Min Sci Geomech Abstr 30:627–631

Gupta HK, Chadha R (eds) (1995) Induced seismicity. Birkhäuser Verlag, Basel

Guyer RA, Johnson PA (1999) Nonlinear mesoscopic elasticity: evidence for a new class of materials. Phys Today 52(4):30–36

Guyer RA, McCall KR, Boinott GN (1995) Hysteresis, discrete memory, and nonlinear-wave propagation in rock—a new paradigm. Phys Rev Lett 74:3491–3494

Hager BH, O'Connell RJ (1981) A simple global model for plate dynamics and mantle convection. J Geophys Res 86:4843–4867

Hahn HG (1985) Elastizitätstheorie. BG Teubner-Verlag, Stuttgart (in German)

Haimson BC (1973) Earthquake related stresses at Rangely, Colorado. In: Proceedings 14th US symposium of rock mechanics, University Park, ASCE, pp 689–708

Haimson BC (1975) The state of stress in the earth's crust. Rev Geophys Space Phys 13:350–352

Haimson BC (1977) Recent in-situ stress measurements using the hydrofracturing technique. In: Proceedings 18th US symposium of rock mechanics, Johnson Publisher, Golden, pp 4C2-1–4C2-6

Haimson BC (1978) The hydrofracturing stress measuring method and recent field results. Int J Rock Mech Min Sci Geomech Abstr 15:167–178

Haimson BC (1980) Near surface and deep hydrofracturing stress measurements in the Waterloo quartzite. Int J Rock Mech Min Sci Geomech Abstr 17:81–88

Haimson BC (1990) Scale effects in rock stress measurements. In: Proceedings international workshop on scale effects in rock masses, Loen, AA Balkema, Rotterdam, pp 89–101

Haimson BC (1997) Borehole breakouts and core disking as tools for estimating in situ stress in deep holes. In: Sugawara K, Obara Y (eds) Proceedings of the international symposium on rock stress, AA Balkema, Rotterdam, pp 35–42

Haimson BC (2006) True triaxial stresses and the brittle fracture of rock. Pure Appl Geophys 163(5–6):1101–1130

Haimson BC (2007) Micromechanisms of borehole instability leading to breakouts in rock. Int J Rock Mech Min Sci 44:157–173

Haimson BC, Chang C (2002) True triaxial strength of the KTB amphibolite under borehole wall conditions and its use to estimate the maximum horizontal in situ stress. J Geophys Res 107 (B10):2257. doi:10.1029/2001JB000647

Haimson BC, Cornet FH (2003) ISRM suggested methods for rock stress estimation—Part 3: hydraulic fracturing (HF) and/or hydraulic testing of pre-existing fractures (HTPF). Int J Rock Mech Min Sci 40:1011–1020

Haimson BC, Fairhurst C (1967) Initiation and extension of hydraulic fractures in rocks. Soc Pet Eng J 7(3):310–318 (September)

Haimson BC, Herrick CG (1986) Borehole breakouts—a new tool for estimating in situ stress. In: Stephansson O (ed) Rock stress and rock stress measurements. Centek Publishers, Lulea, pp 271–280

Haimson B, Kovacich J (2003) Borehole instability in high-porosity Berea sandstone and factors affecting dimensions and shape of fracture-like breakouts. Eng Geol 69:219–231

Haimson BC, Lee MY (1995) Estimating in situ stress conditions from borehole breakouts and core disking—experimental results in granite. In: Proceedings international workshop on rock stress mesurement at great depth, Tokyo, Japan, 8th ISRM congress, AA Balkema, Rotterdam, pp 19–24

Haimson BC, Rummel F (1982) Hydrofracturing stress measurements in the Iceland drilling project drillhole at Reydasfjordur, Iceland. J Geophys Res 87:6631–6649

Haimson BC, Voight B (1977) Crustal stress in Iceland. Pure Appl Geophys 115:153–190

Haimson BC, Lin W, Oku H, Hung J-H, Song S-R (2009) Integrating borehole breakout dimensions, strength criteria, and leak-off test results, to constrain the state of stress across the Chelungpu Fault, Taiwan, Tectonophysics. doi:10.1016/j.tecto.2009.05.016

Hakala M (1999) Numerical study on core damage and interpretation of in situ state of stress. Posiva Report 99-25, Posiva Oy, Olkiluoto, p 234
Hakala M, Sjöberg J (2006) A methodology for interpretation of overcoring stress measurements in anisotropic rock. Working report 2006-99, Posiva Oy, Olkiluoto, 207 pp
Hakala M, Hudson JA, Christiansson R (2003) Quality control of overcoring stress measurement data. Int J Rock Mech Min Sci 40:1141–1159
Hakala M, Kuula H, Hudson JA (2007) Estimating the transversely isotropic elastic intact properties for in-situ stress measurement data reduction: a case study of the Olkiluoto mica gneiss, Finland. Int J Rock Mech Min Sci 44:14–46
Hakami E, Hakami H, Cosgrove J (2002) Strategy for a rock mechanics site descriptive model—development and testing of an approach to modelling the state of stress. SKB research report R-02-03. Swedish Nuclear Fuel and Waste Management Co., Stockholm, p 129
Hamiel Y, Liu Y, Lyakhovsky V, Ben-Zion Y, Lockner D (2004) A viscoelastic damage model with applications to stable and unstable fracturing. Geophys J Int 159:1155–1165
Handin J, Hager RV, Friedman JR, Feather JN (1963) Experimental deformation of sedimentary rocks under confining pressure: pore pressure tests. Bull Am Assoc Pet Geol 47:717–755
Handin J, Heard HC, Magourik JN (1967) Effects of the intermediate principal stress on the failure of limestone, dolomite and glass at different temperatures and strain rates. J Geophys Res 72:611–640
Hansen KM, Mount VS (1990) Smoothing and extrapolation of crustal stress orientation measurements. J Geophys Res 95(B2):1155–1165
Hardebeck JL, Hauksson, E (2001) Crustal stress field in southern California and its implications for fault mechanics. J Geophys Res 106:21859–21882
Hardebeck JL, Michael AJ (2004) Stress orientations at intermediate angles to the San Andreas fault, California. J Geophys Res 109, B11303. doi:10.1029/2004JB003239
Hardy MP, Fairhurst C (1974) Analysis of fracture in rock and rock masses. In: Proceedings 14th annual symposium New Mexico section of ASME, Engineering for the Materials/Energy Challenge, New Mexico, pp 73–80
Hardy HR Jr, Zhang D, Zelanko JC (1989) Recent studies of the Kaiser effect in geologic materials. In: Proceedings of the 4th conference AE/MA in geologic structures and materials. Clausthal-Zellerfeld, Trans Tech Publications, pp 27–55
Harper TR, Szymaski JS (1991) The nature and determination of stress in the accessible lithosphere. Philos Trans R Soc Lond A 337:5–24
Harrison JP, Hudson JA (2000) Engineering rock mechanics: Part 2 illustrative worked examples. Elsevier Science Ltd, Kidlington
Hart R (2003) Enhancing rock stress understanding through numerical analysis. Int J Rock Mech Min Sci 40:1089–1097
Hashin Z (1962) The elastic moduli of heterogeneous materials. J Appl Mech 29:143–150
Hast N (1958) The measurement of rock pressure in mines. Sveriges Geologiska Undersokning, Series C Arsbok 52, No. 3
Hast N (1969) The state of stress in the upper part of the Earth's crust. Tectonophysics 8:169–211
Hauk V (1997) Structural and residual stress analysis by nondestructive methods: evaluation, application, assessment. Elsevier, Amsterdam
Hawkes I (1971) Photoelastic strain gauges and in situ stress measurements. In: Proceedings international symposium on the determination of stresses in rock masses, Lab. Nac. de Eng. Civil, Lisbon, pp 359–375
Hawkes I, Moxon S (1965) The measurement of rock in situ stress using the photoelastic biaxial gauge with the core-relief technique. Int J Rock Mech Min Sci 2:405–419
Hawkes CD, Bachu S, Haug K, Thompson AW (2005) Analysis of in-situ stress regime in the Alberta Basin, Canada, for performance assessment of CO2 geological sequestration sites. In: 4th annual conference on carbon capture and sequestration DOE/NETL
Hayashi K, Masuoka M (1995) Estimation of tectonic stress from slip data from fractures in core samples. In: Proceedings international workshop on rock stress measurement at great depth, Tokyo, Japan, 8th ISRM congress, AA Balkema, Rotterdam, pp 35–39

Hazzard JF, Young RP (2000) Simulating acoustic emissions in bonded-particle models of rock. Int J Rock Mech Min Sci 37(5):867–872

Healy JH, Rubey WW, Griggs DT, Raleigh CB (1968) The Denver earthquakes. Science 161:1301–1310

Hecht E, Zajac A (1979) Optics. Addison-Wesley Publishing Company, Meno Park, CA

Hecht E, Zajac A (2003) Optics. Addison-Wesley Publishing Company, 4th edn. Meno Park, CA

Heidbach O, Drewes H (2003) 3D finite element model of major tectonic processes in the Eastern Mediterranean. In: Nieuwland D (ed) New insights in structural interpretation and modelling. Special Publication Series 212, Geological Society of London, pp 259–272

Heidbach O, Höhne J (2008) CASMI—a visualization tool for the World Stress Map database. Comput Geosci 34:783–791

Heidbach O, Reinecker J, Tingay M, Müller B, Sperner B, Fuchs K, Wenzel F (2007) Plate boundary forces are not enough: Second- and third-order stress patterns highlighted in the World Stress Map database. Tectonics 26, TC6014. doi:10.1029/2007TC002133

Heidbach O, Tingay M, Barth A, Reinecker J, Kurfeß D, Müller B (2008) The 2008 release of the World Stress Map. available online at www.world-stress-map.org

Heidbach O, Tingay M, Barth A, Reinecker J, Kurfeß D, Müller B (2009) Global crustal stress pattern based on the World Stress Map database release 2008. Tectonophysics, doi:10.1016/j.tecto.2009.07.023

Heim A (1878) Untersuchungen über den Mechanismus der Gebirgsbildung. Im Anschluss an die Geologische Monografie der Tödi-Windgällen-Gruppe, Bd 2, Benno Schwabe Verlagsbuchhandlung, Basel. II. Teil Allgemeine Untersuchungen über den Mechanismus der Gebirgsbildung. Abschn 1. Die mechanische Gesteinsumformung bei der Gebirgsbildung. D: Theorie der mechanischen Gebirgsumformung, pp 79–99

Heinze WD, Goetze C (1974) Numerical simulation of stress concentrations in rock. Int J Rock Mech Min Sci Geomech Abstr 11:151–155

Herget G (1974) Ground stress determinations in Canada. Rock Mech 6:53–74

Herget G (1986) Changes of ground stresses with depth in the Canadian Shield. In: Stephansson O (ed) Rock stress and rock stress measurements, Centek Publishers, Lulea, pp 61–68

Herget G (1987) Stress assumptions for underground excavations in the Canadian shield. Int J Rock Mech Min Sci Geomech Abstr 24:95–97

Herget G (1993) Rock stresses and rock stress monitoring in Canada. In: Hudson JA (ed) Comprehensive rock engineering, vol 3. Rock testing and site characterization. Pergamon Press, Oxford, pp 473–496

Hertz HR (1882) Über die Berührung fester elastischer Körper. Journal für die reine und angewandte Mathematik 92:156–171

Hess HH (1962) History of ocean basins. In: Engel AEJ, James HL, Leonard BF (eds) Petrologic studies: a volume to honor AF Buddington. Geol Soc Am, pp 599–620

Hickman, SH (1991) Stress in the lithosphere and the strength of active faults. Rev Geophys 29:759–775

Hickman SH, Zoback MD (1983) The interpretation of hydraulic fracturing pressure time data for in situ stress determination. In: Zoback MD, Haimson BC (eds) Hydraulic fracturing stress measurements. National Academic Press, Washington DC, pp 44–54

Hickman SH, Zoback MD (2004) Stress orientations and magnitudes in the SAFOD pilot hole. Geophys Res Lett 31, L15S12. doi:10.1029/2004GL020043

Hickman SH, Zoback MD, Ellsworth W (2004) Introduction to special section: preparing for the San Andreas fault Observatory at Depth. Geophys Res Lett, 31, L12S01. doi:10.1029/2004GL020688

Hill R (1965) A self-consistent mechanics of composite materials. J Mech Phys Solids 13:213–222

Hillers G, Ben-Zion Y, Mai PM (2006) Seismicity on a fault controlled by rate- and state-dependent friction with spatial variations of the critical slip distance. J Geophys Res 111(B1). doi:10.1029/2005JB003859

Hillis RR, Reynolds SD (2000) The Australian stress map. J Geol Soc Lond 157(5):915–921

Hirth G, Kohlstedt DL (2003) Rheology of the upper mantle and the mantle wedge: a view from the experimentalists. In: Eiler J (ed) The subduction factory. Geophysical monograph series 138. American Geophysical Union, pp 83–105

Hirth G, Teyssier C, Dunlap WJ (2001) An evaluation of quartzite flow laws based on comparisons between experimentally and naturally deformed rocks. Int J Earth Sci 90:77–87

Hoagland RG, Embury JD (1979) A treatment of inelastic deformation around a crack tip due to microcracking, J Am Ceram Soc 63:404–410

Hoagland RG, Hahn GT, Rosenfield AR (1973) Influence of microstructure on fracture propagation in rock. Rock Mech 5:77–106

Hoek E (1968) Brittle failure in rock. In: Stagg KG, Zienkiewicz OC (eds) Rock mechanics in engineering practice. Wiley, London, pp 99–112

Hoek E, Brown ET (1980) Underground excavations in rock. Institution of Mining and Metallurgy, London

Hoerning A (1979) Electrical conductivities of a cracked solid. Pure Appl Geophys 117:690–709

Holcomb DJ (1981) Memory, relaxation, and microfracturing in dilatant rock. J Geophys Res 86:6235–6248

Holcomb DJ (1983) Using acoustic emissions to determine in-situ stress: problems and promise. Geomechanics ASME, AMD, 57:11–21

Holcomb DJ (1992) Localization studies under triaxial conditions. In: Tillerson JR, Warwersik WR (eds) Rock mechanics (Proceedings 33rd US symposium), AA Balkema, Rotterdam, pp 661–670

Holcomb DJ (1993a) General theory of the Kaiser effect. Int J Rock Mech Min Sci Geomech Abstr 30(7):929–935

Holcomb DJ (1993b) Observations of the Kaiser effect under multiaxial stress states: Implications for its use in determining in situ stress. Geophys Res Lett 20(19):2119–2122

Holcomb DJ, Costin LS (1986a) Detecting damage surfaces in brittle materials using acoustic emissions. J Appl Mech 108:536–544

Holcomb DJ, Costin LS (1986b) Damage in brittle material: experimental methods. In: Lamb JP (ed) Proceedings of the 10th US national congress of applied mechanics, pp 107–113

Holcomb DJ, Martin RJ III (1985) Determining peak stress history using acoustic emissions. In: Proceedings of the 26th US symposium on rock mechanics, AA Balkema Rotterdam, pp 715–722

Holcomb DJ, Rudnicki JW (1986) Detecting damage surfaces in brittle material using acoustic emissions. J Appl Mech 53:536–544

Holcomb DJ, Stevens JL (1980) The reversible Griffith crack—a viable model of dilatancy. J Geophys Res 85:7101–7107

Holmes A (1928) Radioactivity and continental drift. Geol Mag 65:236–238

Holt RM, Kenter C (1992) Laboratory simulation of core damage induced by stress release. In: Tillerson JR, Warwersik WR (eds) Rock mechanics (Proceedings 33rd US symposium), AA Balkema, Rotterdam, pp 959–968

Holt RM, Brignoli M, Kenter CJ (2000a) Core quality: quantification of coring-induced rock alteration. Int J Rock Mech Min Sci 37:889–907

Holt RM, Brandshaug T, Cundall PA (2000b) Discrete particle and laboratory modelling of core mechanics. In: Girard J, Liebman M, Breeds C, Doe T (eds) Proceedings 4th North American rock mechanics symposium, Seattle, AA Balkema, Rotterdam, pp 1217–1224

Holt RM, Pestman BJ, Kenter CJ (2001) Use of a discrete particle model to assess feasibility of core based stress determination. In: Elsworth D, Tinucci JP, Heasley KA (eds) Rock mechanics in the national interest, vol 2 (Proceedings of the 38th US symposium of rock mechanics, Washington DC), Swets & Zeitlinger, Lisse, pp 1361–1366

Holt RM, Doornhof D, Kenter CJ (2003) Use of discrete particle modelling to understand stress-release effects on mechanical and petrophysical behaviour of granular rocks. In: Konietsky H (ed) Numerical modeling in micromechanics via particle methods. Swets & Zeitlinger, Lisse, pp 269–276

Holzhausen GR, Johnson AM (1979) The concept of residual stress in rock. Tectonophysics 58:237–267

Hori M, Kameda T (2001) Inversion of stress from strain without full knowledge of constitutive relations. J Mech Phys Solids 49:1621–1638

Horii H, Nemat-Nasser (1985) Compression-induced microcrack growth in brittle solids: axial splitting and shear failure. J Geophys Res 90:3105–3125

Horn HM, Deere DU (1962) Frictional characteristics of minerals. Geotechnique 12:319–335

Hoskin ER, Russell JE (1981) The origin of the measured residual strains in crystalline rocks. In: Friedman M, Logan J, Stearns DW (eds) Mechanical behaviour of crustal rocks. Geophysical monograph 24. American Geophysical Union, Washington DC, pp 187–198

Huang SL, Aughenbaugh NB, Rockaway JD (1986) Swelling pressure studies of shales. Int J Rock Mech Min Sci 23:371–377

Hubbert KM, Willis DG (1957) Mechanics of hydraulic fracturing. Petroleum Transactions, AIME T.P. 4597 210:153–166

Hudson JA (1981) Wave speeds and attenuation of elastic waves in materials containing cracks. Geophys J R Astron Soc 64:133–155

Hudson JA (1986) A higher order approximation to the wave propagation constants for a cracked solid. Geophys J R Astron Soc 87:265–274

Hudson JA (1991) Crack distributions which account for a given seismic anisotropy. Geophys J Int 104:517–521

Hudson JA, Cooling CM (1988) In situ rock stresses and their measurement in the UK—Part 1. The current state of knowledge. Int J Rock Mech Min Sci 25:363–370

Hudson JA, Cornet FH (2003) Special issue on rock stress estimation. Int J Rock Mech Min Sci 40(7–8):955

Hudson JA, Harrison JP (2000) Engineering rock mechanics. Elsevier Science Ltd, Kidlington

Hudson JA, Cornet FH, Christiansson R (2003) ISRM suggested methods for rock stress estimation—Part 1: strategy for rock stress estimation. Int J Rock Mech Min Sci 40(7, 8):991–998

Hughson DR, Crawford AM (1986) Kaiser effect gauging: a new method for determining the pre-existing in-situ stress from an extracted core by acoustic emissions. In: Stephansson O (ed) Rock stress and rock stress measurements. Centek Publishers, Lulea, pp 359–368

Humphreys ED, Coblentz DD (2007) North American dynamics and western US tectonics. Reviews of Geophysics 45, RG3001, American Geophysical Union, p 30

Hutchings MT (1992) Neutron diffraction measurements of residual stress fields: overview and points of discussion. In: Hutchings MT, Krawitz AD (eds) Measurements of Residual and applied stress using neutron diffraction. Kluwer Academic Publishers, Dordrecht, pp 3–18

Hutchings MT, Krawitz AD (eds) (1992) Measurements of residual and applied stress using neutron diffraction. Springer, New York

Hutchings MT, Withers PJ, Holden TM, Lorentzen T (2005) Introduction to characterization of residual stress by neutron diffraction. Taylor & Francis, London

Hyett AJ, Dyke CG, Hudson JA (1986) A critical examination of basic concepts associated with the existence and measurement of in-situ stress. In: Stephansson O (ed) Rock stress and rock stress measurements. Centek Publishers, Lulea, pp 387–391

Ibs-von Seht M, Plenefisch T, Schmedes E (2006) Faulting style and stress field investigations for swarm earthquakes in NE Bavaria/Germany—the transition between Vogtland / NW-Bohemia and the KTB-site. J Seismol 10:197–211

Iinuma T, Kato T, Hori M (2005) Inversion of GPS velocity and seismicity data to yield changes in stress in the Japanese Islands. Geophys J Int 160:417–434

Illies H, Greiner G (1979) Holocene movements and state of stress in the Rhingraben rift system. Tectonophysics 52:349–359

Inglis CE (1913) Stresses in a plate due to the presence of cracks and sharp corners. Trans Inst Nav Archit (London) 60:219–230

Irwin GR (1958) Fracture. In: Flügge S (ed) Handbuch der Physik, vol 6. Springer-Verlag, Berlin, pp 551–590

Irwin WP, Barnes I (1975) Effect of geologic structure and metamorphic fluids on seismic behavior of the San Andreas fault system in central and northern California. Geology 3:713–716

Isacks B, Oliver J, Sykes LR (1968) Seismology and the new global tectonics. J Geophys Res 73:5855–5899

Itasca Consulting Group Inc (2004) PFC2D/3D, particle flow code in 2 and 3 dimensions. Version 3.1 Minneapolis, Minnesota: ICG

Ito H, Nishizawa O, Xue Z, Sano O (1997a) Estimation of in-situ stresses from ASR and DSCA measurements on drill cores in the 1995 Hyogoken-nanbu earthquake source region. In: Suguwara K, Obara Y (eds) Rock stress. Balkema, Rotterdam, pp 355–358

Ito T, Zoback MD (2000) Fracture permeability and in situ stress to 7 km depth in the KTB scientific drillhole. Geophys Res Lett 27:1045–1048

Ito T, Watanabe K, Hayashi K (1997b) Residual stress in rock core samples and stress state at great depths. In: Suguwara K, Obara Y (eds) Rock stress. AA Balkema, Rotterdam, pp 167–170

Ito T, Evans K, Kawai K, Hayashi K (1999) Hydraulic fracture reopening pressure and the estimation of maximum horizontal stress. Int J Rock Mech Min Sci Geomech Abstr 36:811–826

Ito T, Igarashi A, Kato H, Ito H, Sano O (2006) Crucial effect of system compliance on the maximum stress estimation in the hydrofracturing method: theoretical considerations and field-test verification. Earth Planet Space 58:963–971

Jaeger JC, Cook NGW (1979) Fundamentals of rock mechanics, 3rd edn. Chapman & Hall, London

Jaeger JC, Cook NGW, Zimmerman RW (2007) Fundamentals of rock mechanics, 4th edn. Blackwell Publishing, Oxford

Jamison WR, Spang J (1976) Use of calcite twin lamellae to infer differential stresses. Geol Soc Am Bull 87:868–887

Janssen C, Wagner FC, Zang A, Dresen G (2001) Fracture process zone in granite—a microstructural analysis. Int J Earth Sci 90:46–59

Janssen C, Hoffmann-Rothe A, Tauber S, Wilke H (2002) Internal structure of the Precordilleran fault system (Chile)`—insights from structural and geophysical observations. J Struct Geol 24:123–143

Jaoul O, Tullis JA, Kronenberg AK (1984) The effect of varying water content on the creep behaviour of Heavitree quartzite. J Geophys Res 89:4289–4312

Jarosinski M, Beekman F, Bada G, Cloetingh S (2006) Redistribution of recent collision push and ridge push in Central Europe: insights from FEM modelling. Geophys J Int 167:860–880

Jing L (2003) A review of techniques, advances and outstanding issues in numerical modelling for rock mechanics and rock engineering. Int J Rock Mech Min Sci 40:283–353

Jing L, Stephansson O (2007) Fundamentals of discrete element methods for rock engineering. Theory and applications. Developments in geotechnical engineering 85. Elsevier, Amsterdam

Johnson KL (1985) Contact mechanics. Cambridge University Press, Cambridge

Johnson RB (1961) Patterns and origin of radial dike swarms associated with West Spanish Peak and Dike Mountain, South Central Colorado. Geol Soc Am Bull 72:579–590

Jost M, Büsselberg T, Jost Ö, Harjes H-P (1998) Source parameters of injection-induced microearthquakes at 9 km depth at the KTB Deep Drilling Site, Germany. Bull Seismol Soc Am 88:815–832

Judd WR (1964) Rock stress, rock mechanics and research. In: Judd WR (ed) State of stress in the Earth's crust. Elsevier, Amsterdam, pp 5–54

Kachanov LM (1986) Introduction to continuum damage mechanics. Martinus Nijhoff, Dordrecht

Kachanov M, Tsukrov I, Shafiro B (1994) Effective moduli of solids with cavities of various shapes. In: Ostoja-Starzewski M, Jasiuk I (eds) Micromechanics of random media, Part 2. Appl Mech Rev 47(1):S151–S174

Kahle H-G, Straub C, Reilinger R, McClusky S, King R, Hurst K, Veis G, Kastens K, Cross P (1998) The strain rate field in the eastern Mediterranean region, estimated by repeated GPS measurements. Tectonophysics 294:237–252

Kaiser J (1950) Untersuchung über das Auftreten von Geräuschen beim Zugversuch. Dr.-Ing. Dissertation, Technische Universität München (15 Feb. 1950)

Kaiser J (1953) Erkenntnisse und Folgerungen aus der Messung von Geräuschen bei Zugbeanspruchung von metallischen Werkstoffen. Archiv für das Eisenhüttenwesen 24:43–45

Kaiser PK, Guenot A, Morgenstern NR (1985) Deformation of small tunnels. IV. Behaviour during failure. Int J Rock Mech Min Sci Geomech Abstr 22:141–152

Karato S (2008) Deformation of earth materials: an introduction to the rheology of solid earth. Cambridge University Press, Cambridge

Karato S-I, Paterson MS, Fitzgerald JD (1986) Rheology of synthetic olivine aggregates: influence of grain size and water. J Geophys Res 91:8151–8176

Kehle RO (1964) Determination of tectonic stresses through analysis of hydraulic well fracturing. J Geophys Res 69:252–273

Kemeny JM, Cook NGW (1987) Crack models for the failure of rocks in compression. Proc Int Conf Constitutive Laws for Eng Mat 2:879–887

Kemeny JM, Cook NGW (1991) Micromechanics of deformation in rocks. In: Shah SP (ed) Toughening mechanisms in quasi-brittle materials. Kluwer Academic Publishers, Dordrecht, pp 155–188

Kern H (1978) The effect of high temperature and high confining pressure on compressional wave velocities in quartz-bearing and quartz-free igneous and metamorphic rocks. Tectonophysics 44:185–203

Kern H, Richter A (1979) Compressional and shear wave velocities at high temperature and high confining pressure. Tectonophysics 54:231–252

Kern H, Wenk H-R (1990) Fabric-related velocity anisotropy and shear wave splitting in rocks from the Santa Rosa Mylonite Zone, California. J Geophys Res 95(B7):11213–11223

Kim K (1992) In-situ stress in rock engineering projects. In: Lecture notes, short course on modern in-situ stress measurement methods. University of Madison, Wisconsin

Kirby SH (1980) Tectonic stresses in the lithosphere: constraints provided by the experimental deformation of rocks. J Geophys Res 85:6353–6363

Kirby SH (1983) Rheology of the lithosphere. Rev Geophys Space Phys 21:1458–1487

Kirby SH, Kronenberg AK (1987) Rheology of the lithosphere: selected topics. Rev Geophys 25:1219–1244

Kirsch G (1898) Die Theorie der Elastizität und die Bedürfnisse der Festigkeitslehre. Zeitschrift des Vereins Deutscher Ingenieure 42(29):797–807

Klasson H, Leijon B (1990) Rock Stress Measurements in the Deep Boreholes at Kuhmo, Hyrynsalmi, Sievi, Eurajoki and Konginkangas, Report YJT-90-1, Nuclear Waste Commission of Finnish Power Companies, Helsinki, p 112

Klein RJ, Barr MV (1986) Regional state of stress in Western Europe. In: Stephansson O (ed) International symposium on rock stress and rock stress measurements, Centek Publisher, Stockholm, pp 33–45

Knoll P (1990) The fluid-induced tectonic rock burst of March 13, 1989 in Werra potash mining district of the GDR (first results). Gerlands Beiträge zur Geophysik 99:239–245

Kobayashi S, Nishimura N, Matsumoto K (1987) Displacements and strains around non-flat-end borehole. In: Proceedings 2nd international symposium on field measurements in geomechanics vol 2, Kobe, pp 1079–1084

Kobayashi S, Yamaguchi T, Yoshikawa T (1997) Initial stress estimation in rock using ultrasonic. In: Sugawara K, Obara Y (eds) Rock stress. AA Balkema, Rotterdam, pp 161–166

Kocks UF, Tomé CN, Wenk H-R (2000). Texture and anisotropy. Preferred orientations in polycrystals and their effect on materials properties, 2nd paperback edn. Cambridge University Press, Cambridge

Koerner RM, Lord AE (1989) AE detection of prestress in soil and rock. In: Proceedings of the 4th conference on AE/MA in geologic structures and materials. Clausthal-Zellerfeld, Trans Tech Publications, pp 73–86

Kohlstedt DL, Evans B, Mackwell SJ (1995) Strength of the lithosphere: constraints imposed by laboratory experiments. J Geophys Res 100:17587–17602

Kossmat F (1927) Gliederung des varistischen Gebirgsbaus. Abhandlungen sächsische Geologie L.A. 1:1–39 (Leipzig)

Kostrov VV (1974) Seismic moment and energy of earthquakes, and seismic flow of rock. Izv. Acad Sci USSR Phys Solid Earth 1:23–40

Kozlovskaya E, Elo S, Hjelt S-E, Yliniemi J, Pirttijärvi M (SVEKALAPKO Seismic Tomography Working Group) (2004) 3D density model of the crust of southern and central Finland obtained from joint interpretation of the SVEKALAPKO crustal P-wave velocity models and gravity data. Geophys J Int 158:827–848

Kranz RL (1980) The effects of confining pressure and stress difference on static fatigue of granite. J Geophys Res 85:1854–1866

Krantz RW (1989) Orthorhombic fault patterns—the odd axis model and slip vector orientations. Tectonics 8:483–495

Krawitz AD (2001) Introduction to diffraction in materials science and engineering. Wiley, New York

Kreemer C, Holt WE, Haines AJ (2003) An integrated global model of present-day plate motions and plate boundary deformation. Geophys J Int 154:8–34

Kümpel H-J, Erzinger J, Shapiro S (2006) Two massive hydraulic tests completed in deep KTB pilot hole. Sci Drill 3:40–42

Kurita K, Fujii N (1979) Stress memory of crystalline rocks in acoustic emission. Geophys Res Lett 6:9–12

Kusznir NJ (1991) The distribution of stress with depth in the lithosphere: thermo-rheological and geodynamic constraints. Philos Trans R Soc Lond Ser A, 337:95–110

Kutter HK (1993) Influence of drilling method on borehole breakouts and core disking. In: Proceedings 7th International Society Rock Mechanics (ISRM) congress, Aachen, vol 3. Balkema, Rotterdam, pp 1659–1664

Lachenbruch AH, Sass JH (1980) Heat flow and energetics of the San Andreas fault zone. J Geophys Res 85:6185–6223

Lachenbruch AH, Sass JH (1992) Heat flow from Cajon Pass, fault strength, and tectonic implications. J Geophys Res 97(B4):4995–5015

Lacombe O (2007) Comparison of paleostress magnitudes from calcite twins with contemporary stress magnitudes and frictional sliding criteria in the continental crust: mechanical implications. J Struct Geol 29:86–99

La Femina PC, Dixon TH, Malservisi R, Arnadottir T, Sigmundsson F, Sturkell E, Einarsson P (2005) Modeling the secular velocity field across a divergent plate boundary: geodetic GPS and geology, South Iceland. EOS Trans American Geophysical Union 86(18), Jt. Assem. Suppl., Abstract G23C-03

Lajtai EZ (1971) A theoretical and experimental evaluation of the Griffith theory of brittle fracture. Tectonophysics 11:129–156

Längler F, Schneider E (2005) Kontinuumsmechanische Erweiterung der Ultraschallverfahren zur Beschreibung des Spannungszustandes im gesamten Bauteil. DGZfP-Berichtsband 94-CD, Vortrag 49

Lanaro F, Stephansson O (2003) A unified model for characterisation and mechanical behaviour of rock fractures. Pure Appl Geophys 160(5, 6):989–998

Landau L, Lifschitz E (1970) Elastizitätstheorie. Akademie-Verlag, Berlin (in German)

Landolt-Börnstein (1982) Zahlenwerte und Funktionen aus den Naturwissenschaften und Technik. Angenheister G (ed) Band 1b Physikalische Eigenschaften von Gesteinen. Springer, Berlin

Lang PA, Thompson PM, Ng LKW (1986) The effect of residual stress and drill hole size on the in situ stress determined by overcoring. In: O Stephansson (ed), Rock stress and rock stress measurements, Centek Publishers, Lulea, Sweden, pp 687–694

Lavrov AV (1997) Three-dimensional simulation of memory effects in rock samples. In: Rock stress, Proceedings of the international symposium on rock stress, AA Balkema, Rotterdam, pp 197–202

Lavrov AV (2001) Kaiser effect observation in brittle rock cyclically loaded with different loading rates. Mech Mater 33:669–677

Lavrov AV (2003) The Kaiser effect in rocks: principles and stress estimation techniques, journal review article. Int J Rock Mech Min Sci 40:151–171

Lavrov AV (2005) Fracture-induced physical phenomena and memory effects in rock: a review. Strain 41:135–149

Lawn BR, Wilshaw TR (1975) Fracture of brittle solids. Cambridge University Press, Cambridge

Le Pichon X (1968) Sea floor spreading and continental drift. J Geophys Res 73:3661–3697

Lee MY, Haimson BC (1989) Statistical evaluation of hydraulic fracturing stress measurement parameters. Int J Rock Mech Min Sci Geomech Abstr 26:477–482

Lee MY, Haimson BC (1993) Borehole breakouts in Lac du Bonnet granite: a case of extensile failure mechanism. Int J Rock Mech Min Sci Geomech Abstr 30:1039–1045

Lee MF, Mikula P, Kinnersly E (2006) In situ rock stress measurements and stress change monitoring at Mt Charlotte gold mine, Western Australia. In: Lu M, Li CC, Kjorholt H, Dahle H (eds) In-situ rock stress. Taylor & Francis Group, London, pp 259–267

Leeman ER (1959) The measurement of changes in rock stress due to mining. Min Quarry Eng 25(7):300–304

Leeman ER (1964) The measurement of stress in rock—Parts I, II and III. J S Afr Inst Min Metall 65:45–114, 254–284

Leeman ER (1971) The CSIR "Doorstopper" and triaxial rock stress measuring instruments. Rock Mech 3:25–50

Leeman ER, Hayes DJ (1966) A technique for determining the complete state of stress in rock using a single borehole. In: Proceedings 1st congress international society of rock mechanics, vol 2, Lisbon, pp 17–24

Lehtonen A (2005) Evaluation of rock stress estimation by the Kaiser effect. Posiva working report 2005-67, Posiva Oy, Olkiluoto

Lehtonen AV, Särkkä P (2006) Evaluation of rock stress estimation by the Kaiser effect. In: Lu M, Li CC, Kjorholt H, Dahle H (eds) In-situ rock stress. Taylor & Francis Group, London, pp 135–142

Leijon BA (1986) Application of the LUT triaxial overcoring techniques in Swedish mines. In: Stephansson O (ed) Rock stress and rock stress measurements. Centek Publishers, Lulea, pp 569–579

Li C (1998) Theory of the Kaiser effect and its potential applications. In: Proceedings of the 6th conference AE/MA in geologic structures and materials. Pennsylvania, pp 171–185

Li C, Nordlund E (1993) Experimental verification of the Kaiser effect in rocks. Rock Mech Rock Eng 26(4):333–351

Li F-Q et al (1983) Experiments of in-situ stress measurements using stress relief and hydraulic fracturing techniques. In: Zoback MD, Haimson BC (eds) Hydraulic fracturing stress measurements. National Academic Press, Washington DC, pp 130–134

Li Y, Schmitt DR (1998) Drilling-induced core fractures and in situ stress. J Geophys Res 103(B3):5225–5239

Liebowitz H (ed) (1968) Fracture, an advanced treatise. Academic Press, New York

Lin HM, Wu JH, Lee DH (2006a) Evaluating the pre-stress of Mu-Shan sandstone using acoustic emission and deformation rate analysis. In: Lu M, Li CC, Kjorholt H, Dahle H (eds) In-situ rock stress. Taylor & Francis Group, London, pp 215–222

Lin W, Kwasniewski M, Imamura T, Matsuki K (2006b) Determination of three-dimensional in situ stresses from anelastic strain recovery measurements of cores at great depth. Tectonophysics 426:221–238

Lin W, Oikawa Y, Imamura T, Nakamura T, Kwasniewski M, Takahashi M (2006c) Comparisons betwen three-dimensional in-situ stresses determined by anelastic strain recovery and differential strain curve analysis methods. In: Lu M, Li CC, Kjorholt H, Dahle H (eds) In-situ rock stress. Taylor & Francis Group, London, pp 385–391

Lin W, Yeh E-C, Ito H, Hirono T, Soh W (2007a) Preliminary results of stress measurement using drill cores of TCDP Hole-A: an application of anelastic strain recovery method to three-dimensional in-situ stress determination. Terrestrial, Atmospheric and Ocean Sciences 18(2):379–393

Lin W, Yeh E-C, Ito H, Hung J-H, Hirono T, Soh W, Ma K-f, Kinoshita M, Wang C-Y, Song S-R (2007b) Current stress state and principal stress rotations in the vicinity of the Che-

lungpu fault induced by the 1999 Chi-Chi, Taiwan earthquake. Geophys Res Lett 34, L16307. doi:10.1029/2007GL030515
Lindner EN, Halpern JA (1977) In-situ stress: an analysis. In: Proceedings 18th US symposium in rock mechanics. Johnson Publisher, Golden, pp 4C1-1–4C1-7
Lindner EN, Halpern JA (1978) In-situ stress in North America: a compilation. Int J Rock Mech Min Sci Geomech Abstr 15:183–203
Ling CB (1947) On the stresses in a notched plate under tension. J Math Phys 26:284–289
Lithgow-Bertelloni C, Guynn JH (2004) Origin of the lithospheric stress field. J Geophys Res 99, B01408. doi:10.1029/2003JB002467
Lithgow-Bertelloni C, Richards MA (1998) The dynamics of Cenozoic and Mesozoic plate motions. Rev Geophys 36:27–78
Little TA (1995) Brittle deformation adjacent to the Awatere strike-slip fault in New Zealand: faulting patterns, scaling relationships, and displacement partitioning. GSA Bull 107(11):1255–1271
Liu Y, Crampin S, Abercrombie RE (1997) Shear-wave anisotropy and the stress field from borehole recordings at 2.5 km depth at Cajon Pass. Geophys J Int 129:439–449
Liu Z, Bird P (2002) North American plate is driven westward by lower mantle flow. Geophys Res Lett 29(4), 2164. doi:10.1029/2002GL016002
Ljunggren C (1987) Determination of virgin rock stresses by fracturing methods. Licentiate thesis Technical University of Lulea
Ljunggren C, Klasson H (1996) Rock stress measurements at the three investigation sites: Kivetty, Roumavaara and Olkiluoto, Finland. Posiva Work Report PATU-96-26e. Posiva Oy, Olkiluoto
Ljunggren C, Chang Y, Janson T, Christiansson R (2003) An overview of rock stress measurement methods. Int J Rock Mech Min Sci 40:975–989
Lockner DA (1993) Room temperature creep in saturated granite. J Geophys Res 98:475–487
Lockner DA (1995) Rock failure. In: Ahrens TJ (ed) Rock physics and phase relations—a handbook of physical constants, AGU Reference Shelf 3, American Geophysical Union, Washington DC, pp 127–147
Lockner DA, Byerlee JD, Kuksenko V, Ponomarev A, Sidorin A (1991) Quasi-static fault growth and shear fracture energy in granite. Nature 350(6313):39–42
Logan JM, Teufel LW (1986) The effect of normal stress on the real area of contact during frictional sliding in rocks. Pure Appl Geophys 124:471–486
Louchnikov V, Hunt SP, Meyers AG (2004) The use of particle flow code for investigating the stress memory effect in rocks. In: Shimizu Y, Hart R, Cundall P (eds) Numerical modeling in micromechanics via particle methods. Taylor & Francis Group, London, pp 331–340
Louchnikov V, Hunt SP, Meyers AG (2006) Influence of confining pressure on the deformation memory effect in rocks studied by particle flow code, PFC-2D. In: Lu M, Li CC, Kjorholt H, Dahle H (eds) In-situ rock stress. Taylor & Francis Group, London, pp 483–489
Lu J (ed) (1996) Handbook of measurement of residual stresses. Society for experimental mechanics. Prentice-Hall, New Jersey
Lubarda VA, Richmond O (1999) Second-order elastic analysis of dilute distribution of spherical inclusions, Mech Mater 31:1–8
Luthi SM (2001) Geological well logs: their use in reservoir modelling. Springer, Berlin
Lynch JC, Richards MA (2001) Finite element models of stress orientation in well-developed strike-slip fault zones: implications for the distribution of lower crustal strain. J Geophys Res 106:26707–26729
Lynch PA, Tamura N, Lau D, Madsen I, Liang D, Strohschnieder M, Stevenson AW (2007) Application of white-beam X-ray microdiffraction for the study of mineralogical phase identification in ancient Egyptian pigments. J Appl Crystallogr 40:1089–1096
Lyzenga GA, Raefsky A, Mulligan SG (1991) Models of recurrent strike-slip earthquake cycles and the state of crustal stress. J Geophys Res 96(B13):21623–21640
Macherauch E, Hauk V (eds) (1987) Residual stresses in science and technology. Proceedings international conference on residual stresses. Deutsche Gesellschaft für Metallkunde, Informationsgesellschaft, Oberursel

Malkovsky M (1987) The Mesozoic and Tertiary basins of the Bohemian Massif and their evolution. Tectonophysics 137:31–42

Manthei G, Eisenblätter J, Kamlot P, Heusermann S (1998) AE measurements during hydraulic fracturing tests in a salt mine using a special borehole probe. In: Hamstad MA, Kishi T, Ono K (eds) Progress in acoustic emission IX, Proceedings 14th international acoustic emission conference and 5th acoustic emission world meeting, Big Island, Hawaii, pp II60–II69 (9–14 August)

Manthei G, Eisenblätter J, Kamlot P (2003) Stress measurements in salt mines using a special hydraulic fracturing borehole tool. In: Natau O, Fecker E, Pimentel E (eds) Proceedings international symposium on geotechnical measurements and modelling. Karlsruhe, AA Balkema, Rotterdam, pp 355–360

Marder M, Fineberg J (1996) How things break. Phys Today 49:24–29 (September)

Martin CD, Chandler NA (1993) Stress heterogeneity and geological structures. Int J Rock Mech Min Sci Geomech Abstr 30:993–999

Martin CD, Christiansson R (2009) Estimating the potential for spalling around a deep nuclear waste repository in crystalline rock. Int J Rock Mech Min Sci 46:219–228

Martin CD, Simmons GR (1993) The atomic energy of Canada limited underground research laboratory: an overview of geomechanics characterization. In: Hudson JA (ed) Comprehensive rock engineering, vol 3 (Chap 38). Pergamon Press, Oxford, pp 915–950

Martin CD, Read RS, Chandler NA (1990) Does scale influence in situ stress measurements?—Some findings at the underground research laboratory. In: de Cunha P (ed) International symposium on scale effects in rock masses, AA Balkema, Rotterdam, pp 307–316

Martin CD, Martino JB, Dzik EJ (1994) Comparison of borehole breakouts from laboratory and field tests. In: Proceedings Eurock, international symposium on rock mechanics in petroleum engineering, Delft, AA Balkema, Rotterdam, pp 183–190

Martino JB, Thompson PM, Chandler NA, Read RS (1997) The in-situ stress program at AECL's underground research laboratory; 15 years of research (1982–1997). Report No. 06819-REP-01200-0053 ROO (December)

Mastin LG, Heinemann B, Krammer A, Fuchs K, Zoback, MD (1991) Stress orientation in the KTB pilot hole determined from wellbore breakouts. Sci Drill 2:1–12

Matsuki K (1991) Three-dimensional in-situ stress measurement with anelastic strain recovery of a rock core. In: Wittke W (ed) Proceedings 7th international conference on rock mechanics vol 1, Aachen, pp 557–560

Matsuki K, Kaga N, Yokoyama T, Tsuda N (2004) Determination of three dimensional in situ stress from core discing based on analysis of principal tensile stress. Int J Rock Mech Min Sci 41:1167–1190

Matsuki K, Takeuchi K (1993) Three-dimensional in-situ stress determination by anelstic strain recovery of a rock core. Int J Rock Mech Min Sci Geomech Abstr 30(7):1019–1022

Maury V (1987) Observations, researches and resent results about failure mechanisms around single galleries. In: Proceedings 6th congress of the international society of rock mechanics, Montreal, AA Balkema, Rotterdam, pp 1119–1128

Maury V (1991) The role of rock mechanics in oil and gas exploration. Suppl Nat 350:18 (8–10 April)

Maury VM, Sauzay J-M (1987) Borehole instability: case histories, rock mechanics approach, and results. SPE/IADC 16051. In: SPE/IADC drilling conference, New Orleans, LA

Maury V, Santarelli FJ, Henry JP (1988) Core discing: a review. SANGROM symposium: rock mechanics in Africa, pp 221–231

Mavko G, Mukerji T, Dvorkin, J (1998) The rock physics handbook: tools for seismic analysis in porous media. Cambridge University Press, Cambridge

McClintock FA, Walsh JB (1962) Friction of Griffith cracks in rock under pressure. In: Proceedings 4th US national congress. Applied mechanics, vol 2. New York, American Society of Mechanical Engineering, pp 1015–1021

McCutchen WR (1982) Some elements of a theory for in-situ stress. Int J Rock Mech Min Sci Geomech Abstr 19:201–203

McGarr A (1980) Some constraints on levels of shear stress in the crust from observation and theory. J Geophys Res 85:6231–6238

McGarr A (1988) On the state of lithosstatic stress in the absence of applied tectonic forces. J Geophys Res 93(13):609–613

McGarr A, Gay NC (1978) State of stress in the Earth's crust. Annu Rev Earth Planet Sci 6:405–436

McGarr A, Wiebols GA (1977) Influence on mine geometry and closure volume on seismicity in deep level mine. Int J Rock Mech Min Sci Geomech Abstr 14:139–145

McGarr A, Spottiswoode SM, Gay NC (1975) Relationship of mine tremors to induced stresses and to rock properties in the focal region. Bull Seismol Soc Am 65(4):981–993

McKenzie DP (1969) The relation between fault plane solutions for earthquakes and the directions of principal stresses. Bull Seismol Soc Am 59:591–601

McLean MR, Addis MA (1990) Wellbore stability analysis: a review of current methods of analysis and their field application. Paper IADC/SPE 19941, IADC/SPE drilling conference. Houston, Texas (USA)

McPhee DK, Jachens RC, Wentworth CM (2004) Crustal structure across the San Andreas fault at the SAFOD site from potential field and geologic studies. Geophys Res Lett 31, L12S01. doi:10.1029/2003GL019363, 4 pp

Mei S, Kohlstedt DL (2000a) Influence of water on plastic deformation of olivine aggregates 1. Diffusion creep regime. J Geophys Res 105:21457–21469

Mei S, Kohlstedt DL (2000b) Influence of water on plastic deformation of olivine aggregates 2. Dislocation creep regime. J Geophys Res 105:21471–21481

Meissner R (1996) Faults and folds, fact and fiction. Tectonophysics 264:279–293

Mendecki AJ (1997) Seismic monitoring in mines. Chapman & Hall, London

Merkel S, Wenk H-R, Shen G, Gillet P, Mao H-K, Hemley RJ (2002) Deformation of polycrystalline MgO at pressures of the lower mantle. J of Geophys Res 107(B11):2271, doi: 10.1029/2001JB000920

Merrill RH (1967) Three component borehole deformation gage for determining the stress in rock. US bureau of mines report of investigation RI 7015

Michael AJ (1984) Determination of stress from slip data: faults and folds. J Geophys Res 89:11517–11526

Michihiro K, Hata K, Fujiwara T, Yoshioka H, Tanimoto T (1989) Study on estimating initial stress and predicting failure on rock masses by acoustic emission. In: Maury V, Fourmaintraux D (eds) Rock at great depth. AA Balkema, Rotterdam, pp 1025–1032

Michihiro K, Hata K, Yoshioka H, Fujiwara T (1992) Determination of the initial stresses on rock mass using acoustic emission method. J Acoust Emiss 10(1):S63–S76

Mimaki Y, Matsuo K (1986) Investigation of asymmetrical deformation behaviour at the horseshoe-shaped large cavern opening. In: Proceedings international symposium on large rock caverns vol 2, Helsinki, Pergamon Press, Oxford, pp 1337–1348

Mindlin RD (1949) Compliance of elastic bodies in contact. J Appl Mech Trans ASME 71:259–268

Min KB, Jing L (2004) Stress-dependent mechanical properties and bounds of Poisson`s ratio for fractures rock masses investigated by DFN-DEM technique. Int J Rock Mech Min Sci 40(6):795–816

Minster JB, Jordan TH (1978) Present-day plate motions. J Geophys Res 85:5331–5354

Mizuta Y, Sano O, Ishida T, Gang L (2004) A prototypal probe newly developed for stress measurement in the Earth's crust. Chikyu Mon 26(2):97–102 (in Japanese)

Mogi K (1972) Fracture and flow of rocks. Tectonophysics 13(1–4):541–568

Mogi K (2007) Experimental rock mechanics. Taylor & Francis Group, London

Mohr O (1900) Welche Umstände bedingen die Elastizitätsgrenze und den Bruch eines Materials. Zeitschrift des Vereins deutscher Ingenieure 44(45):1524–1530

Mohr HF (1956) Measurement of rock pressure. Min Quarry Eng:178–189

Molnar P (1992) Brace-Goetze strength profiles, the partitioning of strike-slip and thrust faulting at zones of oblique convergence, and the stress-heat flow paradox of the San Andreas fault. In: Evans B, Wong T-f (eds) Fault mechanics and transport properties of rocks: a Festschrift in Honor of WF Brace. Academic, San Diego, California, pp 435–459

Momayez M, Hassani FP (1992) Application of the Kaiser effect to measure in situ stresses in underground mines. In: Tillerson JR Wawersik WR (eds) Proceedings of the 33rd US symposium on rock mechanics, Santa Fe, New Mexico, AA Balkema, Rotterdam, pp 979–987

Momayez M, Hassani FP, Hardy HR Jr (1992) Maximum curvature method: a technique to estimate Kaiser-effect load from acoustic emission data. J Acoust Emiss 10(3, 4):61–65

Mooney WD, Laske G, Guy Masters T (1998) CRUST 5.1: a global crustal model at 5° × 5°. J Geophys Res 103(B1):727–748

Moore D (2005) Review of evidence for a weak San Andreas fault. European Geosciences Union, Geophys Res Abstr 7, 05878, SRef-ID: 1607-7962/gra/EGU05-A-05878

Moore DE, Rymer MJ (2007) Talc-bearing serpentinite and the creeping section of the San Andreas fault. Nature 448:795–797

Moos D, Zoback MD (1990) Utilization of observations of well bore failure to constrain the orientation and magnitude of crustal stresses: application to continental deep sea drilling project and ocean drilling program boreholes. J Geophys Res 95:9305–9325

Mount VS, Suppe J (1987) State of stress near the San Andreas fault: implications for wrench tectonics. Geology 15:1143–1146

Müller W (1987) Experimentelle und numerische Untersuchungen zur Rissausbreitung im anisotropen Gestein in der Nähe von Grenzflächen. Dissertation, University of Bochum, p 199 (in German)

Müller B, Zoback ML, Fuchs K, Mastin L, Gregersen S, Pavoni N, Stephansson O, Ljunggren C (1992) Regional pattern of tectonic stress in Europe. J Geophys Res 97:11783–11803

Müller B, Wehrle V, Zeyen H, Fuchs K (1997) Short-scale variations of tectonic regimes in the Western European stress province north of the Alps and Pyrenees. Tectonophysics 275(1–3):199–220

Müller B, Wehrle V, Hettel S, Sperner B, Fuchs F (2003) A new method for smoothing oriented data and its application to stress data. In: Ameen M (ed) Fracture and in-situ stress characterization of hydrocarbon reservoirs, vol 209. Special Publication Series, Geological Society of London, pp 107–126

Mukhamediev SA, Galybin AN, Brady BHG (2006) Determination of stress fields in elastic lithosphere by methods based on stress orientations. Int J Rock Mech Min Sci 43(1):66–88

Murnaghan FD (1951) Finite deformation of an elastic solid. Dover Publications, New York

Murray MH, Segall P (2001) Modeling broadscale deformation in northern California and Nevada from plate mations and elastic strain accumulation. Geophys Res Lett 28(22):4315–4318

Nadai A (1950) Theory of flow and fracture of solids, vol 1, 2nd edn. McGraw-Hill, New York

Natau O, Röckel Th (1993) Orientation of the in-situ stress field determined from drilling induced fractures, centerline fractures, core disking at the KTB drill site. In: Emmerman R, Lauterjung J, Umsonst T (eds) KTB-report 93-2, pp 199–202

Nemat-Nasser S, Horii H (1982) Compression-induced nonplanar crack extension with application to splitting, exfoliation, and rockburst. J Geophys Res 87(B8):6805–6821

Neuber H (1937) Kerbspannungslehre, 1st edn. Springer-Verlag, Berlin

Neuber H (1984) Kerbspannungslehre, 3rd edn. Springer-Verlag, Berlin

Nichols TC (1975) Deformation associated with relaxation of residual stresses in a sample of Barre granite from Vermont, Professional Paper US Geological Survey 875: 1–32

Nicholson C, Wesson RL (1992) Triggered earthquakes and deep well activities. Pure Appl Geophys 139(3, 4):561–578

Noyan IC, Cohen JB (1987) Residual stress measurements by diffraction and interpretation. Springer, New York

Nur A, Byerlee JD (1971) An exact effective stress law for elastic deformation of rocks with fluids. J Geophys Res 76:6414–6419

O'Brian PJ, Duyster J, Grauert B, Schreyer W, Stöckert B, Weber K (1997) Crustal evolution of the KTB drill site: from oldest relics to the late Hercynian granites. J Geophys Res 102(B8):18203–18220

O'Connell RJ, Budiansky B (1974) Seismic velocities in dry and saturated cracked solids. J Geophys Res 79(35):5412–5426
Obert L (1972) Brittle fracture of rock. In: Liebowitz H (ed) Fracture—an advanced treatise, vol 7. Academic Press, New York, pp 93–155
Obert L, Merrill RH, Morgan TA (1962) In situ determination of stress in rock. Min Eng 51–58
Oka Y, Kameoka Y, Saito T, Hiramatsu Y (1979) Investigations on a new method of determining rock stress by stress relief technique and applications of this method. Rock Mech Jpn 3:68–70
Oreskes N (2001) From continental drift to plate tectonics. In: Oreskes N, Le Grand H (eds) Plate tectonics—an insider's history of the modern theory of the Earth. Westview, Oxford, pp 3–27
Orowan E (1949) Fracture and strength of solids. Rep Prog Phys 12:48–74
Ouchterlony F (1980) Review of fracture toughness testing of rocks, DS 1980 15. Swedish Detonic Research Foundation, Stockholm
Paris PC, Sih GC (1965) Stress analysis of cracks. In: Fracture toughness testing and it's application, ASTM Publication 381, Philadelphia, pp 30–83
Pasic B, Gaurina-Medimurec N, Matanovic D (2007) Wellbore instability: causes and consequences. Rudarsko-geolosko-naftni zbornik 19:87–98
Passchier CW, Trouw RAJ (1996) Microtectonics. Springer-Verlag, Berlin
Paterson MS (1970) A high pressure, high temperature apparatus for rock deformation. Int J Rock Mech Min Sci 7:517–526
Paterson MS (1978) Experimental rock deformation—the brittle field, 1st edn. Springer-Verlag, Berlin
Paterson MS (1987) Problems in the extrapolation of laboratory rheological data. Tectonophysics 133:33–43
Paterson MS (1989) The interaction of water with quartz and its influence in dislocation flow—an overview. In: Karato S-I, Toriumi M (eds) Rheology of solids and of the Earth. Oxford University Press, New York, pp 107–142
Paterson MS, Luan FC (1990) Quartzite rheology under geological conditions. In: Knipe RJ, Rutter EH (eds) Deformation mechanisms, rheology and tectonics. Geological Society Special Publication, pp 299–307
Paterson MS, Wong T-f (2005) Experimental rock deformation—the brittle field, 2nd edn. Springer-Verlag, Berlin
Paulsen TS, Wilson TJ (2009) New criteria for systematic mapping and reliability assessment of monogetic volcanic vent alignments and elongate volcanic vents for crustal stress analyses. Tectonophysics, doi: 10.1016/j.tecto.2009.08.025
Peacock S, Crampin S, Booth DC, Fletcher JB (1988) Shear wave splitting in the ANZA seismic gap southern California: temporal variations as possible precursors. J Geophys Res 93:3339–3356
Peiter A (1992) Handbuch Spannungsmeßpraxis: experimentelle Ermittlung mechanischer Spannungen. Vieweg & Sohn Verlagsgesellschaft mbH, Braunschweig
Pendexter C, Rohn RE (1954) Fracture induced during drilling. J Pet Technol 6:15–49
Peng S, Zhang J (2007) Engineering geology for underground rocks. Springer
Pestman BJ, van Munster JG (1996) An acoustic emission study of damage development and stress memory effects in sandstone. Int J Rock Mech Min Sci Geomech Abstr 33(6):585–593
Pestman BJ, Kenter CJ, van Munster JG (2001) Core-based determination of in-situ stress magnitudes. In: Elsworth D, Tinucci JP, Heasley KA (eds) Rock mechanics in the national interest, vol 2 (Proceedings of the 38th US symposium of rock mechanics, Washington DC), Swets & Zeitlinger, Lisse, pp 1353–1360
Pestman BJ, Holt RM, Kenter CJ, van Munster JG (2002) Field application of a novel core-based in-situ stress estimation technique. Society of Petroleum Engineers Inc. SPE/IRSM 78158 Rock Mechanics Conference Oil Rock, 7 pp
Peterek A, Schröder B (1997) Neogene fault activity and morphogenesis in the basement area north of the KTB drill site (Fichtelgebirge and Steinwald). Geol Rundsch 86:185–190
Pfiffner OA (1982) Deformation mechanisms and flow regimes in limestones from the Helvetic zone of the Swiss Alps. J Struct Geol 4:429–442

Pine RJ, Batchelor AS (1984) Downward migration of shearing in jointed rock during hydraulic injections. Int J Rock Mech Min Sci Geomech Abstr 21:249–263

Pine RJ, Ledingham P, Merrifield M (1983) In-situ stress measurement in the Carnmenellis granite 2—hydrofracture tests at Rosemanowes Quarry to depths of 2000 m. Int J Rock Mech Min Sci Geomech Abstr 20:63–72

Pintschovius L, Prem M, Frischbutter A (2000). High-precision neutron-diffraction measurements for the determination of low-level residual stresses in a sandstone. J Struct Geol 22:1581–1585

Plag HP, Pearlman M (2009) The future geodetic reference frame. In: Global Geodetic Observing System. Springer, Heidelberg, pp 225–236

Plumb RA (1994) Variations of the least horizontal stress magnitude in sedimentary rocks. In: Proceedings 1st North American rock mechanics symposium, Austin, AA Balkema, Rotterdam, pp 71–78

Plumb RA, Hickman SH (1985) Stress-induced borehole elongation: a comparison between four-arm dipmeter and the borehole televiewer in the Auburn geothermal well. J Geophys Res 90:5513–5521

Poirier JP (1985) Creep of crystals. Cambridge University Press, New York

Pollard DD, Saltzer SD, Rubin AM (1993) Stress inversion methods—are they based on fault assumptions. J Struct Geol 15:1045–1054

Pollock AA (1989) Acoustic emission inspection. In: Metals handbook, 9th edn. vol 17. Materials Park, Ohio, ASM International, pp 278–294

Popp T, Kern H (1994) The influence of dry and water saturated cracks on seismic velocities of crustal rocks—a comparison of experimental data with theoretical model. Surv Geophys 15:443–465

Pratt HR, Black AD, Brown WS, Brace WF (1972) The effect of specimen size on the strength of unjointed diorite. Int J Rock Mech Min Sci 9:513–529

Prensky S (1992) Borehole breakouts and in-situ rock stress—a review. Log Anal 33(3):304–312

Prioul R, Donald A, Koepsell R, El Marzouki Z, Bratton T (2007) Foreward modelling of fracture–induced sonic anisotropy using a combination of borehole image and sonic logs. Geophysics 72(4):E135–E147

Provost A-S, Houston H (2001) Orientation of the stress field surrounding the creeping section of the San Andreas fault: evidence for a narrow mechanically weak fault zone. J Geophys Res 106:11373–11386

Pyrak-Nolte LJ (2007) Fracture anisotropy: the role of fracture-stiffness gradients. The Leading Edge 26(9):1124–1127

Pyrak-Nolte LJ, Morris JP (2000) Single fractures under normal stress: the relation between fracture specific stiffness and fluid flow. Int J Rock Mech Min Sci 37:245–262

Qin S, Wang S, Long H, Liu J (1999) A new approach to estimating geo-stresses from laboratory Kaiser effect measurements. Int J Rock Mech Min Sci 36:1073–1077

Ramsay JG, Lisle R (2000) The techniques of modern structural geology, vol 3: applications of continuum mechanics in structural geology. Elsevier, Amsterdam

Ramsey JM, Chester FM (2004) Hybrid fracture and the transition from extension fracture to shear fracture. Nature 428:63–66

Ranalli G (1995) Rheology of the Earth, 2nd edn. Chapman & Hall, London

Ratigan JL (1992) The use of fracture reopening pressure in hydraulic fracturing stress measurements. Rock Mech Rock Eng 25:225–236

Reches Z (1987a) Faulting of rocks in three-dimensional strain fields. II. Theoretical analysis. Tectonophysics 95:133–156

Reches Z (1987b) Determination of the tectonic stress tensor from slip along faults that obey the Coulomb yield criterion. Tectonics 6:849–861

Reinecker J, Tingay M, Müller B (2006) The use of the WSM database for rock engineers. In: Lu M, Li CC, Kjorholt H, Dahle H (eds) In-situ rock stress. Taylor & Francis Group, London, pp 505–510

Reinecker J, Stephansson O, Zang A (2008a) Stress analysis from overcoring data. World stress map project, Guidelines: Overcoring (see DVD)

Reinecker J, Tingay M, Müller B (2008b) Borehole breakout analysis from four-arm caliper logs. World stress map project, Guidelines: Four-arm Caliper Logs (see DVD)

Ren NK, Hudson PJ (1985) Predicting the in-situ state of stress using differential wave velocity analysis. In: Ashworth E (ed) Proceedings 26th US symposium on rock mechanics, Rapid City, South Dakota, pp 1235–1244

Ren NK, Roegiers JC (1983) Differential strain curve analysis—a new method for determining the pre-existing in situ stress state from rock core measurements. Proceedings 5th international congress rock mechanics, pp F117–F127

Reynolds M, Shaw J (2006) FRAC TIP: what controls fracture geometry—Part II. Adam Person Associates Inc. (APA) Petroleum Engineering, Issue Q1: 2, 3

Reynolds SD, Coblentz DD, Hillis RR (2002) Tectonic forces controlling the regional intraplate stress field in continental Australia: results from new finite element modeling. J Geophys Res 107(B7):2131. doi:10.1029/2001JB000408

Ricard Y, Fleitout L, Froidevaux (1984) Geoid heights and lithospheric stresses for a dynamic earth. Ann Geophys 2:267–286

Ricard Y, Richards MA, Lithgow-Bertelloni C, LeStunff Y (1993) A geodynamical model of mantle density heterogeneity. J Geophys Res 98:21895–21909

Rice JR (1980) The mechanics of earthquake rupture. In: Dziewonski A, Boschi E (eds) Physics of the Earth's interior. North-Holland, New York, pp 555–649

Rice JR (1992) Fault stress states, pore pressure distributions, and the weakness of the San Andreas fault. In: Evans B, Wong T-f (eds) Fault mechanics and transport properties of rocks. Academic, San Diego, pp 475–503

Rice JR, Cleary MP (1976) Some basic stress diffusion solutions for fluid saturated elastic porous media with compressible constituents. Rev Geophys Space Phys 14:227–291

Richards MA, Ricard Y, Lithgow-Bertelloni C, Spada G, Sabadini R (1997) An explanation for Earth's long-term rotational stability. Science 275:372–375

Richards MA, Gordon RG, van der Hilst RD (eds) (2000) The history and dynamics of global plate motions. Geophysical monograph series 121. American Geophysical Union, Washington DC

Richardson RM (1992) Ridge forces, absolute plate motions, and the intraplate stress field. J Geophys Res 97:11739–11749

Rindschwentner J (2001) Estimating the global permeability tensor using hydraulically induced seismicity—implementation of a new algorithm and case studies. Master thesis, Freie Universität Berlin

Rossmanith H-P (1983) Modelling of fracture process zones and singularity dominated zones. Eng Fract Mech 17:509–525

Rossmanith H-P (1995) Dynamics of fractured rock. Lecture notes of special workshop during the 2nd international conference on mechanics of jointed and faulted rock, Vienna, p 386

Roth F, Fleckenstein P (2001) Stress orientations found in North-east Germany differ from the West European trend. Terra Nova 13(4):289–296

Rothert E, Shapiro SA, Buske S, Bohnhoff M (2003) Mutual relationship between microseismicity and seismic reflectivity: Case study at the German Continental Deep Drilling Site (KTB). Geophys Res Lett 30(17):1893. doi:10.1029/2003GL017848

Rowe KJ, Rutter EH (1990) Paleostress estimation using calcite twinnings: experimental calibration and application to nature. J Struct Geol 12(1):1–17

Ruina AL (1985) Constitutive relations for frictional slip. In: Bazant ZP II (ed) Mechanics of geomaterials: rocks, Concrete, soils. Wiley, New York, pp 169–188

Rummel F (1979) Stresses in the upper crust as derived from in situ stress measurements—a review. Progress in earthquake prediction research. Vieweg, Braunschweig, pp 391–405

Rummel F (1986) Stresses and tectonics of the upper continental crust, a review. In: Stephansson O (ed) Rock stress and rock stress measurements. Centek Publishers, Lulea, pp 177–186

Rummel F (1987) Fracture mechanics approach to hydraulic fracturing stress measurements. In: Atkinson BK (ed) Fracture mechanics of rocks. Academic Press, London, pp 217–239

Rummel F (1989) Technical annexe: hydraulic fracturing stress measurements theory and practice. In: Louwrier K, Strarose E, Garnish JD, Karkoulias V (eds) European geothermal update. Kluwer Academic Publishers, London, pp 335–344

Rummel F (1991) Hydrofrac- Spannungs- und Permeabilitätsmessungen in der KTB-Vorbohrung Oberpfalz. Abschlussbericht zum Forschungsvorhaben Ru 225/10-3, Ruhr-Universität Bochum, 54 pages, plus Appendix (in German)

Rummel F (ed) (2005) Rock mechanics with emphasis on stress. AA Balkema, Leiden

Rummel J, Baumgärtner J (1985) Hydraulic fracturing in-situ stress and permeability measurements in the research borehole Konzen, Hohes Venn (Germany). Neues Jahrbuch Geologisch Paläontologische Abhandlungen 171:183–193

Rummel F, Jung R (1975) Hydraulic fracturing stress measurements near the Hohenzollern-Graben-structure, SW-Germany. Pure Appl Geophys 113:321–330

Rummel F, Winter RB (1983) Fracture mechanics as applied to hydraulic fracturing stress measurements. Earthq Prediction Res 2:33–45

Rummel F, Zoback MD (1993) Hydrofrac stress profile in the KTB boreholes VB and HB. In: Emmerman R, Lauterjung J, Umsonst T (eds) KTB-Report 93-2. pp 191–194

Rummel F, Möhring-Erdmann G, Baumgärtner J (1986) Stress constraints and hydro-fracturing stress data for the continental crust. Pure Appl Geophys 124(4, 5):875–895

Runcorn SK (1955) Core motion and reversals of the geomagnetic field. Ann Geophys 11:73–79

Rutqvist J, Digby P, Stephansson O, Singh U (1990) Simulation of borehole breakouts with damage material model. In: Maury V, Fourmaintraux D (eds) Proceedings international symposium of rock at great depth, Pau, AA Balkema, Rotterdam, vol 3, pp 1439–1445

Rutqvist J, Tsang C-F, Stephansson O (2000) Uncertainity in the maximum principal stress estimated from hydraulic fracturing measurements due to the presence of the induced fracture. Int J Rock Mech Min Sci Geomech Abstr 37:107–120

Rutter EH (1986) On the nomenclature of mode of failure transitions in rocks. Tectonophysics 122:381–387

Rutter EH, Brodie KH (2004a) Experimental intracrystalline plastic flow in hot-pressed synthetic quartzite prepared from Brazilian quartz crystals. J Struct Geol 26(2):259–270

Rutter EH, Brodie KH (2004b) Experimental grain size-sensitive flow of hot-pressed Brazilian quartz aggregates. J Struct Geol 26(11):2011–2023

Rutter EH, Mainprice DH (1978) The effect of water on the stress relaxation of faulted and unfaulted sandstone. Pure Appl Geophys 116:634–654

Rybacki E, Dresen G (2000) Dislocation and diffusion creep of synthetic anorthite aggregates. J Geophys Res 105(B11):26017–26036

Rybacki E, Dresen G (2004) Deformation mechanism maps for feldspar rocks. Tectonophysics 382:173–187

Rybacki E, Gottschalk M, Wirth R, Dresen G (2006) Influence of water fugacity and activation volume on the flow properties of fine-grained anorthite aggregates. J Geophys Res 111, B03203. doi:10.1029/2005JB003663

Sabato Ceraldi T, Chester FM, Kronenberg AK, Chester J (2006) Experimental propagation of mode II and III shear fractures in sandstones. EOS Trans Am Geophys Union, Fall Meeting, abstract # T43A–1632

Sakurai S, Akutagawa S (1994) Back analysis of in-situ stresses in a rock mass taking into account its non-elastic behaviour. In: Proceedings ISRM international symposium of integral approach to applied rock mechanics, Santiago, Chile, 1: pp 135–143

Sakurai S, Shimizu N (1986) Initial stress back analyzed from displacements due to underground excavations. In: Stephansson O (ed) Rock stress and rock stress measurements. Centek Publishers, Lulea, pp 679–686

Salomon JC (1968) Elastic moduli of a stratified rock mass. Int J Rock Mech Min Sci 5:519–527

Sammis CG, Ashby MF (1986) The failure of brittle porous solids under compressive stress states. Acta Metall 34(3):511–526

Sano O, Ito H, Hirata A, Mizuta Y (2005) Review of methods of measuring stress and its variations. Bull Earthq Res Inst Univ Tokyo 80:87–103

Santarelli FJ, Dusseault MB (1991) Core quality control in petroleum engineering. In: Roegiers J-C (ed) Rock mechanics as a multidisciplinary science. AA Balkema, Rotterdam, pp 111–121
Santisteban JR, Daymond MR, James JA, Edwards L (2006) ENGIN-X: a third generation neutron strain scanner. J Appl Crystallopgr 39:812–825
Savage WZ (1994) Gravity induced stresses in finite slopes. Int J Rock Mech Min Sci Geomech Abstr 31:471–483
Savage WZ, Swolfs HS, Amadei B (1992) On the state of stress in the near surface of the Earth's crust. Pure Appl Geophys 138:207–228
Sayers CM (1988) Inversion of ultrasonic wave velocity measurements to obtain the microcrack orientation distribution function in rocks. Ultrasonics 26:73–77
Sayers CM, van Munster, JG (1991) Microcrack-induced seismic anisotropy of sedimentary rocks. J Geophys Res 96(B10):16529–16533
Sayers CM, van Munster JG, King MS (1990) Stress-induced ultrasonic anisotropy in Berea sand stone. Int J Rock Mech Min Sci Geomech Abstr 27:429–436
Scheffzück Ch, Frischbutter A, Mikula P, Vrana M (1998) Using a double bent-crystal neutron diffractometer for quartz strain measurements on sandstones from the Elbezone. Schriftenreihe für Geowissenschaften 6:9–18
Scheidegger AE (1962) Stresses in the Earth's crust as determined from hydraulic fracturing data. Geologie und Bauwesen 27:45–53
Scheidegger AE (1982) Principles of Geodynamics, 3rd edn. Springer-Verlag, Berlin
Schellart WP (2004) Quantifying the net slab pull force as a driving mechanism for plate tectonics. Geophys Res Lett 31, L07611. doi:10.1029/2004GL019528
Schmidt RA (1980) A microcrack model and its significance to hydraulic fracturing and fracture toughness testing. In: Summers DA (ed) Comp Rock Mechanics. Proceedings 21st US symposium on rock mechanics, pp 581–590
Schmitt DR, Zoback MD (1989) Poroelastic effects in the determination of the maximum horizontal principal stress in hydraulic fracturing tests—a proposed breakdown equation employing a modified effective stress relation for tensile failure. Int J Rock Mech Min Sci Geomech Abstr 26:499–506
Schmitt DR, Diallo MS, Weichman F (2006) Quantitative determination of stress by inversion of speckle interferometer fringe patterns: experimental laboratory tests. Geophys J Int 167(3):1425–1438
Scholz CH (1987) Wear and gouge formation in brittle faulting. Geology 15:493–495
Scholz CH (1990) The mechanics of earthquakes and faulting. Cambridge University Press, New York
Scholz CH (2000) Evidence for a strong San Andreas fault. Geology 28:163–166
Scholz CH (2002) The mechanics of earthquakes and faulting, 2nd edn. Cambridge University Press, New York
Schwarzbach M (1980) Alfred Wegener und die Drift der Kontinente. Wissenschaftliche Verlagsgesellschaft mbH, Stuttgart
Seiki T, Aydan Ö, Kawamoto T (1997) The relation between geological features and the stress state of the Earth's crust in Central Japan. In: Sugawara K, Obara Y (eds) Proceedings international symposium on rock stress, AA Balkema, Rotterdam, pp 385–390
Sekine K, Hayashi K (2009) Residual stress measurements on a quartz vein: a constraint on paleostress magnitude. J Geophys Res 114, B01404. doi:10.1029/2007JB005295
Selsing J (1961) Internal stresses in ceramics. J Am Ceram Soc 44:419
Sen Z, Sadagah BH (2002) Probabilistic horizontal stress ratios in rock. Math Geol 34(7):845–855
Serata S, Sakuma S, Kikuchi S, Mizuta Y (1992) Double fracture method for in situ stress measurement in brittle rock. Rock Mech Rock Eng 25:89–108
Serdengecti S, Boozer G, Hiller KH (1962) Effects of pore fluids on the deformation behaviour of rocks subjected to triaxial compression. Proceedings 5th symposium on rock mechanics, University of Minnesota, Pergamon, Oxford, pp 579–625
Seto M, Nag DK, Vutukuri VS (1999) In-situ rock stress measurement from rock cores using the acoustic emission and deformation rate analysis. Geotech Geol Eng 17(3, 4):1–26

Shapiro SA, Dinske C (2007) Violation of the Kaiser effect by hydraulic-fracturing-related microseismicity. J Geophys Eng 4:378–383

Shapiro SA, Huenges E, Borm G (1997) Estimating the crust permeability from fluid-injection-induced seismic emission at the KTB site. Geophys J Int 137:207–213

Shapiro SA, Audiage P, Royer J-J (1999) Large-scale permeability tensor of rocks from induced microseismicity. Geophys J Int 131:F15–F18

Shapiro SA, Royer J-J, Audiage P (2000) Reply to comment by HF Cornet on "large-scale permeability tensor of rocks from induced microseismicity". Geophys J Int 140:470–473

Shapiro SA, Kummerow J, Dinske C, Asch G, Rothert E, Erzinger J, Kümpel H-J, Kind R (2006) Fluid induced seismicity guided by a continental fault—injection experiment of 2004/2005 at the German deep drilling site (KTB). Geophys Res Lett 33, L01309.doi:10.1029/2005GL024659

Shea WT Jr, Kronenberg AK (1993) Strength and anisotropy of foliated rocks with varied mica content. J Struct Geol 15(9, 10):1097–1121

Shen B (2008) Borehole breakouts and in situ stress. In: Y Potvin, J Carter, R Jeffrey (eds). SHIRMS, Australian Centre for Geomechanics, Perth, pp 407-418 ISBN 978-0-9804185-5-2

Shen B, Stephansson O (1993a) Modification of the G-criterion of crack propagation in compression. Int J Eng Fract Mech 47(2):177–189

Shen B, Stephansson O (1993b) Numerical analysis of mixed mode I and mode II fracture propagation. Int J Rock Mech Min Sci 30:861–867

Shen B, Stephansson O, Rinne M (2002) Simulation of borehole breakouts using FRACOD^{2D}. Oil Gas Sci Technol—Rev IFP 57(5):579–590

Shen B, Rinne M, Stephansson O (2006) FRACOD^{2D}. User's manual version 2.21. Fracom Ltd, Finland

Shen HW (1995) Objective Kaiser stress evaluation in rock. In: Proceedings of the 5th conference AE/MA in geologic structures and materials. Trans Tech Publications, Clausthal-Zellerfeld, pp 177–196

Sheorey PR (1994) A theory for in-situ stresses in isotropic and transversly isotropic rock. Int J Rock Mech Min Sci Geomech Abstr 31:23–34

Sheorey PR (1997) Empirical rock failure criteria. AA Balkema, Rotterdam

Shimada M (1992) Confirmation of two types of fracture in granite deformed at temperatures to 300°C. Tectonophysics 211:259–268

Shimada M (1993) Lithospheric strength inferred from fracture strength of rocks at high confining pressures and temperatures. Tectonophysics 217:55–64

Shimamoto T (1986) A transition between frictional slip and ductile flow undergoing large shearing deformation at room temperature. Science 231:711–714

Shin K, Kanagawa T (1995) Kaiser effect of rock in acousto-elasticity, AE and DR. In: Proceedings of the 5th Conference on AE/MA in geologic structures and materials. Trans Tech Publications, Clausthal-Zellerfeld, pp 197–204

Shkuratnik VL, Kuchurin SV, Vinnikov VA (2007) Regularities of acoustic emission and thermoemission memory effect in coal specimens under varying thermal conditions. J Min Sci 43(4):394–403

Simmons G (1964a) Velocity of shear waves in rocks to 10 kilobars. J Geophys Res 69(6): 1123–1130

Simmons G (1964b) Velocity of compressional waves in various minerals at pressures to 10 kilobars. J Geophys Res 69(6):1117–1121

Simmons G, Siegfried RW, Feves M (1974) Differential strain analysis: a new method for examining cracks in rocks. J Geophys Res 79:4383–4385

Simmons G, Todd T, Baldridge WS (1975) Towards a quantitative relationship between elastic properties and cracks in low porosity rocks. Am J Sci 275:318–345

Simpson DW, Leith W (1985) The 1976 and 1984 Gazli, USSR Earthquakes—were they induced? Bull Seismol Soc Am 75:1465–1468

Sjöberg J (2003) Overcoring rock stress measurements in borehole OL-KR24, Olkiluoto. Posiva working report 2003–60, Posiva Oy, Helsinki

Sjöberg J, Christiansson R, Hudson JA (2003) ISRM suggested methods for rock stress estimation—Part 2: overcoring methods. Int J Rock Mech Min Sci 40:999–1010
Skempton AW (1960) Effective stress in soils, concrete and rock. In: Conference on pore pressure and suction in Soils, Butterworths, London, pp 4–16
Skinner B (1966) Thermal expansion. In: Clark S (ed) Handbook of physical constants. the Geological Society of America Inc., New York, Memoir 97, p 587
Smith M (2007) Showing the way: new large-area survey technology narrows the search of oil and gas. New Technology Magazine (April/May)
Smith WHF, Sandwell DT (1997) Global sea floor topography from satellite altimetry and ship depth soundings. Science 277:1956–1962
Sobolev SV, Zoback MD, Babeyko AY (2003) A thermo-mechanical model of the San Andreas fault system in Central and Northern California, Eos Trans AGU, Fall Meeting Supplement 84(46), Abstract T51A-01
Sonder LJ (1990) Effects of density contrasts on the orientation of stresses in the lithosphere: Relation to principal stress dierections in the Transverse Ranges, California, Tectonics 9(4):761–771
Sone H, Zoback MD (2008) Intraplate stress, strain and earthquake potential in central Japan. Abstracts of the 3rd world stress map conference, Potsdam, Germany, p 35
Spence W (1987) Slab pull and the seismotectonics of subducting lithosphere. Rev Geophys 25:55–69
Sperner B, Müller B, Heidbach O, Delvaux D, Reinecker J, Fuchs K (2003) Tectonic stress in the Earth's crust: advances in the world stress map project. In: Nieuwland DA (ed) New insights into structural interpretation and modelling. Geological Society, London, Special Publications 212:101–116
Spetzler H (1987) Rock fracture and frictional sliding. In: Methods of experimental physics, vol 24, Part A, Academic Press Inc., pp 131–183
Spetzler H, Sobolev G, Koltsov A, Zang A, Getting IC (1991) Some properties of unstable slip on irregular surfaces. Pure Appl Geophys 137:95–112
Spudich P, Guatteri M, Otsuki K, Minagawa J (1998) Use of fault striations and dislocation models to infer tectonic shear stress during the 1995 Hyogo-Ken Nanbu (Kobe) earthquake. Bull Seimol Soc Am, 88(2):413–427
Stein R (1999) The role of stress transfer in earthquake occurrence. Nature 402:605–609
Steinberger B (2000) Plumes in a convecting mantle: models and observations for individual hotspots. J Geophys Res 105:11127–11152
Steinberger B, Torsvik TH (2008) Absolute plate motions and true polar wander in the absence of hotspot tracks. Nature 452:620–623
Steinberger B, Schmeling H, Marquart G (2001) Large-scale lithospheric stress field and topography induced by global mantle circulation. Earth Planet Sci Lett 186:75–91
Stephansson O (1983) Rock stress measurement by sleeve fracturing. In: Proceedings 5th congress international society of rock mechanics (ISRM), Melbourne, Balkema, Rotterdam, pp F129–F137
Stephansson O (1993) Rock stress in the Fennoscandian shield. In: Hudson JA (ed) comprehensive rock engineering, vol 3. Rock testing and site characterization. Pergamon Press, Oxford, pp 445–459
Stephansson O (1997) The importance of rock stress measurement and it's interpretation for rock disposal of hazardous waste. In: Sugawara K (ed) Proceedings of the international symposium on rock stress. Kumamoto, Japan. AA Balkema, Rotterdam, pp 3–13
Stephansson O (2003) Estimation of virgin state of stress and determination of final rock stress model. In: Sugawara K, Obara Y, Sato A (eds) Proceedings international symposium on rock stress, Kumamoto, Japan. Balkema, Lisse, pp 49–60
Stephansson O, Särkkä P, Myrvang A (1986) State of stress in Fennoscandia. In: Proceedings international symposium on rock stress and rock stress measurements, Stockholm, Centek Publisher, Lulea, pp 21–32

Stesky R, Brace W, Riley D, Robin P-Y (1974) Friction in faulted rock at high temperature and pressure. Tectonophysics 23:177–203

Stevens JL, Holcomb DJ (1980) A theoretical investigation of the sliding crack model of dilatancy. J Geophys Res 85:7091–7100

Stopinski W, Dmowska R (1984) Rock resistivity in the Lubin (Poland) copper mine and its relation to variations of strain field and occurrence of rock bursts. In: Gay NC, Wainwright EH (eds) Rock bursts and seismicity in mines, the South African Institute of Mining and Metallury, Johannesburg, Symposium Ser. No. 6, pp 297–307

Stuart CE, Meredith PG, Murrell SAF, van Munster JG (1993) Anisotropic crack damage and stress-memory effects in rocks under triaxial loading. Int J Rock Mech Min Sci Geomech Abstr 30(7):937–941

Su S, Stephansson O (1999) Effect of a fault on in situ stresses studied by the distinct element method. Int J Rock mech Min Sci 36:1051–1056

Sugawara K, Obara Y (1995) Rock stress and rock stress measurements in Japan. In: Proceedings international workshop on rock stress measurement at Great Depth, Tokyo, pp 1–8

Sugawara K, Obara Y, Kaneko K, Aoki T (1986) Hemispherical-ended borehole technique for measurement of absolute rock stress. In: Stephansson O (ed) Rock stress and rock stress measurements. Centek Publishers, Lulea, pp 207–216

Sugawara K, Obara Y (1993) Measuring rock stress. In: Hudson JA (ed) Comprehensive rock engineering, vol 3. Rock testing and site characterization. Pergamon Press, Oxford, pp 533–552

Suzuki (1969) Theory and practise of rock stress measurement by borehole deformation method. In: Proceedings international symposium on determination of stresses in rock masses vol 4, Lisbon, pp 173–182

Swolfs HS, Savage WZ (1985) Topography, stresses, and stability at Yucca mountain, Nevada. In: Proceedings of the 26th US symposium on rock mechanics, Rapid City, AA Balkema, Rotterdam, pp 1121–1129

Swolfs HS, Walsh JB (1990) The theory and prototype development of a stress-monitoring system. Seimol Soc Am Bull 80:197–208

Swolfs HS, Handin J, Pratt HR (1974) Field measurements of residual strain in granitic rock masses. In: (ed) Advances in rock mechanics, Proceedings 3rd Congress. ISRM, vol 2, US Committee for Rock Mechanics, National Academy of Sciences, Washington DC, pp 563–568

Tackley PJ (2000) Mantle convection and plate tectonics: toward an integrated physical and chemical theory. Science 288:2002–2007

Tamura N, MacDowell AA, Celestre RS, Padmore HA, Valek B, Bravman JC, Spolenak R, Brown WL, Marieb T, Fujimoto H, Batterman BW, Patel JR (2002) High spatial resolution grain orientation and strain mapping in thin films using polychromatic submicron X-ray diffraction. Appl Phys Lett 80:3724–3727

Tang CA, Chen ZH, Xu XH, Li C (1997) A theoretical model for Kaiser effect in rock. Pure Appl Geophys 50(2):203–215

Tarantola A, Valette B (1982) Generalized non-linear inverse problem solved using the least squares criterion. Rev Geophys Space Phys 20:219–232

te Kamp L, Rummel F, Zoback, MD (1995a) Hydrofrac stress profile to 9 km depth at the German KTB drill site. In: Emmermann R, Lauterjung J, Umsonst T (eds) Contribution to the 8th annual KTB colloquium. Giessen, pp 19–22

te Kamp L, Rummel F, Zoback, MD (1995b) Hydrofrac stress profile to 9 km depth at the German KTB drill site. In: Proceedings workshop on rock stress in the North Sea, Trondheim, Norway, NTH and SINTEF Publications, Trondheim, pp 147–153

Terzaghi K, Richart FE (1952) Stresses in rock about cavities. Geotechnique3:57–90

Tesauro M, Kaban MK, Cloetingh SAPL (2008) EuCRUST–07: a new reference model for the European crust. Geophys Res Lett 35, L05313. doi:10.1029/2007GL032244

Teufel LW (1982) Prediction of hydraulic fracture azimuth from anelastic strain recovery measurements of oriented core. In: Goodman RE, Heuze E (eds) Issues in rock mechanics, 23rd symposium on rock mechanics, Berkeley, California, pp 238–245

Teufel LW (1983) Determination of in-situ stress from anelastic strain recovery measurements of oriented core. Symposium on low permeability gas reservoirs, Denver, Colorado, SPE/DOE 11649: 421–430

Teufel LW (1986) In situ stress and natural fracture distribution at depth in the Piceance Basin, Colorado: implications to stimulation and production of low permeability gas reservoirs. In: Proceedings 27th US symposium on rock mechanics, Tuscaloosa, SME/AIME, pp 702–708

Teufel LW (1989) Acoustic emissions during anelastic strain recovery of cores from deep boreholes. In: Khair AW (ed) Rock Mechanics as a guide for the efficient utilization of natural resources, Proceedings of the 32rd US symposium, AA Balkema, Rotterdam, pp 269–276

Teufel LW (1993) Determination of in situ stress from partial anelastic strain recovery measurements of oriented cores from deep boreholes. In: Lecture notes of the short course in modern in situ stress measurement methods at the 34th US symposium of rock mechanics, Madison Wisconsin, p 19

Teufel LW, Rhett DW, Farrell HE (1991) Effect of reservoir depletion and pore pressure drawdown on in-situ stress and deformation in the Ekofisk Field, North Sea. In: Proceedings 32nd US symposium on rock mechanics, Norman, AA Balkema, Rotterdam, pp 63–72

Theiner WA (1997) Micromagnetic Techniques. In: Hauk V (ed) Structural and residual stress analysis by nondestructive methods. Elsevier Science, Amsterdam, pp 564–589

Thompson PM, Corthesy R, Leite MH (1997) Rock stress measurements at great depth using the modified doorstopper gauge. In: Sugawara K, Obara Y (eds) Proceedings of the international symposium rock stress, Japan, AA Balkema, Rotterdam, pp 59–64

Thompson PM, Martino J (2000) Application of the Doorstopper gauge system to deep in situ rock stress determinations. Report No. 06819-REP-01200-10019-R00, Atomic Energy of Canada Limited

Timoshenko SP, Goodier JN (1970) Theory of elasticity. McGraw-Hill Book Company, New York

Tincelin ME (1951) Research on rock pressure in the iron mines of Lorraine. In: Proceedings conference international sur les Pressions de Terrains et le Soutenement dans les Chantiers d' Exploitation. 24–28 Avril, Liege, pp 158–175

Tingay M, Reinecker J, Müller B (2008) Borehole breakout and drilling-induced fracture analysis from image logs. World stress map project, Guidelines: Image Logs (see DVD)

Tonon F, Amadei B (2003) Stresses in anisotropic rock masses: an engineering perspective building on geological knowledge. Int J Rock Mech Min Sci 40:1099–1120

Torsvik TH, Müller RD, van der Voo R, Steinberger B, Gaina C (2008) Global plate motion frames: toward a unified model. Rev Geophys 46, RG3004. doi:10.1029/2007RG000227

Townend J (2006) What do faults feel? Observational constraints on the stresses acting on seismogenic faults. In: Abercrombie R, McGarr A, Di Toro G, Kanamori H (eds) Earthquakes: radiated energy and the physics of faulting, geophysical monograph series 170, American Geophysical Union, pp 313–327. doi:10.1029/170GM31

Townend J, Zoback MD (2000) How faulting keeps the crust strong. Geology 28:399–402

Townend J, Zoback MD (2004) Regional tectonic stress near the San Andreas fault in central and southern California. Geophys Res Lett 31, L15S11.doi:10.1029/2003GL018918

Trifu CI (ed) (2002) The mechanism of induced seismicity. Pure Appl Geophys 159(1–3):617 (Topical Issue, Birkhäuser, Basel)

Trifu C-I (ed) (2010) Monitoring induced seismicity. Pure Appl Geophys, Topical Issue, Birkhäuser, Basel, ISBN 978-3-0346-0305-8

Tsenn MC, Carter NL (1987) Upper limits of power law creep of rocks. Tectonophysics 136:1–26

Tsukahara H (1983) Stress measurements utilizing the hydraulic fracturing technique in the Kanto-Tokai area, Japan. In: ZobackMD, Haimson BC (eds) Hydraulic fracturing stress measurements. National Academic Press, Washington DC, pp 18–27

Tullis TE (1977) Reflections on measurement of residual stress in rock. Pure Appl Geophys 115:57–68

Tunbridge LW (1989) Interpretation of the shut-in pressure from the rate of pressure decay. Int J Rock Mech Min Sci Geomech Abstr 26:457–459

Tvergaard V, Hutchinson JW (1988) Microcracking in ceramics induced by thermal or elastic anisotropy. J Am Ceram Soc 71:157–166
Twiss RJ (1977) Theory and applicability of a recrystallized grain size paleopiezometer. Pure Appl Geophys 115:227–234
Udias A (1999) Principles of seismology. Cambridge University Press, Cambridge
Valkó P, Economides MJ (1995) Hydraulic fracture mechanics. Wiley, New York
van Heerden WL (1976) Practical application of the CSIR triaxial strain cell for rock stress measurements. In: Proceedings ISRM symposium on investigation of stress in rock, advances in stress measurements, The Institution of Engineers, Sydney, pp 1–6
Varnes DJ, Lee FT (1972) Hypothesis of mobilization of residual stress in rock. Geol Soc Am Bull 83:2863–2866
Vásárhelyi B, Bobet A (2000) Modeling of crack coalescence in uniaxial compression. Rock Mech Rock Eng 33(2):119–139
Vermilye JM, Scholz CH (1998) The process zone: a microstructural view of fault growth. J Geophys Res 103(B6):12223–12237
Vernik L, Zoback MD, Brudy M (1992) Methodology and application of the wellbore breakout analysis in estimating the maximum horizontal stress magnitude in the KTB pilot hole. Sci Drill 3:161–169
Vervoort A, Govaerts A (2006) Kaiser effect in tri-axial tests of limestone samples. In: Lu M, Li CC, Kjorholt H, Dahle H (eds) In-situ rock stress. Taylor & Francis Group, London, pp 143–149
Villaescusa E, Seto M, Baird G (2002) Stress measurements from oriented core. Int J Rock Mech Min Sci 39:603–615
Villaescusa E, Windsor CR, Li J, Baird G (2003) Experimental verification of AE in situ stress measurements. In: Suguwara K, Obara Y, Sato A (eds) Rock stress. Swets & Zeitlinger, Lisse, pp 395–402
Villaescusa E, Li J, Windsor CR, Seto M (2006) A comparison of overcoring and AE stress profiles with depth in Western Australian Mines. In: Lu M, Li CC, KjorholtH, Dahle H (eds) In-situ rock stress. Taylor & Francis Group, London, pp 223–228
Vine FJ, Matthews DH (1963) Magnetic anomalies over oceanic ridges. Nature 199:947–949
Voight B (1966a) Restspannungen im Gestein. In: Proceedings 1st congress international society of rock mechanics, Lisbon, pp 45–50
Voight B (1966b) Interpretation of in-situ stress measurements. Panel report on Theme IV. In: Proceedings 1st congress international society of rock mechanics, Lisbon, vol 3, pp 332–348
Voight B (1968) Determination of the virgin state of stress in the vicinity of a borehole from measurements of a partial anelastic strain tensor in drill cores. Felsmechanik und Ingenieurgeololgie 6:201–215
Vollbrecht A, Dürrast H, Kraus J, Weber K (1994) Paleostress directions deduced from microcrack fabrics in the KTB core samples and granites from the surrounding area. Sci Drill 4:233–241
Von Schonfeldt H, Fairhurst C (1970) Field experiments on hydraulic fracturing. Soc Petrol Eng J, Am Inst Min Eng 1234–1239
Wallace RE (1951) Geometry of shearing stress and relationship to faulting. J Geol 59:118–130
Walsh JB (1965) The effect of cracks on the compressibility of rock. J Geophys Res 70:381–389
Walsh JB (1980) Static deformation of rock. J Eng Mech Div 106(5):1005–1019
Wang HF (2000) Theory of linear poroelasticity with applications to geomechanics and hydrogeology. Princton series in geophysics. Princton University Press, Princeton
Wang DF, Davies PJ, Yassir N, Enever J (1997a) Laboratory investigation of controls of stress history on ASR response. In: Suguwara K, Obara Y (eds) Rock stress. AA Balkema, Rotterdam, pp 181–186
Wang K, He J, Davis E (1997b) Transform push, oblique subduction resistance, and intraplate stress of the Juan de Fuca plate. J Geophys Res 102:661–674

Warpinski NR, Teufel LW (1986) A viscoelastic constitutive model for determining in-situ stress magnitudes from anelastic strain recovery of core. 61st Annual technical conference and exhibition of SPE, New Orleans, Lousiana, SPE 15368:1–12

Watson GS (1985) Interpolation and smoothing of directed and undirected line data. In: Krishnaiah PR (ed) Multivariate analysis-VI. Elsevier Sciences, New York, pp 613–625

Watt JP, Davies GF, O'Connell RJ (1976) The elastic properties of composite materials. Rev Geophys Space Phys 14(4):541–563

Wawersik WR, Rudnicki JW, Olsson WA, Holcomb DJ, Chau KT (1990) Localization of deformation in brittle rock: theoretical and experimental investigations. In: Shah SP, Swartz SE, Wang ML (eds) Micromechanics of failure of Quasi-Brittle materials. Elsevier Applied Science, pp 115–124

Weber K (1990) Observations on the ductile deformation path of the paragneisses of the KTB pilot hole. In: Emmermann R, Dietrich H-G, Lauterjung J, Woehrl Th (eds) KTB-report 90-8, Niedersaechsisches Landesamt für Bodenforschung, Hannover, J1–J19

Weertman J (1978) Creep laws for the mantle of the Earth. Philos Trans R Soc Lond A 288:9–26

Wegener A (1911) Brief vom 6. November an seinen Mentor Wladimir Köppen

Wegener A (1915) Die Entstehung der Kontinente und Ozeane. Vieweg-Verlag, Braunschweig (in German)

Wegener A (1924) The origin of continents and oceans. Methuen & Co. Ltd., London

Wehrle V (1998) Analytische Untersuchung intralithosphärischer Deformationen und numerische Methoden zur Bestimmung krustaler Spannungsdomänen. Dissertation Universität Karlsruhe (in German)

Wendt I, Carl C, Kreuzer H, Müller P, Stettner G (1992) Ergänzende Messungen zum Friedenfelser Granit (Steinwald) und radiometrische Datierungen der Ganggesteine im Falkenberger Granit. Geol Jahrb A 137:3–24

Wenk H-R, Lonardelli I, Merkel S, Miyagi L, Pehl J, Speziale S, Tommaseo CE (2006) Deformation textures produced in diamond anvil experiments, analysed in radial diffraction geometry. J Phys: Condens Matter 18:S933–S947

Wenzel U (1990) Der Kaiser-Effekt und die Frage nach seiner Verwendbarkeit zur in-situ Spannungsbestimmung an Tiefbohrkernen. Diploma thesis, University of Frankfurt/M

Wenzel F, Heidbach O, Müller B, Reinecker J (2004) Die Weltkarte der tektonischen Spannungen: Methoden—Ergebnisse—Anwendungen. Fridericiana: 19–34.

White JE (1962) Elastic waves along a cylindrical bore. Geophysics 27:327–333

Whittaker BN, Singh RN, Sun G (1992) Rock fracture mechanics. Elsevier, New York

Whyatt JK, Williams TJ, White BG (2000) Ground conditions and the May 13, 1994, rock burst, Coeur d` Alene Mining District, northern Idoho. In: Girard J, Leibman M, Breeds C, Doe T (eds) Proceedings of the 4th North American rock mechanics symposium, AA Balkema, Rotterdam, pp 313–318

Wibberley C (2007) Talc at fault. Nature 448:756–757

Wibberley CAJ, Yielding G, di Toro G (2008) Recent advances in the understanding of fault zone internal structure: a review. In: Wibberley CAJ, Kurz W, Imber J, Holdsworth RE, Collettini C (eds) The internal structure of fault zones: Implications for mechanical and fluid-flow properties 299:5–33, doi: 10.1144/SP299.2 The Geological Society of London

Wieghardt K (1907) Über das Spalten und Zerreiben elastischer Körper. Ztg Math Phys 55:60–103

Wilks KR, Carter NL (1990) Rheology of some continental lower crustal rocks. Tectonophysics 182:57–77

Williams CF, Grubb FV, Galanis SP Jr (2004) Heat flow in the SAFOD pilot hole and implications for the strength of the San Andreas fault. Geophys Res Lett 31, L15S14. doi:10.1029/2003GL019352

Wilson JT (1965) A new class of faults and their bearing on continental drift. Nature 207:343–347

Wilson M (1993) Plate-moving mechanisms—constraints and controversies. J Geol Soc 150:923–926

Wolter KE (1987) Untersuchungen der in situ Spannung, Residualspannung und der Mikrorißsysteme in Graniten Süddeutschlands. Dissertation University of Karlsruhe (in German)

Wolter KE, Berckhemer H (1989) Time dependent strain recovery of cores from the KTB-Deep drill hole. Rock Mech Rock Eng 22:273–287

Wolter KE, Röckel Th, Bücker Ch, Dietrich HG, Berckhemer H (1990) Core disking in the KTB drill cores and the determination of the in situ stress orientation. In: Emmerman R, Dietrich HG, Lauterjung J, Wöhrl Th (eds) KTB-report 90-8, NLfB Hannover, pp G1–G12

Wong IG (1993) The role of geological discontinuities and tectonic stresses in mine seismicity. In: Hudson JA (ed) Comprehensive rock engineering, vol 5. Pergamon Press, Oxford, pp 393–410 (Chap 15)

Wong T-f (1990) Mechanical compaction and the brittle-ductile transition in porous sandstones. In: Knipe RJ, Rutter EH (eds) Deformation mechanisms, rheology and tectonics. Geological Society Special Publication No. 54: 111–122

Wong T-f, David C, Zhu W (1997) The transition from brittle faulting to cataclastic flow in porous sandstones: mechanical deformation. J Geophys Res 102:3009–3025

Wong T-f, Szeto H, Zhang J (1992) Effect of loading path and porosity on the failure mode of porous rock. Appl Mech Rev 45(8):281–293

Worotnicki G (1993) CSIRO triaxial stress measurement cell. In: Hudson JA (ed) Comprehensive rock engineering, vol 3. Oxford, Pergamon, pp 329–394

Worotnicki G, Walton RJ (1976) Triaxial hollow inclusion gauges for determination of rock stresses in-situ. In: Supplement to proceedings ISRM symposium on investigation of stress in rock, advances in stress measurement, Sydney, The Institution of Engineers, Australia, pp 1–8 (Suppl)

Wu S, Zoback M (2008) Observations and modelling of co-seismic stress changes in the M7.6 Chi-Chi earthquake. Abstracts of the 3rd world stress map conference, Potsdam, p 73

Wu XY, Baud P, Wong T-f (2000) Micromechanics of compressive failre and spatial evolution of anisotropic damage in Darley Dale sandstone. Int J Rock Mech Min Sci 37:143–160

Wutzke U (1997) Durch die weiße Wüste. Justus Perthes Verlag Gotha GmbH (in German)

Yamaguchi H, Sakaguchi K, Sato K, Matsuki K, Hashida T (2008) Possibility of shear type fracture in visco-plastic material under confining pressure. In: Tohji K, Tsuchiya N, Jeyadevan B (eds) Water dynamics, 5th international workshop on water dynamics. AIP conference proceedings, vol 987, American Institute of Physics, pp 79–82

Yamaji A, Sato K (2006) Distances for the solutions of stress tensor inversion in relation to misfit angles that accompany the solutions. Geophys J Int 167:933–942

Yamamoto K, Kuwahara Y, Kato N, Hirasawa T (1990) Deformation rate analysis: a new method for in situ stress estimation from inelastic deformation of rock sample under uniaxial compressions. Tohoku Geophys J33(2):127–147

Yardley B (1997) Probe of a plate interior. Nature 389:792–793

Yong C, Wang C (1980) Thermally induced acoustic-emission in Westerly granite. Geophys Res Lett 7(12):1089–1092

Yoshikawa S, Mogi K (1981) A new method for estimation of the crustal stress from cored rock samples: laboratory study in the case of uniaxial compression. Tectonophysics 74:323–339

Yoshikawa S, Mogi K (1989) Experimental studies on the effect of stress history on acoustic emission activity—a possibility for estimation of rock stress. J Acoust Emiss 8(4):113–123

Yuan R-f, Li Y-h, Zhao X-d (2007) Experimental verification and theoretical analysis on felicity effect of acoustic emission in concrete. Key Eng Mater 353–358:2333–2336

Zang A (1991) Theoretische Aspekte der Mikrorißbildung in Gesteinen. Dissertation Universität Frankfurt/Main, p 209. ISBN 3-922540-38-4 (in German)

Zang A (1993a) Finite element study on the closure of thermal microcracks in feldspar/quartz-rocks—Part 1. Grain boundary cracks. Geophys J Int 113:17–31

Zang A (1993b) Finite element study on the closure of thermal microcracks in feldspar/quartz-rocks—Part 2. Intra-, transgranular and mixed cracks. Geophys J Int 113:32–44

Zang A, Berckhemer H (1989) Residual stress features in drill cores. Geophys J Int 99:621–626

Zang A, Berckhemer H (1993) Classification of crystalline drill cores from the KTB deep well based on strain, velocity and fracture experiments. Int J Rock Mech Min Sci Geomech Abstr 30(4):331–342

Zang A, Stöckl H (1991) Numerical study on crack evolution in diagenetically consolidated sandstones. J Geophys Res 96(B6):10039–10048

Zang A, Wong T-f (1995) Elastic stiffness and stress concentration in cemented granular material. Int J Rock Mech Min Sci Geomech Abstr 32(6):563–574

Zang A, Wolter K, Berckhemer H (1989) Strain recovery, microcracks and elastic anisotropy of drill cores from KTB deep well. Sci Drill 1:115–126

Zang A, Berckhemer H, Wolter K (1990) Inferring the in situ state of stress from stress relief microcracking in drill cores. In: Emmermann R, Dietrich H-G, Lauterjung J, Woehrl Th (eds) KTB-Report 90–8, NLfB, Hannover, F1-F20

Zang A, Berckhemer H, Lienert M (1996a) Crack closure pressures inferred from ultrasonic drill core measurements to 8 km depth in the KTB wells. Geophys J Int 124:657–674

Zang A, Lienert M, Zinke J, Berckhemer H (1996b) Residual strain, wave speed and crack analysis of crystalline cores from KTB-VB well. Tectonophysics 263(1–4):219–234

Zang A, Wagner FC, Dresen G (1996c) Acoustic emission, microstructure, and damage model of dry and wet sandstone stressed to failure. J Geophys Res 101 (B8):17507–17521

Zang A, Wagner FC, Stanchits S, Dresen G, Andresen R, Haidekker M (1998) Source analysis of acoustic emissions in Aue granite cores under symmetric and asymmetric compressive loads. Geophys J Int 135:1113–1130

Zang A, Wagner FC, Stanchits S, Janssen C, Dresen G (2000) Fracture process zone in granite. J Geophys Res 105(B10):23651–23661

Zang A, Stanchits S, Dresen G (2002) Acoustic emission controlled triaxial rock fracture and friction tests. In: Dyskin AV, Hu X, Sahouryeh E (eds) Structural integrity and fracture. Swets & Zeitlinger, Lisse, pp 289–294

Zbinden P (2006) New Gotthard rail link. Int Soc Rock Mech News J 6(2):33–36

Zemanek JE, Glen E Jr, Norton LJ, Cardwell RL (1970) Formation evaluation by inspection with the borehole televiewer. Geophysics 35:254–269

Zhang J, Wong T-f, Davis DM (1990) Micromechanics of pressure-induced grain crushing in porous rocks. J Geophys Res 95(B1):341–352

Zheng Z, Kemeny J, Cook NGW (1989) Analysis of borehole breakouts. J Geophys Res 94:7171–7182

Zimmerman RW (1991) Compressibility of sandstones. Elsevier, Amsterdam

Zimmerman RW, Meyer LR, Cook NGW (1994) Grain and void compression in fractured and porous rock. Int J Rock Mech Min Sci Geomech Abstr 31(2):179–184

Zoback MD (1991) State of stress and crustal deformation along weak transform faults. Phils Trans R Soc Lond A 337:141–150

Zoback MD (2000) Strength of the San Andreas. Nature 405:31–32

Zoback MD (2007) Reservoir geomechanics. Cambridge University Press, New York

Zoback MD, Harjes H-P (1997) Injection-induced earthquakes and crustal stress at 9 km depth at the KTB deep drilling site, Germany. J Geophys Res 102(B8):18477–18491

Zoback MD, Zoback, ML (2002a) State of stress in the Earth's lithosphere. In: Lee WHK (ed) International handbook of earthquake and engineering seismology, vol 81A, pp 559–568

Zoback MD, Zoback ML (2002b) Stress in the Earth's lithosphere, In encyclopedia of physical science and technology, 3rd edn. Academic Press, pp 16, 143–154

Zoback MD, Tsukahara H, Hickman S (1980) Stress measurements at depth in the vicinity of the San Andreas fault: implications for the magnitude of shear stress with depth. J Geophys Res 85:6157–6173

Zoback MD, Moos DL, Mastin L, Anderson RN (1985) Wellbore breakouts and in situ stress. J Geophys Res 90:5523–5530

Zoback MD, Mastin LG, Barton C (1986) In-situ stress measurements in deep boreholes using hydraulic fracturing, wellbore breakouts, and Stonely wave polarization. In:

Stephansson O (ed) Rock stress and rock stress measurements. Centek Publishers, Lulea, pp 289–299

Zoback MD, Zoback ML, Mount VS, Suppe J, Eaton JP, Healy JH, Oppenheimer DH, Reasenberg P, Jones L, Raleigh CB, Wong IG, Scotti O, Wentworth C (1987) New evidence on the state of stress on the San Andreas fault system. Science 238:1105–1111

Zoback MD, Apel R, Baumgärtner J, Brudy M, Emmerman R, Engeser B, Fuchs K, Kessels W, Rischmüller H, Rummel F, Vernik L (1993) Upper-crustal strength inferred from stress measurements to 6 km depth in the KTB borehole. Nature 365:633–635

Zoback MD, Hickman S, Ellsworth W (2008) State of stress and shear localization in and near the San Andreas fault in central California: results from the SAFOD project. Proceedings 3rd world stress map conference, Potsdam

Zoback MD, Barton CA, Brudy M, Castillo DA, Finkbeiner T, Grollimund BR, Moos, DB, Peka P, Ward CD, Wiprut DJ (2003) Determination of stress orientation and magnitude in deep wells. Int J Rock Mech Min Sci 40:1049–1076

Zoback ML (1992) First- and second-order patterns of stress in the lithosphere: the world stress map project. J Geophys Res 97:11703–11728

Zoback ML, Zoback MD (1980) State of stress in conterminous United States. J Geophys Res 85:6113–6156

Zoback ML, Zoback MD, Adams J, Assumpcao M, Bell S, Bergman EA, Bluemling P, Brereton NR, Denham D, Ding J, Fuchs K, Gay N, Gregersen S, Gupta HK, Gvishiani A, Jacob K, Klein R, Knoll P, Magee M, Mercier JL, Mueller BC, Paquin C, Rajendran K, Stephansson O, Suarez G, Suter M, Udias A, Xu ZH, Zhizhin M (1989) Global patterns of tectonic stress. Review article, Nature 341:291–298

Zogala B, Zuberek WM, Goroskiewicz A (1992) Acoustic emission in Carboniferous sandstone and mudstone samples subjected to cyclic heating. Acta Mont SerA 3(89)

Zulauf G (1993) Brittle deformation events at the western margin of the Bohemian Massif. Geol Rundsch 82:489–504

World Wide Webs

Crustal Thickness, United States Geological Survey, Menlo Park, CA (USA)
http://quake.wr.usgs.gov
Plate Velocity Vectors, UNAVCO Boulder, CO (USA)
http://sps.unavco.org
World Stress Map, GFZ German Research Centre for Geosciences (Germany)
http://www.world-stress-map.org

Stress Movies Content

Movie	Book chapter and movie title	Master tape recorded	Playing time, min	Disk MB
M1	Chap. 1 Introduction by Priv.-Doz. Dr. Arno Zang, Stresses in a Body	Aug. 2007	2:40	123
M2	Chap. 2 Introduction by Prof. Ove Stephansson, Book Content	Dec. 2008	11:46	542
M3.1	Chap. 3 Laboratory Fracture Test, GFZ Equipment	Dec. 2008	3:42	172
M3.2	Chap. 3 Triaxial Test Equipment, Dr. Sergei Stanchits, GFZ	Apr. 2009	12:31	575
M4	Chap. 4 Rock Stress Terminology, Priv.-Doz. Dr. Arno Zang and Prof. O. Stephansson	Feb. 2009	7:21	276
M5.1	Chap. 5 Introduction Paterson Laboratory by Dipl-Ing. Michael Naumann, GFZ	Apr. 2009	7:32	347
M5.2	Chap. 5 Interview with Dr. Erik Rybacki, Paterson Laboratory, GFZ	Dec. 2008	15:45	724
M6.1	Chap. 6 Overcoring Borre Probe, MSc Civil Eng. Ulf Lindfors, Dr. Daniel Ask and Mrs. Sofi Berg, Vattenfall Power Consultant AB, Lulea, Sweden	Nov. 2007	15:43	720
M6.2	Chap. 6 Overcoring Data Analysis, MSc Civil Eng. Ulf Lindfors, Vattenfall Power Consultant AB, Lulea, Sweden	Nov. 2007	10:16	472
M7.0	Chap. 7 Interview with Prof. Bezalel C. Haimson (University of Wisconsin-Madison) at World Stress Map Conference, Potsdam	Oct. 2008	17:51	823
M7.1	Chap. 7 Conventional Hydraulic Fracturing (HF) and Quadruple Packer Testing (QPT), Dr. Daniel Ask, Vattenfall Power Consultant AB, Lulea, Sweden	Nov. 2007	11:37	533
M7.2	Chap. 7 Sleeve Fracturing, Dr. Daniel Ask, Vattenfall Power Consultant AB, Lulea, Sweden	Nov. 2007	2:23	108

Movie	Book chapter and movie title	Master tape recorded	Playing time, min	Disk MB
M7.3	Chap. 7 Hydraulic Testing on Pre-Existing Fractures (HTPF). Dr. Daniel Ask, Vattenfall Power Consultant AB, Lulea, Sweden	Nov. 2007	3:44	169
M7.4	Chap. 7 Interview with Dr. Daniel Ask, Vattenfall Power Consultant AB, Lulea, Sweden	Nov. 2007	18:25	846
M8	Chap. 8 Anelastic Strain Recovery. Dr. Klaus Wolter, DCM, Walluf, Germany	Aug. 2007	2:51	131
M9.1	Chap. 9 Interview with Prof. Mark D. Zoback (Stanford University) at GFZ Potsdam	Jan. 2008	11:42	539
M9.2	Chap. 9 Interview with Dr. Jonny Sjöberg, Vattenfall Power Consultant AB, Lulea, Sweden	Nov. 2007	7:19	336
Data	Chaps. 10–11 World Stress Map data set Release 2008 provided by Dr. Oliver Heidbach, GFZ Potsdam	Oct. 2008	–	206
Total			2:34:08	7642

Index

MIX
Papier aus verantwortungsvollen Quellen
Paper from responsible sources
FSC® C105338

If you have any concerns about our products,
you can contact us on
ProductSafety@springernature.com

In case Publisher is established outside the EU,
the EU authorized representative is:
Springer Nature Customer Service Center GmbH
Europaplatz 3, 69115 Heidelberg, Germany

Printed by Libri Plureos GmbH
in Hamburg, Germany